Chemical Soil Stabilization

Chemical Soil Stabilization

Contributors

Elmira Saljnikov, Dragan Cakmak et al.

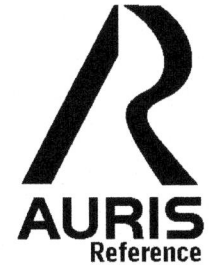

AURIS
Reference

www.aurisreference.com

Chemical Soil Stabilization

Contributors: Elmira Saljnikov, Dragan Cakmak et al.

Published by Auris Reference Limited

www.aurisreference.com

United Kingdom

Printed in 2017 for Sale in the Indian Subcontinent

Chemical Soil Stabilization

ISBN: 978-1-78154-973-5

British Library Cataloguing in Publication Data
A CIP record for this book is available from the British Library

Printed in the United Kingdom

Exclusively distributed by CBS Publishers & Distributors Pvt. Ltd.

Sales & Distribution Rights only for India, Pakistan, Bangladesh, Sri Lanka, Nepal and Bhutan.This book is not to be sold outside these territories.

Contents

List of Abbreviations

ANOVA	Analysis of variance
BNF	Biological nitrogen fixation
CDR	Clay dispersion ratio
CFI	Clay flocculation index
CRD	Completely randomized design
CF	Continuous fallowing
CTT	Conventional tillage and transplanting
DSS	Decision support system
DR	Dispersion ratio
DOC	Dissolved organic carbon
DOM	Dissolved organic matter
ECSA	Eastern, central and southern africa
FTIR	Fourier transform infrared spectroscopy
GHG	Greenhouse gases
ISFM	Integrated soil fertility management
MWD	Mean-weight diameter
NT	No-tillage
OC	Organic carbon
OM	Organic matter
PMC	Potentially mineralizable carbon
PMN	Potentially mineralizable nitrogen
RUSLE	Revised universal soil loss equation
SOC	soil organic carbon
SOM	Soil organic matter
SOM	Soil organic matter
SABRN	Southern africa bean research network
WSOM	Water soluble organic matter
WDS	Water-dispersible silt
WAP	Weeks after planting

List of Contributors

Elmira Saljnikov
Institute of Soil Science, Belgrade, Serbia

Dragan Cakmak
Institute of Soil Science, Belgrade, Serbia

Saule Rahimgalieva
West-Kazakhstan Agrarian-Technical University named after Zhangir-Khan, Uralsk, Kazakhstan

C.A. Igwe
Department of Soil Science, University of Nigeria, Nsukka, Nigeria

S.E. Obalum
Department of Soil Science, University of Nigeria, Nsukka, Nigeria

M.C. Hernandez-Soriano
Division of Soil and Water Management, KU Leuven, Belgium

B. Kerré
Division of Soil and Water Management, KU Leuven, Belgium

A. Sevilla-Perea
Instituto Andaluz de Ciencias de la Tierra, University of Granada - CSIC, Spain, Spain

M.D. Mingorance
Instituto Andaluz de Ciencias de la Tierra, University of Granada - CSIC, Spain, Spain

Shinya Funakawa
Graduate School of Agriculture, Kyoto University, Japan

Hiroshi Yoshida
Graduate School of Agriculture, Kyoto University, Japan

Tetsuhiro Watanab
Graduate School of Agriculture, Kyoto University, Japan

Soh Sugihara
Graduate School of Agriculture, Kyoto University, Japan

Method Kilasara
Faculty of Agriculture, Sokoine Agricultural University, Tanzania

Takashi Kosaki
Graduate School of Urban Environmental Sciences, Tokyo Metropolitan University Japan

Manoj K. Jha
North Carolina A&T State University USA

Andrea Rubenacker
Departamento de Recursos Naturales, Facultad de Ciencias Agropecuarias, Universidad Nacional de Córdoba, Córdoba Argentina

Paola Campitelli
Departamento de Recursos Naturales, Facultad de Ciencias Agropecuarias, Universidad Nacional de Córdoba, Córdoba Argentina

Manuel Velasco
Departamento de Recursos Naturales, Facultad de Ciencias Agropecuarias, Universidad Nacional de Córdoba, Córdoba Argentina

Silvia Ceppi
Departamento de Recursos Naturales, Facultad de Ciencias Agropecuarias, Universidad Nacional de Córdoba, Córdoba Argentina

Bart Minten
International Food Policy Research Institute, Addis Ababa, Ethiopia

Claude Randrianarisoa
United States Agency for International Development (USAID), Madagascar

Celia Maria Maganhotto de Souza Silva
Embrapa Environment Brazil

Elisabeth Francisconi Fay
Embrapa Environment Brazil

Alberto C. de Campos Bernardi
East region, Sao Carlos - SP, Brazil

Patrícia P. A. Oliveira
East region, Sao Carlos - SP, Brazil

Odo Primavesi
East region, Sao Carlos - SP, Brazil

Ren Wan-Jun
Sichuan Agricultural University, Wenjiang, Sichuan, China

Huang Yun
Sichuan Agricultural University, Wenjiang, Sichuan, China

Yang Wen-Yu
Sichuan Agricultural University, Wenjiang, Sichuan, China

Roland Nuhu Issaka
CSIR-Soil Research Institute, Academy Post Office, Kwadaso-Kumasi Ghana

Moro Mohammed Buri
CSIR-Soil Research Institute, Academy Post Office, Kwadaso-Kumasi Ghana

Eric Owusu-Adjei
CSIR-Soil Research Institute, Academy Post Office, Kwadaso-Kumasi Ghana

Satoshi Tobita
Japan International Research Center for Agricultural Sciences, Ohwashi, Tsukuba, Japan

Satoshi Nakamura
Japan International Research Center for Agricultural Sciences, Ohwashi, Tsukuba, Japan

Lubanga Lunze et al.
Institut National pour l'Etude et la Recherche Agronomiques (INERA), Kinshasa Democratic Republic of Congo

Emmanuel Ibukunoluwa Moyin-Jesu
Agronomy Department, Federal College of Agriculture, Akure, Ondo State Nigeria

D.-G. Kim
Landcare Research, Palmerston North 4442, New Zealand

R. Vargas
Departamento de Biolog´ıa de la Conservacion, Centro de Investigaci ´ on Cient ´ ´ıfica y de Educacion Superior de Ensenada (CICESE), Ensenada, BC, Mexico
Department of Plant and Soil Sciences, Delaware Environmental Institute, University of Delaware, Newark, DE 19717, USA

B. Bond-Lamberty
Pacific Northwest National Laboratory, Joint Global Change Research Institute at the University of Maryland – College Park, College Park, MD 20740, USA

M. R. Turetsky
Department of Integrative Biology, University of Guelph, Guelph, ON, Canada

Preface

Soil stabilization is a general term for any physical, chemical, biological, or combined method of changing a natural soil to meet an engineering purpose. It is a method of improving soil properties by blending and mixing other materials. The text *Chemical Soil Stabilization* highlights new ground improvement techniques as well as recent innovations in soil modification and stabilization procedures. The objectives of first chapter are to examine the effects of summer fallow on the characteristics of soil organic matter (SOM) on a long-term basis as well as on a short-term basis with special reference to readily decomposable fractions. Second chapter focuses on microaggregate stability of tropical soils and its roles on soil erosion hazard prediction. The aim of third chapter is to evaluate the chemical composition of organic matter (OM) in artificial soils obtained from organic wastes combined with the FeM at different ratios. In fourth chapter, the regional trend in soil fertility with respect to the soil mineralogical and chemical properties has been investigated. The main objective of fifth chapter is to use soil and water assessment tool (SWAT) model to quantify soil moisture distribution on a watershed scale and evaluate the impact of applying cover crop conservation practice on soil moisture content. Fire impact on several chemical and physicochemical parameters in a forest soil has been investigated in sixth chapter. Forest preservation, flooding, and soil fertility of Madagascar forests have been studied in seventh chapter. Eighth chapter focuses on effect of salinity on soil microorganisms. Ninth chapter provides information regarding the management of liming and fertilization of intensively managed pastures based on soil analysis and requirement of the grass. The objectives of tenth chapter are to clarify the influence of the new technology on soil fertility and microbial populations of paddy field. Eleventh chapter focuses on indigenous fertilizing materials for enhancing soil productivity in Ghana. Integrated soil fertility management in bean-based cropping systems has been presented in last chapter.

Chapter 1

SOIL ORGANIC MATTER STABILITY AS AFFECTED BY LAND MANAGEMENT IN STEPPE ECOSYSTEMS

Elmira Saljnikov[1], Dragan Cakmak[1], and Saule Rahimgalieva[2]

[1]Institute of Soil Science, Belgrade, Serbia

[2]West-Kazakhstan Agrarian-Technical University named after Zhangir-Khan, Uralsk, Kazakhstan

INTRODUCTION

Soil Organic Matter Status

Soil organic matter (SOM) is most reactive and powerful factor in the formation of soil and in its fertility. Formation of soil and accumulation of organic matter are a function of interactions between biological factors and parent rocks under certain hydrothermal conditions and are one of the sections of a continuous chain of the trophic bounds between different life forms, serving as a first and a last section at the same time. The later is because SOM contain the main nitrogen stock, nearly the half of phosphorus, significant part of sulphur and other macro- and micronutrients for sustaining life and productivity of plants. Although soil organic matter comprise only five percent of total soil structure it has been a major research topic throughout the history of soil science, which is generally regarded to have been ongoing for approximately a century [1, 2].

Discovering the role and fate of soil organic matter has been a great challenge for the scientists. There are many argues about definitions of SOM among soil scientist. One of the most dynamic definitions of the SOM was given by [3]: the amount of organic carbon contained in a particular soil is a function of the balance between the rate of deposition of plant residues in or on soil and the rate of mineralization of the residue carbon by soil biota. In fact organic matter in soil always is in a very dynamic state, where transformations of bio-products occur constantly. The mechanisms through which soil organic C can be biologically stabilized depend on the decomposition of the soil mineral phase and the chemical structure of the organic residues added to the soil.

Climate is the most powerful factor that determines the array of plant species at any given location, the quantity of plant material produced, and the intensity of microbial activity in the soil. Climate influences soil organic carbon (SOC) content primarily through the effects of temperature, moisture, and solar radiation. Related studies found that amounts of SOC were positively correlated with precipitation and, at a given level of precipitation, negatively correlated with temperature [4, 5]. Climatic influences on biologically active fractions of SOM are not well understood. Therefore, one of the focuses in this study was investigation of the dynamics of labile SOM under the different hydrothermal conditions of steppe ecosystems.

Another powerful factor determining SOM reserves is plant biomass inputs and outputs. In agricultural systems, where soil and plant residues are often intensively manipulated, human impact on decomposition is especially pronounced [6]. Management practices like tillage, selection of crops and cropping sequences, and fertilization can alter decomposition rates by their effects on soil moisture, soil temperature, aeration, composition and placement of residues. Many studies confirm that under the similar climatic condition, carbon and nitrogen retention in soil is influenced by crop management systems, such as *crop rotation* [7, 8], *tillage* [5, 9], *residue management* [10] and fertilization and fertility [7, 10, 11]. This Chapter will discuss an impact of different land management practices on the labile (biologically active) pool of soil organic matter.

Decomposition

Decomposition is the progressive break down of organic, ultimately into inorganic constituents. The decomposition process is mediated mainly by soil microorganisms, which derive energy and nutrients from decomposing substrate. Plant litter decomposes very rapidly and although the carbon from plant litter represents only a small fraction of C in soil, about half of the CO_2 output from soil, globally, comes from decomposition of the annual litter fall [12]. Decomposition is central to the biogeochemical cycles in terrestrial, aquatic and atmospheric systems. It releases nutrients and energy associated in organic materials and feeds them back into local and global cycles, thereby affecting land, and air and water quality (Fig. 1).

Three interrelated factors regulate decomposition: the quality of the residue, the physical-chemical environment in which decomposition occurs and the type of organisms in the decomposer community. All organic carbon in soils can serve as potentially suitable as substrate. Vegetation can influence

SOC levels as a result of the amount, placement and biodegradability of plant residues returned to the soil. The fate of surface deposited residues depends on the activity of soil microorganisms and fauna and their ability to mix these residues into surface mineral horizons. Microorganisms are the major contributors to soil respiration and are responsible for 80-95% of the mineralization of carbon. Humans can affect decomposition by altering some of these factors, especially in agricultural systems. The current understanding of decomposition processes, learned from field and laboratory studies, is embodied in simulation models, e.g., the first-order kinetic model [110].

One of the effects of global warming is accelerated decomposition of soil organic matter, thereby releasing CO_2 to the atmosphere, which will further enhance the warming trend [104]. The United Nations Framework Convention on Climate change (Kyoto Protocol of 1997), allows organic carbon stored in arable soils to be included in calculations of net carbon emissions. By altering organic matter production, litter quality, and belowground C allocation, however, changes in vegetation type can influence microbial decomposition [105] and root respiration and therefore soil respiration rates [80]. As a result of global climate change and alterations in land use many ecosystems are currently experiencing concurrent changes in the abiotic and biotic controls on soil respiration. Given the large quantity of CO_2 that soils respire annually and the role CO_2 plays in greenhouse warming, an understanding of SR response to climate change and alterations in vegetation resulting from land use is critical.

Labile Pool of Soil Organic Matter

Labile carbon is the fraction of soil organic carbon with most rapid turnover times and its oxidation drives the flux of CO_2 between soils and atmosphere. Labile organic matter pools are fine indicators of soil quality that influence soil function in specific ways and that are much more sensitive to changes in soil management practice [e.g., 13]. The biggest and main source for labile organic matter is a 'light' fraction organic matter (or particulate OM, or macroorganic matter; [8, 14-17] that consists of partially decomposed plant litter. This 'light' organic matter acts as a substrate for soil microbial activity, a short-term reservoir of nutrients, a food source for soil fauna and loci for formation of water stable macroaggregates.

The biological determination of labile SOM is the carbon decomposed by microorganisms during the microbial growth. This biological definition of labile SOM includes two aspects: labile soil organic carbon chemically and physically assessable; the organic carbon that is chemically decomposable but physically un-assessable due to protection by clay minerals is not considered as a labile organic carbon. Generally, soil organic matter is divided into stable

(70-96%), active (2-30%) and plant litter (0-20%) fractions (Fig.2). The active fraction mainly consists of microbial biomass and their metabolites, the organic substrate in different stages of decomposition and non-humic substances, with turnover time from 0.8 to 5 years. The stabilized or passive fraction of SOM is passive, chemically and physically protected matters. The physically protected OM has turnover time from 20 to 50 years; the chemically protected –from 800 to 1200 years.

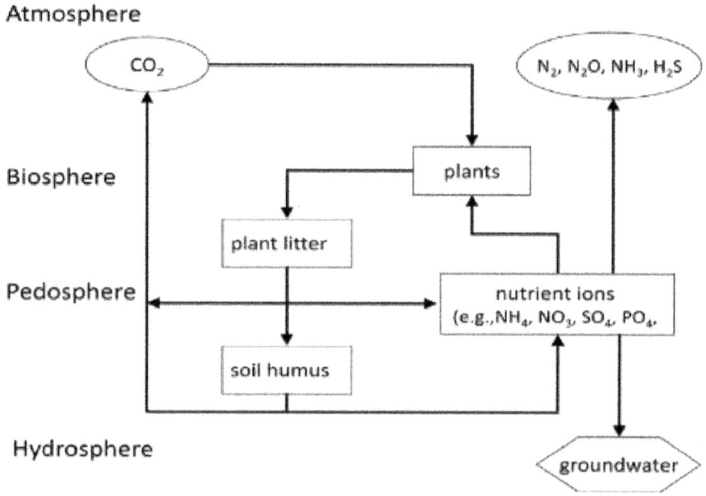

Figure 1: Conceptual model of soil organic matter decomposition (modified from [106]).

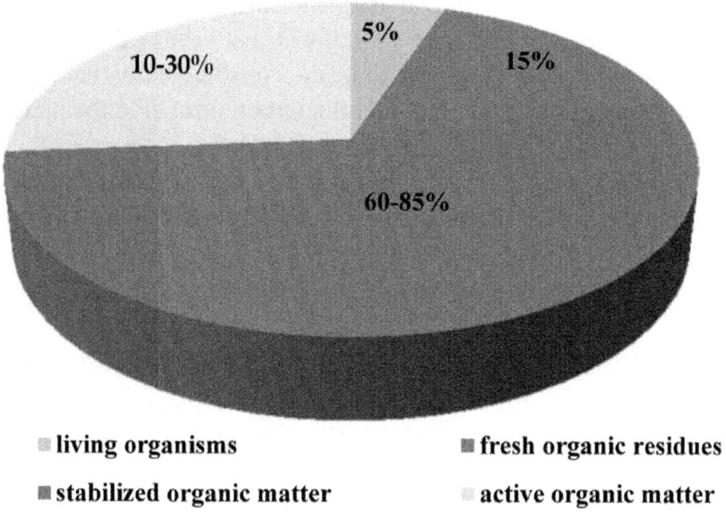

Figure 2: Composition and distribution of fractions of soil organic matter

The 10 to 30% active fraction is responsible for the support of soil microorganisms. This fraction is most sensitive to soil management practices. Although labile OM comprises a small part of total SOM, it is the main source for nutrients and energy for microorganisms and plants, and main source for carbon dioxide flux from soil. The roles of stable and labile SOM differ. An active fraction mainly influences the activity of microorganisms, the stability of macroaggregates, filtration speed, and the speed of nutrient mineralization. Whilst, the stable fraction influences mainly water-holding capacity, soil cation exchangeable capacity and soil microaggregation.

Fresh plant litter decomposes very quickly and the decomposition usually occurs not as a single step, but as a cascade. Fresh material, usually plant residue, undergo hydrolysis and redox reactions and then converted into altered forms. The transformed organic material, so called 'light' fraction (LF), in turn, is susceptible to further decomposition. A small part of LF is utilized for microbial synthesis, which after death contribute back to LF. Greatest part of LF is subjected to further mineralization resulting in mineral products, which is of direct practical interest of soil scientists from agronomical and ecological point of view, because as mentioned above, about half of the CO_2 output from soil, globally, comes from decomposition of the annual litter fall.

Thus, transformations of SOM are generally concentrated within labile pool. The end products of organic matter mineralization (e.g., CO_2, NO_3, NH_4) can give us valuable information about ability of a given soil to supply plants with nutrients and/or ability to stabilize soil organic matter.

Nitrogen (N) is generally the most common growth-limiting nutrient in agricultural production systems Nitrogen taken up by crops is derived from a number of sources, particularly from fertilizer, biological N fixation and mineralization of N from soil organic matter, crop residues, and manures [18]. Large amounts of mineralizable N can accumulate under grassland with the result that crops grown immediately after cultivation of long-term grass may derive much of their N from mineralization. In contrast, soils that have been intensively cropped often mineralize little N, leaving crops heavily dependent on fertilizer nitrogen. This chapter will present the impact of different fertilization experiments on soil labile OM.

Labile fractions of SOM neither have been fully described nor successfully isolated [19, 20]. However, procedurally defined fractions such as carbon and nitrogen mineralized under controlled conditions and "light" fraction organic carbon proved to be good indicators of subtle changes in SOM, because they affects the nutrient dynamics within single growing season, the organic matter content under contrasting management regimes, and C sequestration over extended periods of time. Organic matter quality may also be characterized by estimates of kinetically defined pools obtained by fitting the simulation models to data of carbon and nitrogen mineralization [21, 22]. Although soil labile organic carbon is constituted of amino acids, simple carbohydrates, a fraction of microbial biomass, and other simple organic compounds, a clear chemical or physical definition of soil labile organic carbon is difficult if not impossible. We here present a biological definition of soil labile organic carbon as microbial degradable carbon associated with microbial growth. This biological definition includes two aspects: soil labile organic carbon is both chemically degradable and physically accessible by soil microbes. Organic carbon that is chemically degradable but physically inaccessible by microbes due to clay mineral protection is not regarded here as soil labile organic carbon [13].

Mollisols soils are the most fertile and productive soils and therefore they are often overexploited for agricultural needs. The area under *Mollisols* in Kazakhstan occupies 25.3 million ha and in Ukraine 60.4 million ha. During the Soviet period the political aim was a rapid increase in grain production that was achieved by indiscriminate plowing of as large area of virgin lands as possible. However, such intensive cultivation of these soils resulted in drastic decrease in its humus content. In this Chapter four types of *Mollisols*: *Hupludolls*, *Argiudolls*, *Calciustolls* and *Haplustolls* studied for characterization of the "light fraction organic matter" for the scenarios considered and estimation of relationship between C/N ratio and mineralization rates are presented.

MATERIALS AND METHODS

Description of Study Sites

Four experimental sites from Eurasian steppes were examined for soil organic matter fraction. They are: Kharkov (dry forest-steppe, east Ukraine), Uman (moist forest-steppe, central Ukraine), Kherson (dry steppe, south Ukraine) and Astana (dry steppe, northern Kazakhstan,). The sites are located in different soil-ecological zones and differ in the amount of precipitation, temperature, soil type and vegetation

Soil Sampling and Analysis

Topsoil samples (0-20-cm) were collected in spring-summer. Three sub-samples for chemical and five sub-samples for biological analysis were taken from each sampling point. The soil samples were air-dried followed by grinding, and were passed through a 2-mm sieve for chemical analysis. Samples for biological analysis were stored at fresh condition at 4° C before analysis. The dried soil samples were analyzed for total N concentration using a full automatic analyzer (Shimazu NC-800-13N). Organic C was determined by dichromate oxidation method [23].

Labile Carbon (Potentially Mineralizable Carbon, PMC)

The rate of disappearance of plant residues can be described using a kinetic model. First-order kinetic model is usually used to characterize decomposition of plant residues, assuming that the annual input of plant residues is independent of the rate of their decomposition. Using first-order kinetics to describe decomposition implies that the metabolic potential of the soil microbial biomass exceeds the substrate supply.

Carbon mineralization was determined using laboratory incubation techniques via measuring soil respiration. The fresh soils were brought to 50% of WHC followed by incubation in a square-plastic jar (500-ml) at 30°C for 70 days. The evolved CO_2 was trapped in an alkali solution (10-ml of 1M NaOH) that was replaced every 14 days, and cumulative CO_2 was measured by titration with 0.5*M*HCl. The amount of mineralizable carbon was estimated from the rate of CO_2-C evolved during 70 days of incubation using nonlinear regression according to the following equation [24]: $C_{min}=C_0(1-e^{-kt})$,where, C_{min} is an experimental value of mineralized C (mg kg^{-1} soil) at time *t* (days) that was plotted to fit the equation, C_0 is potentially mineralizable carbon (PMC) (mg kg^{-1} soil) that was calculated after fitting the curve, and *k* is a nonlinear mineralization constant, i.e. the fraction mineralized per day (d^{-1}) [25].

Labile Nitrogen (Potentially Mineralizable Nitrogen, PMN)

Potentially mineralizable N is a measure of the active fraction of soil organic N, which is chiefly responsible for the release of mineral N through microbial action. Mineralizable N is composed of a heterogeneous array of organic substrates including microbial biomass, residues of recent crops, and humus. Despite a continuing research effort [26, 27], chemical tests that are selective for the mineralizable portion of soil N are not available and incubation assays remain the preferred way of estimating mineralizable N.

Mineralized N was determined after incubation of soils for 2-, 4-, 6-, 8-, 10-weeks and analyzed for nitrate and ammonium N content by colorimetric method following extraction with 2*N*KCl solution. Nitrate N was analyzed after reduction of NO_3ions to NO_2 by passing the extraction through cadmium column. Ammonium N was analyzed by salicylate nitroprusside method [18]. The amount of mineralizable N (N_0) was obtained after fitting the data of mineralized N (N_{min}) every 14 days to the first order kinetic model

[25]: $N_{min}=N_0*(1-e^{-kt})$, where, N_{min} is an experimental value of mineralized N at a given time (*t*) that was plotted to fit the equation, N_0 is a potentially mineralizable nitrogen (PMN) that was calculated after fitting the curve, and *k* is nonlinear mineralization constant.

Microbial Biomass

Soil microbial biomass measurements have been used in studies of soil organic matter dynamics and nutrient cycling in a variety of terrestrial ecosystems. They provide a measure of the quantity of living microbial biomass present in the soil, and in arable soils account for ~1%–5% of the total soil organic matter [28, 29]. Measurements of the carbon (C) and nitrogen (N) contained in the soil microbial biomass provide a basis for studies of the formation and turnover of soil organic matter, as the microbial biomass is one of the key definable fractions [30]. The data can be used for assessing changes in soil organic matter caused by soil management [31] and tillage practices [32], for assessing the impact of management on soil strength and porosity, soil structure and aggregate stability [33], and for assessing soil N fertility status [21].

Soil microbial biomass was determined by chloroform fumigation-extraction technique as described by [34]. For each sample, four sub-samples of field-moist soil were placed in flasks, moistened to field capacity and conditioned for 3 days at 25°C. Two sub-samples were fumigated with chloroform in a vacuum chamber for 5 days at 25°C and the other two sub-samples (controls) were incubated without fumigation at the same temperature. All samples were extracted with 0.5 m of K_2SO_4 at ratio 5:1. Microbial biomass C was measured by dissolved organic carbon analyzer (TOC-5000) and microbial biomass N was determined by colorimetric method. The microbial biomass C and N were calculated using an equation relating the increased release of C and N as a result of $CHCl_3$ fumigation and a factor representing the fraction of biomass C and N extracted by $K2SO_4$ [35].

"LIGHT" Fraction Organic Matter (LF)

"Light" fraction organic matter (LFOM) was separated by density separation using reagent-grade NaI solution adjusted to 1.8 g cm^{-3} [36]. 10g of soil was suspended in 40 ml of NaI solution (sp.gr. = 1,7) and the soil dispersed for 30 seconds using a Virtis homogenizer. After centrifugation, the floating material, i.e., the 'light' fraction was transferred directly to a vacuum filtration unit. The LFOM was then washed (three aliquots of 10 ml 0.01M $CaCl_2$ followed by three aliquots of distilled water), dried at 70°C for 15h and weighed. The residue was re-suspended and the procedure was repeated to ensure complete collection of the LF. The composite LF was finely ground and analyzed for N and C concentrations.

Statistics

Descriptive statistical analyses were performed using SYSTAT-8 software [24]. Variability among treatments in each region was within the range of variability among the regions for all the cases. Sigma Plot 8 software [25] was applied for modeling C mineralization pattern and mineralization rate constant.

IMPACT OF SOIL MANAGEMENT PRACTICES ON CONTENT OF TOTAL ORGANIC C AND N IN SOIL

Soil Total C and N in Fertilization Experiment

Mean annual mineralization of humus depends upon many factors. However, in case of unified soil and climatic conditions the limiting factor of soil organic matter mineralization becomes the cultivated plant and the technology of crop cultivation. Time, depth, frequency and intensity of cultivation are directly related to the amount of humus mineralization [37, 38].

The experiment with application of different dozes of mineral and organic fertilizers was conducted on*Mollisols*, in Uman (Table 1). The results of the study confirm the role of manure in contribution to both stable and labile soil organic matter. The content of soil organic carbon was not increased after 36 years application of mineral fertilizer in most of the treatments, compared to the control, while application of high rates of manure (O) alone maintained the higher accumulation of soil organic carbon (Table 2).

Manure contains humic acids [39], which directly contributes to the soil humic acids and favors humification processes [40, 41]. As this experiment has been performing since 1964, the long-term input of high rates of manure contributed to SOM via direct inputs of humic acids into the soil, showing the higher soil organic C than in other treatments. Content of total N in the treatments was not statistically different as indicated by the same letters in Table 2. Insignificant effect of the mineral fertilizers on the accumulation of soil organic C and N is due to quick depletion of mineral fertilizer in the soil either by means of microbial utilization [42] and by plant consumption, or by direct losses via leaching and/or volatilization.

Soil Total C and N in Fallow Frequency Experiment in Astana, Kazakhstan

Under nearly 50 years of monoculture of wheat, summer bare fallow has been practiced in crop rotation in order to retain moisture, to accumulate nutrients through mineralization and to control weed infestation. Fallowed fields are usually cultivated many times to keep the land bare during the whole cropping season. Of great concern is, however, the adverse effect of fallow, that is, the changes in soil organic matter (SOM) quality and quantity in the context of degradation of the fertility of chernozem soils and subsequent agricultural sustainability. The studies of [43-45] have demonstrated that fallowing significantly exacerbates the depletion of SOM. Organic C and N content of soil after 33 years of cropping decreased with increasing frequency of fallow in a rotation on Canadian soils (53).

The objectives of this study were to examine the effects of summer fallow on the characteristics of SOM on a long-term basis (length of crop rotation with a variety of frequencies of fallow) as well as on a short-term basis (pre- and post-fallow phases) with special reference to readily decomposable fractions.

To investigate the impact of bare fallow on soil SOM dynamics the five representatives fallow-spring wheat crop rotation were selected (2-year, 4-year and 6-year with one year of bare fallowing). Soil samples were collected from pre- (2R-pre, 4R-pre and 6R-pre) and post-fallow (2R-post, 4R-post and 6R-post) phases in each rotation. Also, for comparison the continuous cropping of spring wheat (CW) and continuous fallowing (CF) were sampled for comparison.

Table 1: Fertilization treatments in Uman experimental site from 10-year crop rotation in surface soil of*Argiudolls*, Ukraine

Treatments	Fertilization rates kg ha^{-1} N year^{-1}	Fertilizer	N applied per rotation (10 years), kg N ha^{-1} rotation^{-1}	
CON	no	no	no	no
M1	N_{45}	$(NH_4)_2SO_4$	450	450
M3	N_{135}	$(NH_4)_2SO_4$	1350	1350
O	Manure $N_{67.5}$	manure	675	675
MO1	N_{22}+ manure $N_{22.5}$	$(NH_4)_2SO_4$	225	450
		manure	225	
MO3	N_{45}	$(NH_4)_2SO_4$	675	1350
		manure	675	

*Amount of N in manure was calculated as: one ton of cattle manure contains approximately 5 kg of N

Table 2: Effect of fertilization treatments on soil organic C (SOC) and total N (TN) in surface soil of*Argiudolls*, Uman, Ukraine

Fertilization rates kg ha^{-1} year^{-1}	Treatments	Organic C	Total N	C/N ratio
		g kg^{-1} soil		
no	CON	20.4a	1.64a	12
N_{45}	M1	19.6a	1.62a	12
N_{135}	M3	20.8a	1.76a	12
Manure $N_{67.5}$	O	21.9b	1.77a	12
N_{22}+ manure $N_{22.5}$	MO1	20.4a	1.71a	12
N_{22} + manure $N_{67.5}$	MO3	20.1a	1.72a	12

Soil organic carbon (SOC) content was significantly affected by long-term fallowing. The CF system maintained the least SOC, while 6R and CW stored the most SOC (Table 3). SOC was inversely proportional to fallow frequency, indicating the negative effect of fallow on long-term accumulation of SOM. The effect of the rotations on total nitrogen (TN) paralleled that described for SOC. The highest TN concentrations were observed in the 6R and CW systems and lowest concentrations in the CF system.

To protect the field against weeds and to store more moisture and nutrients in the soil, fallowed field are cultivated 4 to 5 times during the vegetative season. Such intensive mechanical disturbance causes enhanced mineralization of SOM in fallow, firstly, due to better aeration of surface soil, and secondly, particular organic matter occluded within aggregates might become exposed to microbial attack after disruption of aggregates. Additionally, bare fallow does not contribute plant residues for the replenishment of SOM.

In general, distributions of SOC and TN among rotations with different fallow frequencies were comparable to those reported by [50-52] for Chernozem soils. Frequently fallowing systems such as 2R showed less SOM than less frequently fallowing systems, such as 6R. Our results confirmed the findings from North American arable systems that frequently fallowing system accelerates mineralization of SOM [51-53]

Table 3: Effects of fallow (F) frequency and rotation phase on soil organic C (SOC) and total N (TN) in surface soil of *Haplustolls*, Astana, Kazakhstan

Rotation phase	Rotation phase sampled	SOC	TN	C-to-N ratio
		kg Mg^{-1} soil		
CF	Cont. Fallow	21.9a[x]	1.97a	11
2R-pre	(F)[y]-W	25.4b	2.26b	11
2R-post	F-(W)	25.1b	2.16b	12
4R-pre	(F)-W-W-W	26.1b	2.26b	12
4R-post	F-(W)-W-W	24.9b	2.19b	11
6R-pre	(F)-W-W-W-W-W	31.0c	2.57c	12
6R-post	F-(W)-W-W-W-W	30.6c	2.50c	12
CW	Cont. W	27.2c	2.38c	13

[X]a-c: values within columns followed by the same letter are not significantly different (*P*=0.05) as determined by *LSD* analysis.

[Y] ( ) denotes rotation phase sampled.

IMPACT OF AGRICULTURAL PRACTICES ON LABILE SOM

Labile SOM fractions such as the "light" fraction C [14],microbial biomass carbon [15], mineralizable C [8, 16] are highly sensitive tochanges in C inputs to the soil and will provide a measurable change before any such change in total organic matter [17].In contrast, the more stable (humified) poolsare probably the more appropriate and representative fractions for C sequestration characterization [54].

Carbon mineralization potentials measured via soil respiration in a cascade measurement of the carbon dioxide efflux produced from soil metabolic processes consists of mainly microbial decomposition of soil organic matter and root respiration [55]. A great deal of research money and effort has been invested in studies of soil respiration in recent years because of the potential impacts of this process on the Greenhouse Effect [55]. Measurement of potentially mineralizable C represents a bioassay of labile organic matter using the indigenous microbial community to release labile organic fractions of C. Mineralizable N is also an important indicator of the capacity of the soil to supply N for crops. Individual labile organic matter fractions, such as easily mineralizable carbon and nitrogen, the microbial biomass and activity, are sensitive to changes in soil management and have specific effects on soil function [8]. Together they reflect the diverse but central effects that organic matter has on soil properties and processes.

Soil Labile OM in Fertilization and Experiment

In the 21st century the mineral fertilizers became determinant for obtaining contented yield of agricultural crops. However, the application of only mineral fertilizer might lead to accelerated mineralization of soil organic matter not mentioning the ecological aspects. Most scientist agree that prolonged application of manure either stabilizes the initial content of humus or increases its content, depending on the rates of applied manure [38, 56-58].

Soil Mineral Nitrogen in Fertilization Experiment

The study of the fertilization experiment showed that amount of soil mineral N (min-N) was in direct correlation with added N (Fig 3). The highest content of min-N was recorded in MO3 and M3 treatments, followed by O, MO1and M1 treatments (Figure 3). As shown in Table 1, M3 and MO3 treatments received the highest rate of N that was 1350 kg of N per ha per rotation that was the reason of the increased amount of min-N. Generally, min-N was distributed proportionally to the amount of applied N. But some difference was observed between application of mineral N alone and combination of N applied with mineral fertilizer and manure. For example, M1 and MO1 treatments received the same amount of N in whole rotation, where M1 treatment received only mineral N, and MO1 treatment received 50% N from the mineral fertilizer and 50% N from the manure. Similar pattern was observed in the case of M3 and MO3. Fertilizer N was quickly utilized by plants and microorganisms, while N of manure was decomposed more slowly supplying the soil with N for longer period.

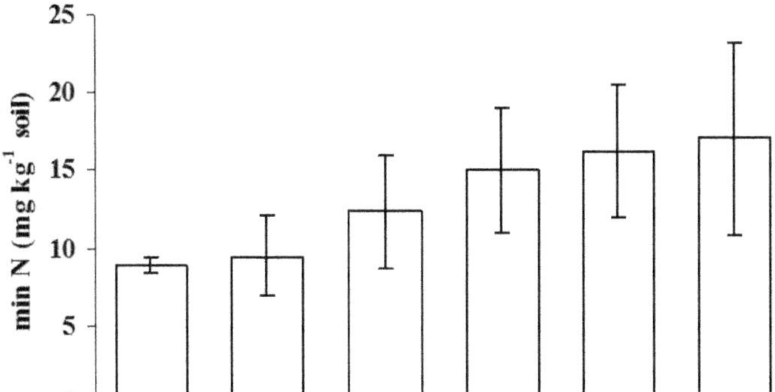

Figure 3: Soil mineral nitrogen at sampling time from fertilization treatments, Uman

The [56] reported that "ready" humic substances applied with cattle manure is thermodynamically non-stable and therefore is subject to faster decomposition and mineralization. Probably, these non-stable organic substances in manure were the main source of mineralized nitrogen resulting in increased accumulation of labile forms of N in manured experiment.

Nitrogen Mineralization Potentials in Fertilization Experiment

Mineralization rate constant (k) among the treatments varied significantly ($P<0.05$). The treatments where the high rates of manure wereapplied showed higher mineralization rate (Figure 4). All treatments but MO3 have lowered their mineralization rate by the end of incubation (56-70 days). MO3is the treatment that received mineral fertilizer and high rate of manure.Nitrogen of the mineral fertilizer serves as an easy available substrate for microorganisms at the beginning of the incubation then, after the available mineral nitrogen was depleted by microbial utilization, the nitrogen of the manure was exposed to microbial attack showing high mineralization rate after 70 days of incubation, while in O treatment, manure was attacked from the beginning because no mineral N was added to the soil. Manure consists of labile and non-labile fractions of organic compounds. After the labile fractions of manure were mineralized, the mineralization rate was slowed down thus showing lowered rate after eight weeks of incubation.

PMN content was the highest in the treatments where high rates of manure were applied that are O and MO3 (Table 4). Manure was applied about 19 months before the soil sampling. During about eight months the soil was frozen and no microbial activity was undergoing. It takes about 275 days to start releasing mineral N from manure, and about 391 days for complete mineralization or for reaching the stabilization point [59]. By the time of sampling manure had been releasing N for about 90 days, therefore, during the laboratory incubation manure continued to release mineral N, showing higher PMN in O and MO3 treatments.

Cattle manure contains "ready" humic substances that can be directly and immediately involved into immobilization processes. Probably, these non-stable organic substances in manure were the main source of mineralized N resulting in increased accumulation of labile forms of N under the manured treatments.

Table 4: Potentially mineralizable nitrogen as influenced by different fertilization, Uman

Treatment	**Fertilization rates**	**PMN**	**C/N**	**Soil total N as PMN, %**
	kg ha^{-1}yea^{r-}1	**mg k^{g-}1**		
CON	No	75.61	12.7	4.6
M1	$_{N4}5$	84.64	8.6	5.2
M3	$_{N13}5$	78.50	15.2	4.5
O	Manure $_{N67.}5$	96.44	12.0	5.4
MO1	N_{22} + manure $_{N22.}5$	88.17	13.7	5.2
MO3	N_{22}+ manure $_{N67.}5$	151.92	6.8	8.8

Table 5: Microbial biomass carbon (MBC) and nitrogen (MBN) as influenced by different fertilization, Uman

Treatment	**Fertilization rate kg ha^{-1} yea^{r-}1**	**MBC**	**MBN**	**C/N**	**Soil organic C and total N (%)**	
		mg kg^{-1} soil			**as MBC**	**as MBN**
CON	no	459a	33.3a	14	2.09	2.02
M1	$_{N4}5$	586b	35.5a	17	2.72	2.19
M3	$_{N13}5$	531c	21.3b	25	2.35	1.21
O3	Manure $_{N67.}5$	566bc	85.3c	7	2.40	4.81
MO1	N_{22}+ manure $_{N22.}5$	585b	71.5d	8	2.64	4.19
MO3	N_{22}+ manure $_{N67.}5$	677d	69.6d	10	3.10	4.05

Microbial Biomass Carbon and Nitrogen in Fertilization Experiment

Microbial biomass N (MBN) in the treatments where manure was applied amounted from 72 to 85 mg kg^{-1} soil (Table 5). In the treatments where no manure was applied MBN amounted from 21 to 36 mg kg^{-1} soil. Moreover, application of the high rate of mineral fertilizer alone decreased the MBN content. The ratio of MBC to MBN also shows distinctive difference between the treatments where high ratio was observed under the mineral fertilization and lower ratio under the manure application. Noticeably, highest C to N ratio was under the highest rate of mineral fertilization (M3) and the lowest was under the high rate of manure application (O3). This is due to intensive utilization of added nitrogen by microorganisms. Addition of biomass substrate with the content N more than 1.5 to 1,7% do not need additional fertilizer by nitrogen, the soil N satisfies the need of microorganisms during the decomposition. Always, the "requirements" of microorganisms are satisfied first of all, disregarding on the need of plants for nitrogen.

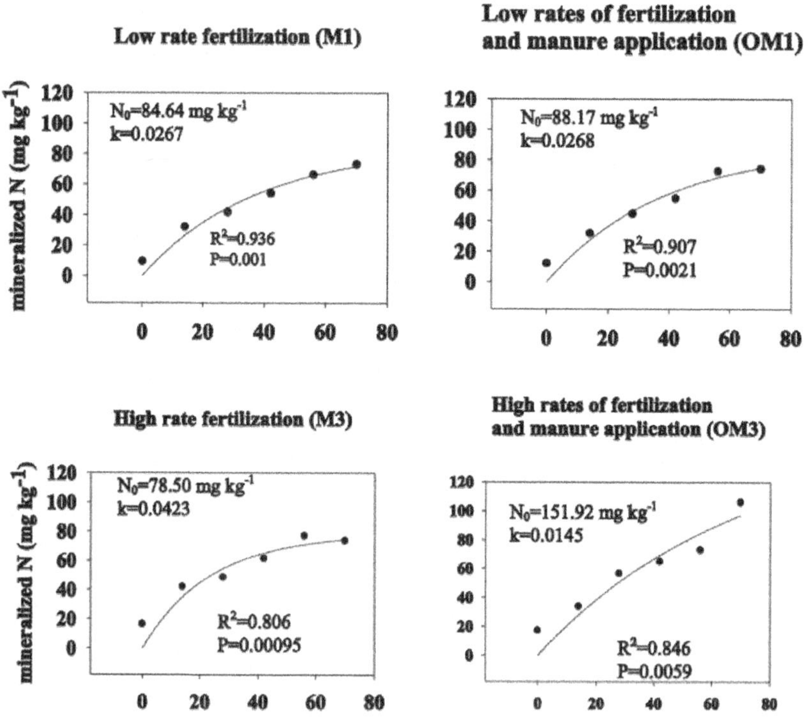

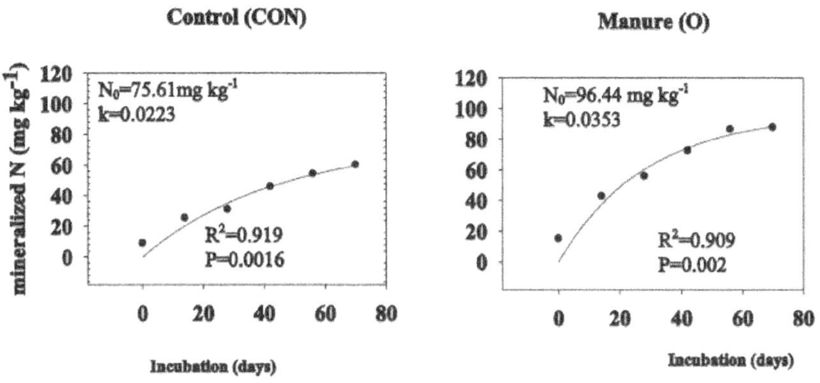

Figure 4: Fitting curves of nitrogen mineralization in fertilization experiment in U_{man}, Ukraine, as described with the first order kinetic model: $N_{min}=N_0(1-e^{-kt})$, where N_{min} is the mineralized N at time t, N_0 is the potentially mineralizable N (PMN), k is the mineralization rate constant.

Impact of Fertilization and Irrigation Practices on Total Soil Organic C and N

This study was conducted in long-term experiments with fertilization and irrigation in Kherson, south Ukraine. Sampling scheme of the Kherson experiments were 1) irrigated plus fertilized treatment (I+F), 2) irrigated only (I); 3) fertilized only (F) and 4) control that was neither fertilized nor irrigated (CON).

Analysis of variance showed that soil organic carbon (SOC) and total nitrogen (TN) were not statistically different among treatments. However, there were observed a different trend in accumulating SOM (Table 6). Contents of SOC and TN were for 7.19% and 9.30% respectively, greater in I+F treatments than in the control CON. Higher accumulation of SOC and TN under I+F treatment is due to higher biomass production under irrigation and fertilization. Consequently, higher plant biomass contributes to SOM. Fertilization (F) alone, or irrigation (I) alone maintained similar amount of organic C and total N. Kherson region is characterized with very small amount of precipitation, around 300 mm annually. In such dry conditions, fertilization does not pay off in terms biomass production because the applied fertilizer cannot be dissolved and be available for plant consumption.

Table 6: Organic carbon and total nitrogen concentration in irrigation experiment, Kherson

Treatments	Applied treatment	Organic C	Total N	C/N
		g kg-1 soil		
I+F	Irrigated and fertilized	16.7a	1.29a	13
F	Fertilized	16.4a	1.25a	13
I	Irrigated	15.8a	1.24a	13
CON	no	15.5a	1.17a	13

*Fertilizer was applied as $N_{120}P_{120}K_{120}$ for every fertilization treatment at rates indicated as subscripted mark.

*Irrigation water was applied at rate of 3200m^3 ha^{-1} for every irrigated treatment.

Soil Mineral Nitrogen in Irrigation Experiment

Content of soil mineral N was different among the treatments at P=0.1 (Fig. 5). As expected, the greatest differences were observed between I+F and non-irrigated treatments (F and CON).

Irrigated treatment accumulated higher min-N than the non-irrigated because irrigation of dry soil resulted in flash in microbial growth. During the long dry period the most of microorganisms dies, and then later upon irrigation those microbial necromass serves as easy source of N for survived microorganisms. The microorganism's body has the narrowest C-to-N ratio that is an indicator of the most easily available N source. Thus the flash in microbial growth accelerates mineralization processes in soil and moistening of dry soil causes disruption of organic compounds and soil particles that may contain organic substances. Subsequently, the disrupted organic material is more sensitive for microbial attack thus contributing to N mineralization.

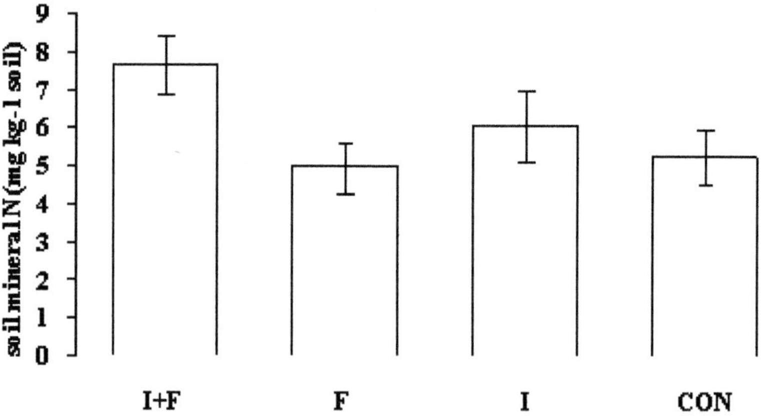

Figure 5: Soil mineral nitrogen in irrigation experiment, Kherson

Nitrogen Mineralization Potentials in Irrigation Experiment

Potentially mineralizable nitrogen (PMN) was significantly different (P=0.01) among the treatments with the highest mineralization rate under I+Ftreatment (k=0.0192) (Figure 6). The highest accumulation of mineralizable N (PMN) was also obtained under the I+Ftreatment (171.73 mg kg^{-1}) (Table 7), while all other treatments maintained statistically not different amounts of PMN.

I+F treatment maintained higher plant biomass returned into soil. Later when the soil was placed under the favourable laboratory conditions, those accumulated residues were subjected to mineralization showing higher PMN.

Fertilization alone (F) had suppressed mineralization on the field because of deficiency of water necessary for microbial activity. But when the soil was placed under favourable laboratory conditions the accumulated organic substrate is mineralized, thus giving nearly the same amount of PMN as the irrigated treatment.

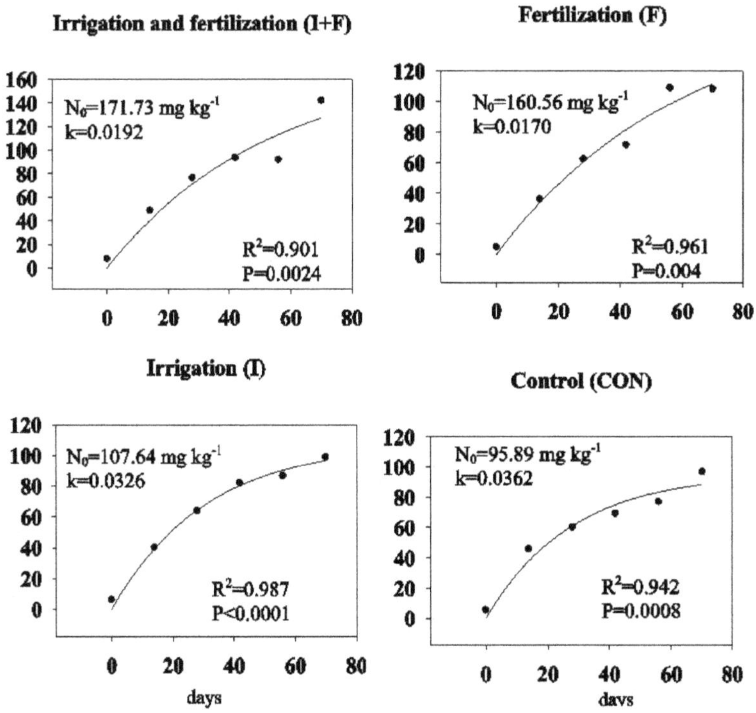

Figure 6: Fitting curves of nitrogen mineralization in fertilization experiment, Kherson, as described by the first order kinetic model: $Nmin=N_0(1-e^{-kt})$, where Nmin is the mineralized N at time t, N_0 is the potentially mineralizable N (PMN), k is the mineralization rate constant.

Carbon Mineralization Potentials in Irrigation Experiment

Irrigated plus fertilized treatment (I+F) showed the highest carbon mineralization rate as well as amount of PMC in 70d (Fig. 7; Table 7). Irrigation of dry soil disrupts soil structure thereby making previously sequestered carbon available for microbial utilization. The [60, 61] found that soil drying destroyed 1/3 to 1/4 of biomass and after remoistening the biomass was progressively restored to approximately the same size as before drying.

In this study, high temperatures and dry conditions have caused death of microorganisms that were immobilized during desiccation via adsorption on clay surfaces and/or transformation into other forms of organic compounds. Then, the following irrigation revived microbial community and disrupted soil clay particles that released stabilized organic matter. In the study conducted by [62] the similar results were obtained, where extra mineralized ^{14}C, due to soil desiccation came from non-living residues, likely to be those that were stabilized by adsorption to clay surfaces.

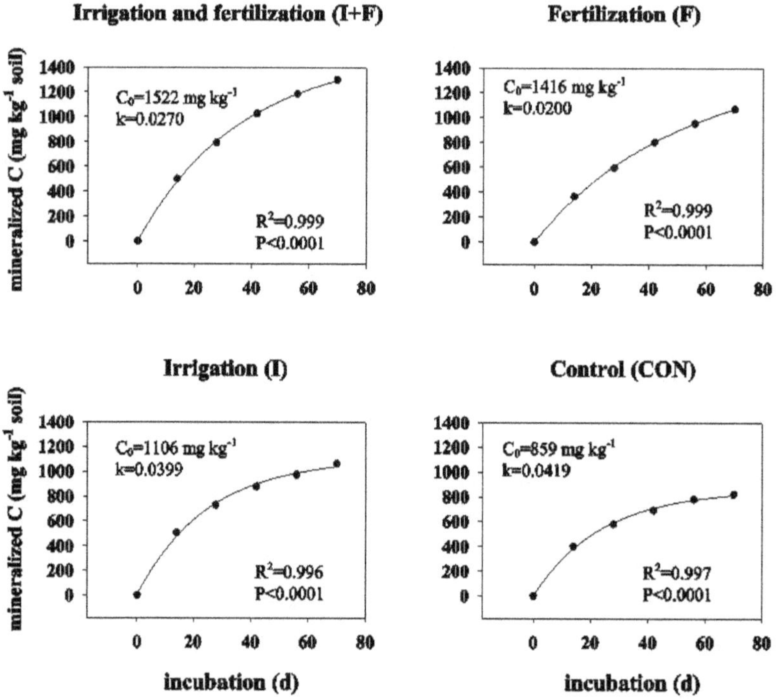

Figure 7: Fitting curves of carbon mineralization in fertilization experiment, Kherson, as described by the first order kinetic model: $C_{min}=C_0(1-e^{-kt})$, where C_{min} is the mineral-

ized C at time t, C_0 is the potentially mineralizable C (PMC), k is the mineralization rate constant.

The highest percentage of PMC and PMN in soil organic C were under I+F treatment (Table 7). This is in accordance with the earlier discussion and confirms the hypothesis that there are at least two reasons responsible for it: firstly, irrigation of dry soil causes enhanced mineralization of soil organic matter, and secondly, fertilization of irrigated soil provides higher plant biomass that contributes to the accumulation of labile organic matter.

Table 7: Mineralizable carbon and nitrogen (PMC and PMN) in irrigation experiment, Kherson

Treatments	**Applied treatment**	**PMC**	**PMN**	C/N	**Soil organic C and total N (%)**	
		mg kg^{-1} soil			**as PMC**	**as PMN**
I+F	Irrigated and fertilized	1522	171.73	8.9	9.11	13.31
F	Fertilized	1416	160.56	8.8	8.63	12.84
I	Irrigated	1105	107.64	10.2	6.99	8.68
CON	no	858	95.89	8.9	5.53	8.20

"Light Fraction" Organic Matter in Irrigation Experiment, Kherson

The I+Ftreatment maintained the highest amount of 'light' fraction dry matter (LFDM), carbon (LFC), nitrogen (LFN) and their proportions in soil organic carbon (SOC) and total nitrogen (TN) (Table 8). One of the reasons is, as discussed earlier, higher biomass production in this treatment, hence higher organic substrate was added with residues. Desiccation that caused the death of a large number of microorganisms, followed by immobilization and condensation of their dead tissues thus increasing the amount of recalcitrant, soluble organic C is another reason [60]. Moreover, irrigation of desiccated soil also causes the death of microorganisms due to the osmoregulatory shock [63] that also could contribute to the LFOM.

Microbial Biomass in Irrigation Experiment, Kherson

Microbial biomass carbon (MBC) and nitrogen (MBN) significantly differed among the treatments (Table 9). The highest MBC and MBN were obtained under the irrigation alone (I) treatment followed by the irrigated plus fertilized (I+F) treatment. The least microbial biomass was obtained under the fertilized alone (F) treatment.

Table 8: "Light fraction" dry matter (LFDM), carbon (LFC) and nitrogen (LFN) in irrigation experiment, Kherson

Treatment	**Applied treatment**	**LFDM**	**LFC**	**LFN**	C/N	**Soil organic C and total N (%)**	
		g mg^{-1} soil	**mg kg^{-1}soil**			**as LFC**	**as LFN**
I+F	Irrigated and fertilized	9.27	1884	122	15	11.3	9.4
F	Fertilized	8.52	1591	104	15	9.7	8.4
I	Irrigated	6.49	1589	104	15	10.0	8.4
CON	no	6.82	1684	108	16	10.8	9.2

Table 9: Microbial biomass in irrigation experiment, Kherson

Treatment	**Applied treatment**	**MBC**	**MBN**	C/N	**Soil organic C and total N (%)**	
		mg kg^{-1} soil			**as MBC**	**as MBN**
I+F	Irrigated and fertilized	618	160	4	3.7	12.4
F	Fertilized	450	76	6	2.7	6.1
I	Irrigated	733	175	4	4.6	14.2
CON	no	636	128	5	4.1	10.9

Such distribution of microbial biomass was expected because moisture conditions are a major factor controlling survival and activity of microorganisms in the soil [64]. Drying and remoistening of soils strongly affects microbial growth and activity [61, 65, 66]. After remoistening of dried soil, available C components were assimilated and transformed partly into new biomass C, and partly involved into CO_2 that evolved into the atmosphere [63].

General Discussion of Fertilization, Manure Application and Irrigation Experiments

Many researchers recorded positive effects of manure application on SOM [38, 57, 67-69]. For example, in Nebraska, annual application of 13.5 t ha^{-1} of manure (dry matter) during 31 years on irrigated land has increased content of humus from 0.98 to 1.67%. The [59] found out that increased application of manure resulted in intensification of C mineralization, especially the C that is included in fulvic acids, and in lesser extent in humic acids. Biological analysis showed that application of high rates of manure activates the biochemical processes, which is controlled by particular microbiological community that has ability for active transformations not of only simple organic substances (e.g. fulvic acids), but also of more complex and hardly decomposable substances (e.g. humic acid).

Increased microbial activity in irrigated treatments in Kherson has been ascribed to the rapid metabolization of biomass-derived substrate resulting from the death of part of the microbial community during drying [61-63, 70] and/or rapid rewetting of the desiccated soil material [100]. Alternate drying and re-moistening increases the mobility of organic matter and results in the release of N as ammonium and amides [42].

Impact of Bare Fallow on Soil Labile Organic Matter in Haplustolls,Astana, Kazakhstan

With increasing cultivation intensity, the SOM of the less stable pools is decomposed, as indicated by decreasing portions of sand-sized SOM (2-0.05 mm) [19, 46, 47], or light fraction C [48, 49). Organic compounds adsorbed to surfaces of clay particles might become exposed to microbial attack after disruption of aggregates due to tillage.

To investigate the impact of bare fallow on soil SOM dynamics the five representatives fallow-spring wheat crop rotation were selected (2-year, 4-year and 6-year with one year of bare fallowing). Soil samples were collected from pre- (2R-pre, 4R-pre and 6R-pre) and post-fallow (2R-post, 4R-post and 6R-post) phases in each rotation. The continuous fallowing (CF) and continuous cropping of wheat (CW) were also sampled to see the effect of fallow impact on SOM.

SOIL Mineral Nitrogen in Fallow Frequency Experiment

On a long-term basis, the CF system accumulated the highest amount of soil mineral nitrogen (min-N). But min-N was strongly affected by summer fallow on a short-term basis as well. Pre- and post-fallow phases showed significant differences with min-N accumulating in post-fallow than in pre-fallow phase (Table 10). Post-fallow phases accumulated 3.0-, 1.9- and 1.9-fold amounts of min-N of pre-fallow phases in 2R, 4R and 6R, respectively

As expected, the CF system maintained the highest amount of soil min-N that was due to enhanced mineralization of SOM compared to the other systems. The short-term effect of fallow on the accumulation of min-N is clearly observed as well. During fallow phase min-N is not subjected to either plant uptake or leaching, thus resulting in a greater accumulation of soil min-N in post-fallow (2R-post, 4R-post and 6R-post) than in pre-fallow phases (2R-pre, 4R-pre and 6R-pre).

Nitrogen Mineralization Potentials in Fallow Frequency Experiment (PMN)

The pattern of N mineralization showed a different trend between pre- and post-fallow phases in all rotations (Fig. 8). Pre-fallow phases (Fig.8. a, c and e) were characterized by a larger value of PMN (N_0), a smaller mineralization rate constant (k), and a shorter initial delay of mineralization (c) than in the post-fallow phases (Fig.8. b, d, and f).

Fallow influenced accumulation of PMN on short-term basis, that is, pre-fallow phases (2R-pre, 4R-pre and 6R-pre) accumulated more PMN than post-fallow (2R-post, 4R-post and 6R-post) phases (Table 10). The lowest PMN was observed under the CF system (69 mg kg^{-1}) and the highest under 6R-pre (124 mg kg^{-1}). Pre-fallow phases accumulated 2.4, 1.3 and 1.5 fold amount of PMN of post-fallow phases in 2R, 4R and 6R, respectively. Larger amounts of mineralized nitrogen (N_0) in the pre-fallow phases indicate larger storage of PMN in these soils than in post-fallow soils. Differences in the rate constant (k) between pre- and post-fallow phases indicate that fallowing has caused changes in the quality of the PMN. Due to multiple cultivations of fallows the soil is subjected to alternating wet-dry cycles. The wet period provided better moisture condition microorganism activity and produced greater biomass than in cropped fields. Then in the subsequent dry period the greater biomass turned into necromass due to drought. This cycle may be repeated several times in a cropping season. Later, during incubation in the laboratory, this microbial necromass as well as living biomass was rapidly mineralized showing a higher mineralization rate constant in the post fallow than in the pre-fallow phases. The soils from the post-fallow phase showed a longer initial delay of mineralization, suggesting that higher concentration of min-N compared to pre-fallow phase probably stimulated microbial activity and resulted in immobilization of mineralized N during the initial stages of incubation.

The long-term effect of fallow was not observed for soil min-N or PMN suggesting that N mineralization is only affected by the substrate added during the previous year or the latest cycle of rotation. Nitrogen in the forms of NO_3^- and NH_4^+ is assimilated by plants and returned into soil whereas C originates from CO_2 in the air and ploughed as organic residue into soil. Nitrogen transformations are closely related to the processes of mineralization of its organic forms in plant-soil system. Therefore, in plant-soil systems N cycling is affected over shorter period than C cycling.

Table 10: Effect of fallow frequency and rotation phase on labile organic matter content, Astana

Rotation phase	**PMC**	**min-N**	**PMN**	**C-to-N ratio**	**Organic C as PMC**	**Total N as PMN**
	mg kg^{-1} soil				%	
CF	794bx	46a	69a	11	3.6	3.5
2R-pre	1194ab	14b	166b	7	4.7	7.4
2R-post	1012b	42a	69a	15	4.0	3.2
4R-pre	1224a	13b	86c	14	4.7	3.8
4R-post	1215a	24c	67a	18	4.9	3.1
6R-pre	1524c	16b	124b	12	4.9	4.8
6R-post	1300ac	30c	82c	16	4.3	3.3
CW	1581c	14b	93c	17	5.8	3.9

Carbon Mineralization Potentials in Fallow Frequency Experiment, Astana

Differences in PMC among the rotation systems ($P<0.001$) were more clearly shown than for SOC. PMC ranged from 3.6 (CF) to 5.8% (CW) of the SOC.

The amount of PMC was more affected by the long-term effect of fallow than by the short-term effect and was inversely proportional to fallow frequency.

Continuous wheat (CW) and 6-y systems (6R) had higher amount of PMC that was inversely proportional to fallow frequency and indicated the long-term effect of fallow. These results corroborate with the study conducted by [73] who found for a silt-loam in southwestern Saskatchewan that mineralized C (measured after 30 days at 21°C) represented 1.06 and 1.45% of SOC in a 2-y fallow-wheat rotation and continuous growing of wheat, respectively. The [74] found that C mineralization was not related to the amount of crop residue from the previous year. In our study PMC was a little higher in the pre-fallow (2R-pre, 4R-pre and 6R-pre) than in the post-fallow (2R-post, 4R-post and 6R-post) phases, probably reflecting the input of crop and weed residues in the preceding year.

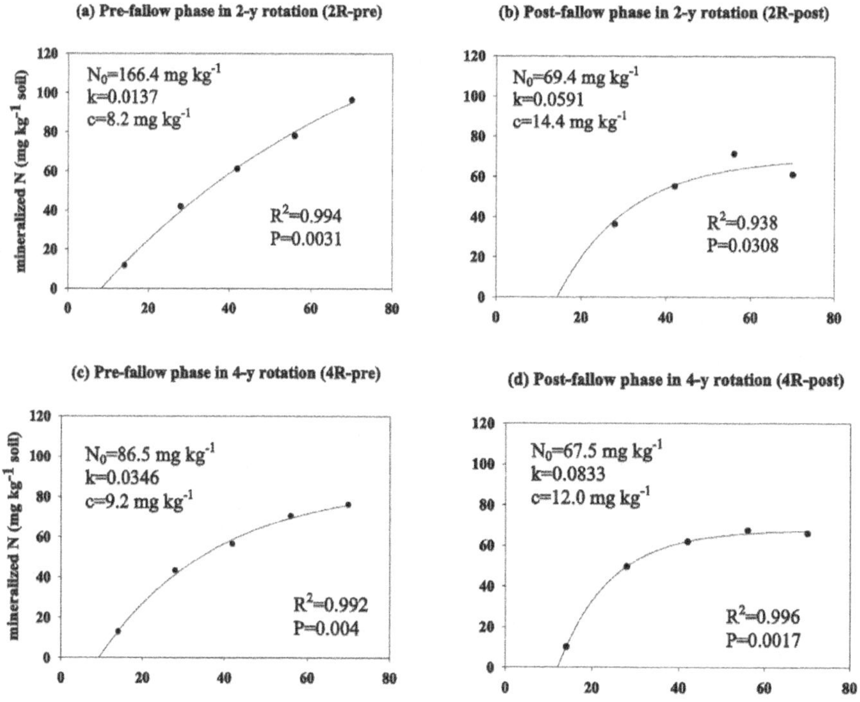

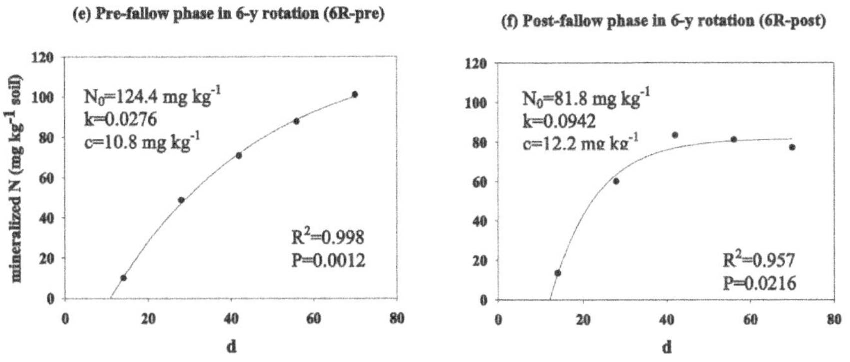

Figure 8: Fitting curves of N mineralization from pre- and post-fallow phases of the 2-, 4-, and 6-year wheat-fallow 40-years rotation experiment, as described by the first order kinetic model with an initial delay of mineralization ($N_{min} = N_0(1\text{-}e^{-kt-c})$, where, N_{min} is mineralized N at time t,;N_0 is value of potentially mineralizable N (PMN), k is a mineralization rate constant, and c is an initial delay in mineralization.

Light Fraction" Organic Matter in Fallow Frequency Experiment, Astana

The amount LF-OM was highly responsive to fallow frequency, accounting for 1.1(CF)-6.3(CW)% of the SOC and 0.8(CF)-4.3(CW)% of the TN (Table 11). LF-OM, as expressed on the basis of dry matter (LF-DM), C (LF-C) or N (LF-N), was inversely related to fallow frequency. For example, the LF-C content of the CW system was 7.2 times higher than that in the CF system. These results agree with those of other studies [e.g. 36, 75], where LF content was highest under continuous cropping and lowest in those with a high frequency of summer fallow. Additionally, LF-C was affected by the rotation phase, showing larger amounts in pre- than in post-fallow phases in 4R and 6R rotations.

"Light fraction" of SOM (LF-OM) consists mainly of plant residues, small animals and microorganisms adhering to plant-derived particulate matter at various stages of decomposition that serves as a readily decomposable substrate for soil microorganisms and also as a short-term reservoir of plant nutrients [76]. The "light fraction' C (LF-C) was positively correlated with PMC (Fig. 9), and confirm the hypothesis that the reduced fallowing system has more potential to supply soil with easily mineralizable C. However, there was no linear correlation between LF-N and PMN, presumably because the high C-to-N ratio of the LF-OM temporary induced N immobilization [36].

The content of labile OM, which is closely related to LF-OM, may be governed by the degree to which temperature and moisture conditions constrain decomposition of accumulated residues [52]. Under the CW system decomposition of residues during periods of favourable soil temperature was retarded by the depleted soil moisture [77, 78,]. Then, when moisture and temperature constraints were removed during laboratory incubations, soil showed a high respiration rate [36]. On the contrary, residues in the 2R system during the fallow phase were always exposed to an extended period with favourable moisture and temperature. Therefore, labile organic matter was rapidly depleted in the field, and in the laboratory respiration rates were much lower in 2R than in CW [52].

Grain Yield and Weed Biomass

The main controller of biological activity in soil is the SOM generated from crop residue, while crop residues are in direct correlation with crop yield and sometimes with yield of weeds. In systems with no pesticide application, such as control treatments, or in case of the CW (continuous wheat cropping), where no break in lifecycle of weeds occur, the weed biomass might significantly contribute to the input of crop residue in soil and affect the crop yield output.

Table 11: Effects of fallow frequency and rotation phase on the amount of "light fraction" dry matter (LF-DM), "light fraction" C and N (LF-C and LF-N) and their proportions in SOC and TN in surface soil of Haplustolls, Astana

Rotation phase	LF-DM	LF-C	LF-N	C-to-N ratio	Soil C as LF-C	Soil N as LF-N
	$g\ kg^{-1}$ soil	$mg\ kg^{-1}$ soil			%	
CF	0.9	240a[x]	15a	16	1.1	0.8
2R-pre	3.6	810b	51b	18	3.2	2.2
2R-post	2.5	660b	38bc	16	2.6	1.7
4R-pre	5.7	1330c	81d	17	5.1	3.6
4R-post	5.3	1250c	73d	17	5.0	3.3
6R-pre	6.4	1560d	74d	20	5.1	3.0
6R-post	6.0	1500d	75d	21	4.9	2.9
CW	7.4	1730e	103e	17	6.3	4.3

[x]a-e: values within columns followed by the same letter are not significantly different ($P<0.005$) as determined by *LSD* analysis.

The highest grain yield was produced in the first year of the rotations (the year after fallow) (Figure 10). However, the yield decreased sharply with in the second and successive years after fallowing. This trend is, firstly, because plants in a post-fallow phase take advantage of higher soil min-N. Secondly, because when a field is in fallow provides the only break for weed infestation; the amount of weeds was generally least in the first year after fallow and reduced competition for nutrients.

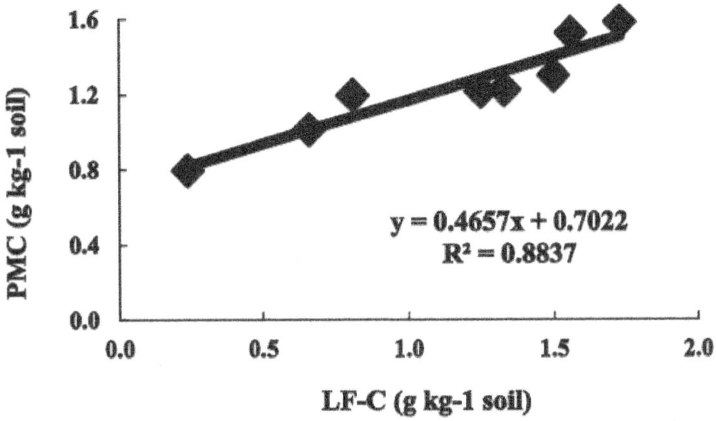

(a)

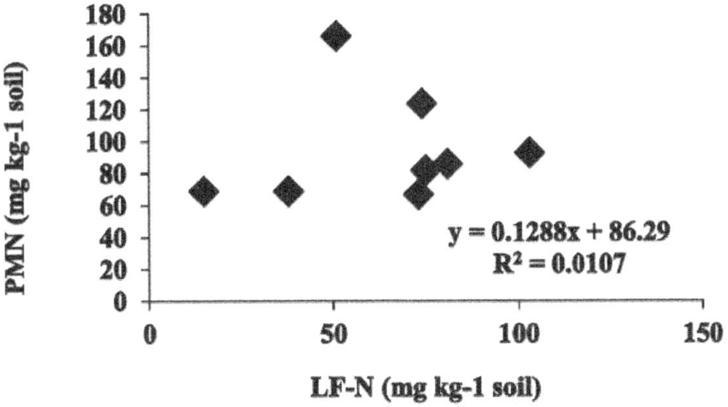

(b)

Figure 9: Correlation of potential mineralizable C and N) with “light” fraction C (LF-C) and N (LF-N)

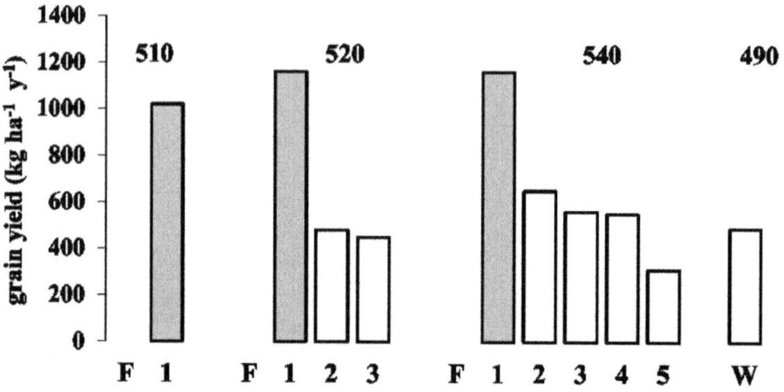

Figure 10: Grain yield (1994-2000) of spring wheat affected by distance from fallowing year. (F is fallow; 1, 2, 3…5 are succession of crops after fallow. (the numbers above bars are average yield per rotation including the fallow year).

In contrast to the grain yield, weed infestation reached its maximum in the second years after fallow in 4R and 6R. Probably, some of the weeds were not destroyed during the fallow and their seeds remained dormant but germinated in the second year after fallow. Correlation between the grain yield and weed infestation (average for ten years) is presented by the following equation of multiple linear regression:

$Y = 20.82 - 0.189X,$

where X total amount of weeds, pieces/m^{-2}. The coefficient of determination was also high ($R^2 = 0.78$) or 78% of changes of the yield depend on weed infestation.

The highest grain output per whole rotation, was obtained by 6-year rotation, being 540 kg ha^{-1} that parallels the distribution of soil labile C (PMC and LF-C). These results indicate that longer rotations with fewer fallows contribute more to the accumulation of SOM than shorter rotations with frequent fallowing, and that SOM is replenished continuously under less bare-fallowing system.

The results on *Mollisols* in North Kazakhstan suggested that N dynamics were closely related to the recent input of substrate added as plant residue while C dynamics were more related to long-term substrate addition.

Yearly input of plant residue in a 6-y wheat-fallow rotation system built up more labile OM, especially LF-C or readily decomposable C, whereas 2-y rotation system with a high frequency of fallow depleted SOM via accelerated mineralization. Therefore, with no fertilizer or pesticides application, in the semi-arid regions of northern Kazakhstan, the inclusion of fallow in wheat monoculture every 6 years is the most appropriate farming system in terms of sustainability in both grain production and soil fertility.

The results of this study may provide prediction of SOM response to fallow frequency in wheat-based rotation systems in Chernozem soils of semi-arid regions: the susceptibility of labile fractions of OM and their relationship to fallow frequency suggest the possibility of managing labile OM through controlling the length of wheat-fallow rotation systems.

CHARACTERIZATION OF SOIL ORGANIC MATTER DYNAMICS FROM DIFFERENT CLIMATIC ZONES

Many studies have shown that climatic factors, namely temperature and precipitation are major determinants of microbial diversity and activity in soil at global [79[, regional [5, 80] and local [81] scales [47, 82]. The great meridian and latitudinal extension of Mollisols [83] determines a wide variety of climatic conditions that influence the main genetic characteristics of soils and their natural growth and agronomic properties. The aim of this study was to find out effects of temperature and moisture on soil organic matter accumulation and decomposition.The study conducted by [109] also showed that in dry and cold conditions of dry steppe, soil respiration was mostly controlled by soil temperature while residue input was a function of moisture conditions.

Four types of Mollisols from four different climatic regions were sampled: *Hupludolls* (southern forest-steppe, Kharkov, Ukraine), *Argiudolls* (northern forest-steppe, Uman, Ukraine), *Calciustolls*(southern steppe, Kherson, Ukraine) and *Haplustolls*(northern steppe, Astana, Kazakhstan) (Table 12). The sampling sites represent the most typical soil type and ecosystem for each given region. The selected geographical regions are characterized as wet-frigid (Kharkov; 6.5°C, mean annual temperature, 542 mm, mean annual precipitation), wet-mesic (Uman; 8.5°C, 660 mm), dry-thermic (Kherson; 11°C, 332 mm) and dry-frigid (Astana; 0°C, 324 mm).

Table 12: Soil Types and Climatic Characteristics of the Study Sites

Site	**Geographical coordinates**	**Precipitation (mm)**	**Mean temperature, (°C)**		**Ecological and climatic region**		**Soil Taxonomy, USDA**
			winter	**summer**			
Kharkov	50°N, 36°E	515-570	-10	+18	southern forest-steppe	wet-frigid	*Hapludolls*

Uman	48.8°N, 30.2°E	550-770	-5	+17	northern forest-steppe	wet-mesic	*Argiudolls*
Kherson	46.6°N, 32.6°E	315-350	0	+22	southern steppe	dry-thermic	*Calciustolls*
Astana	51.3°N, 71.1°E	300-350	-18	+19	northern steppe	dry-frigid	*Haplustolls*

Soil Organic C and Total N in Different Climatic Zones

The highest amount of soil organic carbon (SOC) and total nitrogen (TN) was observed in wet-frigid (Kharkov) region, 26.8 and 2.50 g kg^{-1} soil, respectively, and the lowest in dry-mesic (Kherson) region with 15.3 and 1.24 g kg^{-1} soil, respectively (Table 13). This is mainly due to the inherently higher humus content in comparison with other studied soils. These results agree with previously reported studies where the stock of SOC was generally greater in colder and wetter compared to hotter and drier climates [28, 84]. In our study the TN content was significantly higher in frigid (Kharkov and Astana) than in mesic and thermic (Uman and Kherson) regions. Higher amounts of precipitation generally lead to a higher plant biomass production and organic C input [e.g. 15]. In addition, lower temperatures, especially in winter when it falls below a threshold for biological activity, limits decomposition of SOM resulting in accumulation over time [84].

Lack of water in the dry-frigid (Astana) region retarded mineralization of plant residues [8] while low winter temperatures conserved plant residues that were partially mineralized when the temperature was favorable, partially immobilized and partially accumulated as a labile organic matter [85].

In wet-mesicUman, the relative temperature sensitivity of decomposition was greater than the net primary productivity [3], where the higher amount of precipitation naturally produced a greater plant biomass that was quickly decomposed due to favorable temperature and moisture conditions. Dry and hot condition in Kherson suppressed the production of plant biomass and limited accumulation of SOM.

Table 13: General properties of different types of Mollisols from different

Soil type and hydrothermal regime	Total N	Organic C	C/N	pH	$CaCO_3$	Sand (200-20μm)	Silt (20-2μm)	Clay (<2μm)
	g kg^{-1} soil				%		%	
Hapludolls; Wet-frigid (n=24)	2.50	26.8	11	6.3	0	20.4	37.5	42.1
Argiudolls; Wet-mesic (n=24)	1.70	20.5	12	5.6	0	22.9	37.7	39.4
Calciustolls; Dry-thermic (n=12)	1.24	15.3	12	7.5	5.5	43.4	26.9	29.7
Haplustolls; Dry-frigid (n=24)	2.29	20.0	9	8.2	1.6	25.6	30.6	43.8

Labile C Fractions of Soil Organic Matter from Mollisols in Different Climatic Zones

Potentially Mineralizable C and Rate Constant K from Mollisols in Different Climatic Zones

Drier regions accumulated a higher amount of potentially mineralizable carbon (PMC) that is presented as C_0 in fitting curves (dry-thermic and dry-frigid, 1225 and 1222 mg kg^{-1} soil, respectively) than of wetter regions (wet-frigid and wet-mesic, 754 and 1091 mg kg^{-1} soil, respectively) (Fig. 11). Corresponding rate constant (k) followed a similar trend as PMC (k value of 0.031 and 0.026 in dry-thermic and dry-frigid, respectively; 0.013 and 0.021 in wet-frigid and wet-mesic, respectively). The shape of the fitting curve of Kharkov (wet-frigid) soil greatly differed from others by being the lowest, i.e. reflecting the least amount of mineralized C, and the straightest, i.e. corresponding with the slowest rate of decomposition. On the other extreme, Kherson (dry-thermic) and Astana (dry-frigid) soils showed the highest and the most curved shapes of the fitting curves that reflect the greater amount of mineralized C and faster decomposition rate in 70 days of incubation.

Moisture was the main factor influencing the amounts of soil labile C. In wetter regions (Kharkov and Uman) microbial respiration is always higher [66], and more organic substrate was utilized than in drier regions. In contrast, soil microorganisms from drier regions (Astana and Kherson) experienced moisture deficiency and were unable to use the existing available organic substrate. Consequently, when microbial activity was not limited by moisture during the laboratory incubation there was enough energy substrate to promote a high respiration rate. Additionally, in dry conditions lethal effects contributed "dead biomass" to the organic substrate pool [66, 86, 87]. This easily available substrate was rapidly taken up and utilized by surviving soil microorganisms, thus contributing to the increased soil respiration observed when soils from dry regions were moistened [66]. Severe moisture conditions in Astana and Kherson, firstly, enhanced turnover of MB and condensation of microbial products, thus increasing the amount of soluble C [60], and secondly, caused disruption of soil aggregates that resulted in the liberation of protected organic C [88].

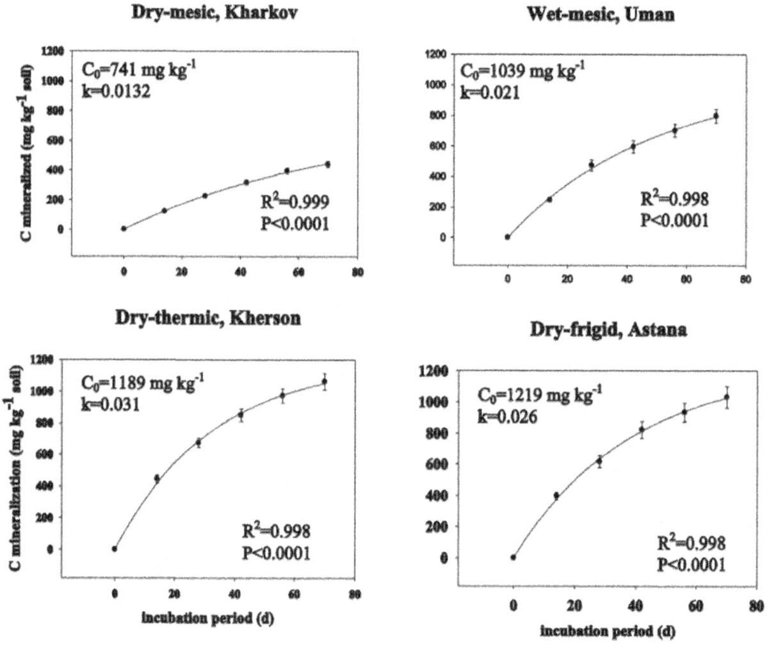

Figure 11: Fitting curves of C mineralization in soils from different climatic regions as described with the first order kinetic model (Cmin= $C_0(1\text{-}e^{-kt})$, where, Cmin is a mineralized C at time *t*, C_0 is a potentially mineralizable C (PMC), k is a mineralization rate constant)

Site variation in Kharkov and Uman were not significant (Figure 12). However, dry regions (Kherson and Astana) showed the biggest variation in the amount of potentially mineralizable C (PMC) and *k*values within the site. This indicates that SOM of dry regions are more sensitive to the imposed agronomic treatments than wetter regions. As discussed in previous sections, in drier regions the microbial activity is suppressed by lack of water. Then, when the water limitation is excluded the flash of mineralization take a place, where rapidly growing microorganisms compete for the N thus involving the more stable SOM into the mineralization process. Therefore, the vulnerability of SOM in drier regions should be considered when designing the agronomic treatments.

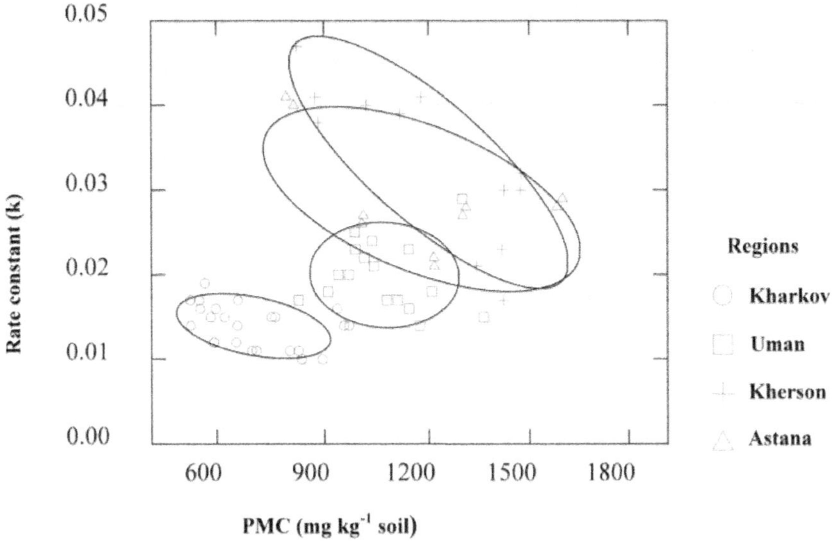

Figure 12: Scatter-plot of mineralized C and decomposition rate constant (k) in *Mollisols* from different climatic regions

"Light" Fraction Organic Matter (Lf-C) From Mollisols in Different Climatic Zones

The highest amount of "light" fraction carbon (LFC) was observed in the dry-thermic region (1687 mg kg^{-1} soil, Kherson), followed by the dry-frigid region (1436 mg kg-1 soil, Astana), and the least amount of LFC was observed in wet regions (1180 and 1105 mg kg-1 soil in Kharkov and Uman, respectively) (Table 14). Obviously, drier conditions in Kherson and Astana retarded decomposition of SOM, contributing to accumulation of LFC. Generally, distribution of LFC among sites was well correlated with PMC (r^2= 0.619). The amount of LFC was more affected by precipitation rather than by temperature. These results correspond to studies of, for example, [77, 78] who reported that decomposition of SOM during the period of favourable soil temperature is inhibited by lack of water. In wetter Kharkov and Uman, favorable moisture and temperatures during vegetation season promoted mineralization of "light fraction" OM resulting in less accumulation of LFC.

Table 14: Labile Carbon Fractions in different types of *Mollisols*

Site	MBC		PMC		LFC	
	mg/kg	**% of SOM**	**mg/kg**	**% of SOM**	**mg/kg**	**% of SOM**
Wet-frigid (n=24)	203±18	0.80	754±39	3.00	1180±31	4.65
Wet-mesic (n=24)	206±16	1.00	1091±34	5.10	1106±49	5.39
Dry-thermic (n=12)	281±31	1.84	1225±56	8.00	1687±67	11.03
Dry-frigid (n=24)	309±36	1.54	1222±50	6.16	1436±58	7.18

Microbial Biomass Carbon in Mollisols from Different Climatic Zones

Microbial biomass C (MBC) was significantly higher under drier (281 and 309 mg kg^{-1} soil in Kherson and Astana, respectively) comparing with wetter (203 and 206 mg kg^{-1} soil in Kharkov and Uman, respectively) regions (Table 14). The effect of temperature on microbial biomass was not clearly observed in this study. As shown by data of LFC, drier regions accumulated a greater amount of organic substrate that favoured accumulation of microbial biomass due to its availability as energy source. However, because microbial activity and survival are in direct physiological dependence on available water [87], the moisture deficiency during the summer period retarded microbial activity and caused death of moisture–sensitive microorganisms. After remoistening of the soils, the inhibition of microbial activity by dry conditions was reactivated and the available necromass was rapidly metabolized by soil microorganisms leading to the higher accumulation of MBC in drier regions.

Changes in the relative contribution of bacteria and fungi to soil respiration occur as soil dries [66] Kharkov and Uman soils normally undergo less severe fluctuations in water potential that Kherson and Astana. [89, 90] have shown that bacterial activity is largely restricted to water films in soil in contrast to fungi activity. Hyphae extension occurs at much lower potentials allowing fungi to bridge air-filled pores and actively explore for nutrients [91].

Relationship between Soil Organic Matter and Clay Content

The inert carbon is strongly correlated with clay content, while most changes in both carbon and nitrogen occur in the readily decomposable fraction [92].

Firstly, clay minerals can absorb large organic molecules directly, reducing their availability to decomposition. Secondly, organic material may be located in pores too small for microorganisms to enter [93-97).

In this study, clay content was highest in Kharkov and Astana regions (43.1% and 48.8%, respectively) versus Uman and Kherson (39.4% and 29.7%, respectively) (Table 15). It is reasonable to conclude that higher clay content and plant biomass production in Kharkov maintained higher SOM. [92] reported that inert carbon was strongly correlated with clay content, while most changes in both carbon and nitrogen occur in the readily decomposable fraction. [98] determined that “light fraction” (LF) of fine silt and coarse clay was more humified and more aromatic than other LF, concluding that LF represents a continuum of undecomposed to highly humified materials. Sites with higher silt fraction (2-0.2µm) that are Kharkov and Uman (37.5% and 37.7%, respectively) might form organo-mineral complexes with large molecules of LF, where those mineral-associated LF probably were not retrieved from these soils during the separation procedure, whereas, Kherson and Astana contained less silt fraction (26.9% and 26.6%, respectively) that could entrap LF, resulting in higher LFOM in these soils. However, although Astana soil showed the highest content of clay the SOM in this soil was less than in Kharkov. This is explained by the lack of water in dry-frigid Astana that produces less plant biomass, and inhibits mineralization processes contributing to the accumulation of labile OM, which explains higher PMC content in this soil. Also, organic compounds adsorbed to surfaces of clay particles become exposed to microbial attack after disruption of aggregates due to severe dry-wet conditions on soil in Astana [34]. The lowest clay content (29.7%) and lack of water in Kherson can explain the lowest SOM content in this soil, since the dry-thermic conditions don’t contribute to high plant biomass.

Table 15: Granulometric composition of studied soil from different climatic regions

Regions	**Siol organic carbon**	**Sand**	**Silt**	**Clay**
	g kg^{-1}soil	**200-20µm**	**20-2µm**	**<2µm**
Kharkov (n=24)	25.4	20.4	37.5	42.1
Uman (n=18)	20.5	22.9	37.7	39.4
Kherson (n=12)	15.3	43.4	26.9	29.7
Astana (n=24)	20.0	25.6	30.6	43.8

General Discussion of Effect of Moisture and Temperature on SOM In Mollisols

Total SOM among the four regions was distributed as follows: dry-thermic < dry-frigid ≤ wet-mesic< wet-frigid. While the labile OM distributed oppositely, as follows: dry-thermic ≥ dry-frigid > wet-mesic> wet-frigid. In Figure 13 the proportions of labile and stable carbon in the studied regions is presented. The highest amount of stable C and the least amount of labile C was found in wet-frigid (Kharkov) region, while the least amount of stable and the greatest amount of labile C was found in dry-thermic (Kherson) region. Because wet-frigid (Kharkov) region maintained the highest amount of total SOC and the least amount of easily mineralizable organic matter (PMC), the suggestion is: in wet-frigid region transformation of organic substrates into more stable humified forms of OM has taken place more actively.

Higher precipitation contributed stronger to the amount of total SOM, while plant biomass production under drier conditions in a lesser degree was subjected to decomposition due to moisture deficiency, thus contributing to the amount of labile SOM. The amount of C reflecting the labile fraction increases with increasing temperature, while the amount of recalcitrant C more controlled by low temperature.

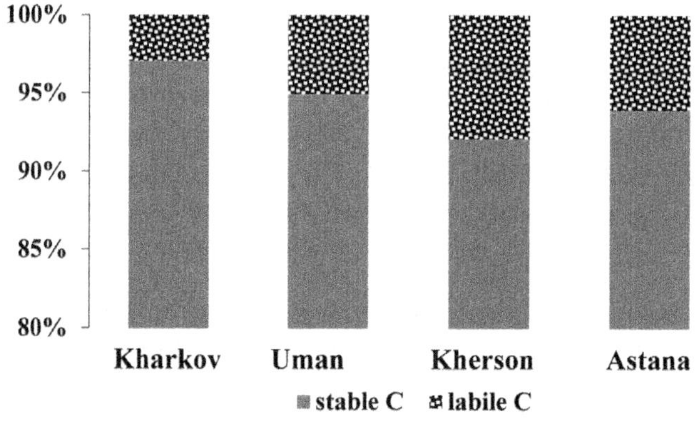

Figure 13: Distribution of labile and stable C in different *Mollisols* from four studied sites

Readily decomposable substrates were also found to originate partially from non-living SOM [62, 63]. This source of non-biomass substrate may become available by aggregate disruption, litter defragmentation and substrate desorption, and redistribution of water, oxygen, substrate and microorganisms resulting from drying and rewetting of soil [62, 63, 65, 99, 100]. Soil drying and rewetting promotes the turnover of carbon derived from added plant material [62, 63]. Drier condition causes more disruptions of entrapped or stabilized organic matter when the soil is rewetted. Also, higher respiration in the soils exposed to wet-dry cycles may have been due to utilization of organic substrate that were gradually built up due to limited microbial activity when the soil was air-dried [66].

CARBON TO NITROGEN RATIO (C TO N) OF PLANT RESIDUES

Quite precise integral index of quality of organic matter, from which the intensity of its decomposition depends, appears the ratio of carbon to nitrogen (C: N). Plant remainders with the wide ratio C to N do not ensure the sufficiency of nitrogen for the metabolism of microorganisms at their high activity. When the rapidly metabolized substrata (carbohydrates) are depleted, the limitation of nourishment is changed from nitrogen to carbon.

"Critical" C to N ratio, which characterizes accessibility to the microorganisms of the nutrients contained in the remainders, and their influence on the soil fertility, varies from 15 to 30 depending on the pool of mineral nitrogen in the soil, the quality of organic matter, duration of their of the decomposition. With an increase in the value indicated the processes of decomposition slow down and the immobilization of nitrogen occurs. With the smaller C:N values the intensive mineralization of plant litter occurs as a result of the invigoration of the activities of the microflora. The studies on the*Mollisols* showed that among the studied crops the maximum speed of decomposition was for the postharvest remainders of alfalfa and pea, which have the narrowest C to N ratio, 19,4 and 26,8 respectively. The percentage of the decomposition of pea in the first two months was 47,2, in seven months - 52,8, in a year - 82,4. The lowest intensity of decomposition process was recorded under the residues of winter wheat. The changes of the ratio in the crop residues occurred in essence due to the content of nitrogen, which increased more rapidly than carbon. In more easily decomposing plant biomass the proteins of microbial synthesis are formed more rapidly thus contributing to the development of humification processes.

During the decomposition process, 55-70% of the carbon in the residue is released to the atmosphere as CO_2, 5-15% of the C is incorporated into the soil microbial biomass, and the remaining 15-40% of C is partially stabilized as new soil humus [101, 102]. Crop residues contain lignin, which is resistant to decomposition and becomes a substrate for soil humus formation. The more lignin in a residue type the slower the decomposition and the larger its contribution to the soil humus formation. Older, mature residues will tend to have more lignin content than young, non-mature residue from the same crop. The simple sugars, amino acids, polysaccharides, proteins, and lipids decompose first in the decomposition process.

Nitrogen (N) is necessary for the decomposition process. The less N the residue contains relative to the C, the slower the decomposition rate. If the residue N content is low, or the C:N ratio high, the decomposition process will require the input of N from either available soil inorganic N or fertilizer. Generally, residues with N concentrations less than 1.5% or C:N ratios greater than 30 will require N from sources outside the residue itself; it will immobilize soil N [107]. Residues with greater N concentration or lower C:N ratios, as is frequently the case with legume residues or non-mature residues, tend to decompose at a more rapid rate and will release or mineralize N. The actual decomposition rate will depend on N content and chemical composition of the residue and the environmental conditions such as soil moisture and temperature. Crop residues such as corn and wheat residue have large C:N ratios and soil N will be immobilized during the decomposition process. However, mineralization of N will generally start to occur after 50-60% of the residue has been decomposed [108] when enough C has been volatilized as CO_2 and N immobilized such that the remaining residue C:N ratio is below 30. Nitrogen immobilization reduces N availability to the growing crop and N mineralization increases N availability. The addition of a large quantity of oxidizable carbon from litter with the content of N less than 1,5% creates microbiological demand on N, thus immobilizing N of plant litter and inorganic N of soil. Effect of substrate composition on decomposition is described by C to N ration, based on the fact that N is the most limiting factor for decomposition. With ongoing decomposition the C to N ratio is narrowing thus the carbon energy supply is decreasing.

CONCLUSIONS

The studies confirm that the transformations of SOM are generally concentrated within labile pool. The process of mineralization of organic matter in soil is controlled predominantly, by the climatic factors (moisture and temperature) and plant litter quality (mainly by content of N), then by land management.

In the 21st century the mineral fertilizers became determinant for obtaining contented yield of agricultural crops. However, the application of only mineral fertilizer might lead to accelerated mineralization of the more stable soil organic matter. Instead, application of manure in combination with mineral fertilizer [38, 56-58, 103] might contribute in both obtaining the content crop yield and sustaining the soil fertility.

Yearly input of plant residue in less fallowed rotation system built up more labile OM, especially readily decomposable C, whereas frequent fallowing depletes SOM via accelerated mineralization. Therefore, in the semi-arid regions the inclusion of fallow in wheat monoculture every 6 years is the most appropriate farming system in terms of sustainability in both grain production and soil fertility.

Higher precipitation produced higher plant biomass contributing to the amount of SOM with further decomposition upon temperatures and soil texture. While less plant biomass production in drier regions in a lesser degree was subjected to decomposition due to moisture deficiency, thus contributing to the amount of labile SOM. Because wet-frigid region maintained the highest amount of total organic carbon and the least amount of easily mineralizable organic matter (PMC), transformation of organic substrates into more stable humified forms of organic matter might have taken place more actively in this region.

Understanding the processes of SOM changes under the impact of land management and different moisture and temperature regimes would greatly contribute to most ecological problem of C sequestration supplying with valuable information about land use management.

REFERENCES

1. Sollins, P., Homann, P. and Caldwell, B.A. 1996. Stabilization and destabilization of soil organic matter – mechanisms and controls. Geoderma 74: 65-105
2. Six J, Bossuyt H, De Gryze S, Denef K. 2004. A history of research on the link between (micro) aggregates, soil biota, and soil organic matter dynamics. Soil Till Res 79:7–31
3. Baldock, J.A., Nelson, P.N, 2000. Soil organic matter. In Sumner, M.E., (Ed.), Handbook of Soil Science. CRC Press, Boca Raton, pp. B25-B84
4. Post, W.M., Emmanuel, W.R., Zinke, P.J. and Stangenberger, A.G., 1982. Soil carbon pools and world life zones. Nature 298: 156-159.
5. Saljnikov E., Cakmak, D., Kostic L., and Maksimovic, S. 2009. Labile fractions of soil organic carbon in Mollisols from different climatic regions. Agrochimica. Vol. LIII, 6
6. Campbell, C.A. 1978. Soil organic carbon, nitrogen, and fertility. In: Schnitzer M., Khan, S.U., (Eds.) Soil organic matter. Elsevier, New York, pp 173-271
7. Beiderbeck, V.O., Campbell, C.A. and Zentner, R.P., 1984. Effect of crop rotation and fertilization on some biological properties of a loam in southwestern Saskatchewan. Can. J. Soil Sci. 64: 355-367.

8. Saljnikov-Karbozova, E., Funakawa, S., Akhmetov, K., and Kosaki, T. 2004. Soil organic matter status of Mollisols soil in North Kazakhstan: effects of summer fallow. Soil Biol.Biochem. 36: 1373-1381
9. Campbell, C.A. and Souster, W., 1982. Loss of organic matter and potentially mineralizable nitrogen from Saskatchewan soils due to cropping. Can. J. Soil Sci. 62: 651-656.
10. Rasmussen, P.E., Allmaras, R.R., Rohde, R.R. andRoager, N.C., 1980. Crop residue influences on soil carbon and nitrogen in a wheat-fallow system. Soil Sci. Soc. Am. J. 44: 596-600
11. Saljnikov E., Hospodarenko H., Funakawa S and Kosaki T. 2005. Effect of fertilization and manure application on nitrogen mineralization potentials in Ukraine. Zemljiste i biljka, vol.54, 3: 221-230.
12. Couteaux, M.M., Bottner, P. And Berg, B. 1995. Litter decomposition, climate and litter quality. Tree 10: 63-66.
13. Zou, X.M.; H.H. Ruan, Y. Fu, X.D. Yang and L.Q. Sha.Estimating soil labile organic carbon and potential turnover rates using a sequential fumigation–incubation procedure. Soil Biol. Biochem.37, 10, 2005, pp. 1923-1928
14. Christensen, B.T. 1986. Barley straw decomposition under field conditions: Effect of placement and initial nitrogen content on weight loss and nitrogen dynamics. Soil Biol. Biochem. 18: 523-529
15. Sparling, G.P., 1992. Ratio of microbial biomass to soil organic carbon as a sensitive indicator of changes in soil organic matter. Aust. J. Soil Res. 30: 195-207
16. Franzluebbers, K., Weaver, R.W., Juo, A.S.R. and Franzluebbers, A.J., 1994. Carbon and nitrogen mineralization from cowpea plant parts decomposing in moist and in repeatedly dried and rewetted soil. Soil Biol. Biochem. 26: 1379-1387.
17. Gregorich, E. G. and Janzen, H.H. 1996. Storage of soil carbon in the light fraction and macroorganic matter. P. 167-190. In M.R. Carter and B.A. Stewart (ed.) Structure and organic matter storage in agricultural soils. CRC press, Boca Raton, FL
18. Keeney, D.R. 1982. Nitrogen—availability indices. In A.L. Page et al., Eds. Methods of Soil Analysis. Part 2, 2nd ed. Chemical and Microbiological Properties, Agronomy 9. SSSA and ASA, Madison, WI, pp. 711–733
19. Christensen, B.T., 1996. Carbon in primary and secondary organo-mineral complexes. Advances in Soil Science 24, 97-165.

20. Elliott, E.T., Paustain, K., Frey, S.D., 1996a. Modeling the measurable or measuring the modelable: A hierarchical approach to isolating meaningful soil organic matter. In D.S.

21. Elliot, L.F., Lynch, J.M., and Papendick, R.I. 1996b. The microbial component of soil quality. In G. Stotsky and J.M.Bollag, Eds. Soil Biochemistry, Vol. 9. Marcel Dekker, New York, NY, pp. 1–20.

22. Breland, T.A. and Eltun, R., 1999. Soil microbial biomass and mineralization of carbon and nitrogen in ecological, integrated and conventional forage and arable cropping systems. Biol. Fert. Soils, 30: 193-201.

23. Nelson, D.W. and Sommers, L.E. 1996. Total carbon, organic carbon, and organic matter. In Page, A.L., Miller, R.H., Keeney, D.R. (eds.) Methods of Soil Analysis, Part 3, SSSA, Madison, WI.

24. SPSS Inc., SYSTAT 1998. version 8.0, Statistics, Chicago, IL.

25. SPSS Inc., 2007. Sigma Plot version 6.0, Programming guide, Chicago, IL.

26. Jalil, A., Campbell, C.A., Schoenau, J., Henry, J.L., Jame, Y.W., and Lafond, G.P. 1996. Assessment of two chemical extraction methods as indices of available nitrogen. Soil Sci. Soc. Am. J. 60: 1954–1960.

27. Picone, L.I., Cabrera, M.L., and Franzluebbers, A.J. 2002. A rapid method to estimate potentially mineralizable nitrogen in soil. Soil Sci. Soc. Am. J. 66: 1843–1847.

28. Jenkinson, D.S. 1988. Soil organic matter and its dynamics.. In A. Wild (ed.) Soil Condition and Plant Growth. Longman, New York. pp. 564-607.

29. Smith, J.L. and Paul, E.A. 1990. The significance of soil microbial biomass estimations. In J.-M. Bollag and G. Stotzky, Eds. Soil Biochemistry, Vol. 6. Marcel Dekker, New York, NY, pp. 357–396

30. Brookes, P.C., Ocio, J.A., and Wu, J. 1990. The soil microbial biomass: its measurement, properties and role in soil nitrogen and carbon dynamics following substrate incorporation. Soil Microorganisms. 35: 39–51

31. Powlson, D.S., Brookes, P.C., and Christensen, B.T. 1987. Measurement of soil microbial biomass provides an early indication of changes in total soil organic matter due to straw incorporation. Soil Biol. Biochem. 19: 159–164

32. Spedding T.A., Hamel C., Mehuys G.R., Madramootoo C.A. (2004): Soil microbial dynamics in maize-growing soil under different tillage and residue management systems. Soil Biology and Biochemistry, 36:

499–512.

33. Herna′ndez-Herna′ndez, R.M.andLo′pez-Herna′ndez,D. 2002. Microbial biomass, mineral nitrogen and carbon content in savanna soil aggregates under conventional and no-tillage. Soil Biol. Biochem. 34: 1563–1570.
34. Jenkinson D.S. and Powlson D.S. 1976. The effects of biocidal treatment on metabolism in soil: V. A method for measuring soil biomass. Soil Biol. Biochem. 8: 209-213
35. Sparling, G.P and West, A. W. 1998. A direct extraction method to estimate soil microbial biomass C: calibration in situ using microbial respiration and 14C labeled cells. Soil Biol. Biochem. 20: 337-343
36. Janzen, H.H., Campbell, C.A, Brandt, S.A., LaFond, G.P. and L.Townley-Smith. 1992. Light-Fraction Organic Matter in Soils from Long-Term Crop Rotations. Soil Sci.Soc.Am.J. 56: 1799-1806.
37. Nosko, B.S., 1987. Changes of humus status of Typical Chernozem upon fertilization. Pochvovedenie, 5: 26-31 (in Russian)
38. Chesnyak, G. Ya., 1981. Changes of humus quantity and quality and the ways of providing of positive humus balance in Typical Chernozems of Ukraine under intensive agricultural use. In publications of VI Meeting of Soil Science Society of USSR, V.2, pp.42-43. (in Russian).
39. Aleksandrova, L.N., 1980. Soil organic matter and its transformation. Nauka, Moscow, p. 288. (in Russian).
40. Kononova, M.M., 1956. Humus of the main soils of SSSR, its nature and ways of forming. Pochvovedenie, 3: 18-30. (in Russian).
41. Mamontov, V.T., 1971. Effect of agricultural use on the agronomical properties of thick Chernozem in west forest-steppes of Ukraine. In doctoral thesis of the author, p. 234. (in Russian).
42. Ilyaletdinov, A., 1988. Microbiological conversion of nitrogen compounds in the soil. Nauka, Moscow, 119-154pp.
43. Ferguson, W.S., Gorby, B.J., 1971. Effect of various periods of seed-down to alfalfa and bromegrass on soil nitrogen. Canadian Journal of Soil Science 51. 65-73.
44. Clarke, A.L., Russell, J.S., 1977. Crop sequential practices. In: J.S. Russell, E.L. Greacen, (Eds), Soil Factors in Crop Production in a Semi-arid Environment. University of Queensland Press, St. Lucia, pp. 279-300.
45. Dormaar, J.F., 1983. Chemical properties of soil and water-stable aggregates after sixty-seven years of cropping to spring wheat. Plant and

Soil 75, 51-61.

46. Bird, M.I., Chivas, A.R., Head J., 1996. A latitudinal gradient in carbon turnover times in forest soils. Nature 381, 143-146.
47. Amelung W., Flach, K.W., Zech W., 1998. Climatic effects on soil organic matter composition in the Great Plains. Soil Science Society of America Journal 61, 115-123.
48. Christensen, B.T., 1992. Physical fractionation of soil and organic matter in primary particles and density separates. Advances in Agriculture 20, 2-90.
49. Trumbore, S.E., Chadwick, O.A., Amundson R., 1996. Rapid exchange between soil carbon and atmospheric carbon dioxide driven by temperature change. Science 272, 393-396.
50. Collins, H.P., Rasmussen, P.E., Douglas, C.L., 1992. Crop rotation and residue management effects on soil carbon and microbial dynamics. Soil Science Society of America Journal 56, 783-788.
51. Campbell, C.A., Zentner, R.P., 1993. Soil organic matter as influenced by crop rotations and fertilization. Soil Science Society of America Journal 57, 1034-1040.
52. Biederbeck, V.O., Janzen, H.H., Campbell, C.A., Zentner, R.P., 1994. Labile soil organic matter as influence by cropping practices in an arid environment. Soil Biology & Biochemistry 12, 1647-1656.
53. Janzen, H.H., 1987. Soil organic matter characteristics after long-term cropping to various spring wheat rotations. Canadian Journal of Soil Science 67, 845-856.
54. Cheng, HH, & Kimble, JM. 2001. Characterization of soil organic carbon pools. In R. Lal, JM Kimble, RF Follett, & BA Stewart (Eds.), Assessment methods for Soil Carbon. CRC/Lewis Press, Boca Raton, FL, 676 pp.
55. Singh J.S. and Gupta S.R. 1977. Plant decomposition and soil respiration in terrestrial ecosystems. Bot. Rev. 43:449-528.
56. Chesnyak, G. Ya., 1986. Modification to the determination of coefficient of humification of plant residues in Typical Chernozems of forest-steppe Ukraine in grain-beet crop rotation. Agrochemistry and soil science, Kiev, 49: 77-92. (in Russian).
57. Voroney, R.P., 1988. Loss of organic matter in Ontario soils. Highlights. V.11, No.3, pp. 25-29.
58. Pare, T., Dinel, H., Moulin, A.P. and Townley-Smith, L., 1999. Organic matter quality and structural stability of a Black Chernozemic soil under different manure and tillage practices. Geoderma 91: 311-326

59. Kharin, S.V., 1993. Humification and regulation of humus status of different cropping systems in Typical Chernozems of west forest-steppe of Ukraine. PhD thesis, Institute of Soil Science and Agrochemistry after Sokolovski, Kharkov, Ukraine (in Russian).

60. Lundquist, E.J., Jackson, L.E. and Scow, K.M. 1999. Wet-dry cycles affect dissolved organic carbon in two California agricultural soils. Soil Biol. Biochem. 31: 1031-1038.

61. Bottner, P., 1985. Response of microbial biomass to alternate moist and dry conditions in a soil incubated with 14C- and 15N-labelled plant material. Soil Biol. Biochem. 17: 329-337

62. Van Gestel, M., Merckx, R. and Vlassak, K., 1993b. Microbial biomass responses to soil drying and rewetting: the fate of fast- and slow-growing microorganisms in soils from different climates. Soil Biol. Biochem. 25: 125-134.

63. Van Gestel, M., Merckx, R. and Vlassak, K., 1993a. Microbial biomass responses to soil drying and rewetting: the fate of fast- and slow-growing microorganisms in soils from different climates. Soil Biol. Biochem. 25: 109-123.

64. Pulleman, M. and Tietama, A., 1999. Microbial C and N transformations during drying and rewetting of coniferous forest floor material. Soil Biol. Biochem. 31: 275-285.

65. Lund, V. and Goksoyr, J., 1980. Effects of water fluctuations on microbial mass and activity in soil. Microbial Ecology 6: 115-123

66. Orchard, V.A. and Cook, F.J. 1983. Relationship between soil respiration and soil moisture. Soil Biol. Biochem. 15: 447-453.

67. Kononova, M.M., Pankova, N.A. and Belchikova N.P., 1949. Changes in quality and quantity of soil organic matter under cultivation. Pochvovedenie, 1: 28-37. (in Russian).

68. Beauchamp, E.G., 1980. Nitrogen from liquid dairy cattle manure for corn. Highlights Agr. Res. In Ontario, V.3, No.4, pp.10-12.

69. Anderson, D.W., Joug, E de, Verity, G.E. and Gregorich, E.G., 1986. The effect of cultivation on the organic matter of soils of the Canada prairies. Transact. 13 Cong. Int. Soc.Soil Sci., Hamburg, 13-20 Aug., V. 4, SI, S.9, pp.1344-1345.

70. Van Gestel, M., Ladd, J.N., and Amato, M., 1991. Carbon and nitrogen mineralization from two soils of contrasting texture and microaggregate stability: influence of sequential fumigation, drying and storage. Soil Biol. Biochem. Vol.23, No.4, 313-322

71. Rubinstein, M.I., 1959. Decomposition rate of organic matter of virgin Chernozem in Northern Kazakhstan during their cultivation. Soviet Soil Science 11: 1332-1335

72. Dzhalankuzov, T D. and Redkov, V.V., 1993. Changes in morphological and agrochemical properties of calcareous Southern Chernozems of North Kazakhstan due to long-term cultivation. In: Proceedings of Academy of Sciences of Republic of Kazakhstan, Biology Series 1: 53-58 (in Russian)

73. Campbell, C.A., Biederbeck, V.O., McConkey, B.G., Curtin, D., Zentner, R.P., 1999. Soil quality-effect of tillage and fallow frequency. Soil organic matter quality as influenced by tillage and fallow frequency in a silt loam in southwestern Saskatchewan. Soil Biology & Biochemistry 31, 1-7.

74. Campbell, C.A., Moulin, A.P., Bowren, K.E., Janzen, H.H., Townly-Smith, L., Biederbeck, V.O., 1992. Effect of crop rotation on microbial biomass, specific respiratory activity and mineralizable nitrogen in a Black Chernozemic soil. Canadian Journal of Soil Science 72, 417-427.

75. Haynes, R.J., 2000. Labile organic matter as an indicator of organic matter quality in arable and pastoral soils in New Zealand. Soil Biology & Biochemistry 32, 211-219

76. Gregorich, E.G., Carter, M.R., Angers, D.A., Monreal, C.M., Ellert, B.H., 1994. Towards a minimum data set to access soil organic matter quality in agricultural soils. Canadian Journal of Soil Science 74, 367-385

77. Shields, J.A. and Paul, E.A. 1973. Decomposition of 14C-laballed plant material under field conditions. Can. J. Soil Sci. 53: 297-306, 1973.

78. Douglas, C.L. Jr., Rickman, R.W., Klepper, B.L., Zuzel, J.F. and Wysocki, D.J. 1992. Agroclimatic zones for dryland winter wheat producing areas of Idaho, Washington, and Oregon. Northwest Science 66: 26-34

79. Raich, J. W. and C. S. Potter. 1995. Global Patterns of Carbon Dioxide Emissions from Soils. Global Biogeochemical Cycles 9(1)23-36. 10.3334/CDIAC/lue.db1015

80. Fierer N, B Colman, JP Schimel, RB Jackson. 2006. Predicting the temperature dependence of microbial respiration in soil: A continental-scale analysis Global Biogeochemical Cycles 20, GB3026c

81. Davidson, E.A., S.E. Trumbore, and R. Amundson. 2000. Biogeochemistry: Soil warming and organic carbon content. Nature. 408:789-790.

82. Janzen, H.H., 2004. Carbon cycling in earth systems-a soil science perspective. Agriculture, Ecosystems and Environment, 104, 399-417

83. Soil Taxonomy 2nd ed., 1999. USDA, Washington, DC. pp.555-655.

84. Franzluebbers, A.J., Haney, R.L., Honeycutt, C.W., Arshad, M.A., Schomberg, H.H. and Hons, F.M. 2001. Climatic influences on active fractions of soil organic matter. Soil Biol.Biochem. 33: 1103-1111

85. Aleksandrova, L.N. 1972. Study of the humification of plant residues and of the nature of newly formed humic acids. Pochvovedenie 7: 37-45.

86. Marumoto, T., Kai, H., Yoshida, T. and Harada T. 1977. Drying effect on mineralization of microbial cell walls as a source of decomposable soil organic matter due to drying. Soil Sci. Plant Nutr. 23: 9-19

87. Mikha, M.M., Rice, C.W. and Milliken, G.A. 2004. Carbon and nitrogen mineralization as affected by drying and wetting cycles. Soil Biol. Biochem. 37: 339-347

88. Wu, J. and Brookes, P.C. 2005. The proportional mineralization of microbial biomass and organic matter caused by air-drying and rewetting of a grassland soil. Soil Biol. Biochem. 37: 507-515.

89. Wong, P.T.W. and Griffin, D.M., 1976a. Bacterial movement at high matric potentials-I. In artificial and natural soils. Soil Biol. Biochem. 8: 215-218.

90. Wong, P.T.W. and Griffin, D.M., 1976b. Bacterial movement at high matric potentials-II. In fungal colonies. Soil Biol. Biochem. 8: 219-223.

91. Griffin, D.M., 1969. Soil water in the ecology of fungi. Annual Review of Phytopathology, 7: 289-310

92. Körschens, M., Weigel, A. and Schulz, E., 1998. Turnover of soil organic matter (SOM) and long-term balances-tools for evaluating sustainable productivity of soils. Z. Pflanzenernähr. Bodenk., 161: 409-424

93. Juma, N.G., 1993. Interrelationship between soil structure /texture, soil biota/soil organic matter and crop production. Geoderma 57: 3-30

94. Elliott, E.T., 1986. Aggregate structure and carbon, nitrogen and phosphorous in native and cultivated soils. Soil Sci.Soc. Am. J., 50: 627-633

95. Gupta, V.V.S.R. and Germida, J.J., 1988. Distribution of microbial biomass and its activity in different soil aggregate size classes as affected by cultivation. Soil Biol. Biochem. 20: 777-786

96. Amelung, W. and Zech, W., 1996. Organic species in ped surface and core fractions along a climosequence in the prairie, North America. Geoderma 74: 193-206

97. Gregorich, E.G., Kachanoski, R.G. and Voroney, R.P., 1989. Carbon mineralization in soil size fractions after various amounts of aggregate disruption. Journal of Soil Sci. 40: 649-659

98. Turchenek, L.W. and Oades, J.M., 1979. Fractionation of organo-mineral complexes by sedimentation and density techniques. Geoderma 21: 311-343

99. Sommers, L.E., Gilmour, C.M., Wildung, R.E. and Beck, S.M., 1981. The effect of water potential on decomposition processes in soil. In: Parr, J.F., Gardner, W.R., Elliot, L.F. (Eds.), Water Potential Relations in Soil Microbiology. Soil Sci. Soc. Am., Madison, pp.97-117.

100. Kieft, L.T., Soroker, E. and Firestone, M.K., 1987. Microbial biomass response to a rapid increase in water potential when dry soil is wetted. Soil Biol. Biochem. 19: 119-126.

101. Jenkinson, D.S. 1971. The Accumulation of Organic Matter in Soil Left Uncultivated. Rothamsted Experimental Station Report for 1970, part 2: 113-137

102. Stott, D.E., and J.P. Marten, 1989. Organic matter decomposition and retention in arid soils. Arid Soil Res. Rehab. 3:115

103. Kononova, M.M., 1951. Problems of soil humus and contemporary methods of their study. Moscow, pp.390. (in Russian).

104. Jenkinson, D.S., Adams, D.E. and Wild A. 1991. Model estimation of CO2 emissions from soil in response to global warming. Nature 351: 304-306

105. Zak, J.C., M.R. Willig, D.L. Moorhead and H.G. Wildman, 1994. Functional diversity of microbial communities: a quantitative approach, Soil Biol. Biochem. 26, 1101–1108.

106. Gregorich, E.G. and Janzen, H.H., 2000. Decomposition. In Handbook of Soil Science. Sumner M. E. (Ed.), pp. C107-C120

107. Schomberg, H.H., Steiner J.L., Unger, P.W. 1994. Decomposition and Nitrogen Dynamics of Crop residues: Residue Quality and Water effects. Soil Sci.Soc.Am.J. 58:372-381.

108. Douglas CL, Allmaras RR, Rassmussen PE, Ramig RE, Roager NC. 1980. Wheat straw decomposition and placement effects on decomposition in dryland agriculture of the Pacific Northwest. Soil Sci.Soc.Am.J. 44:833-837.

109. Funakawa S., Yanai J., Takata Yu., Karbozova-Saljnikov E., and Kosaki T. 2007. Dynamics of water and soil organic matter under grain farming in Northern Kazakhstan – Toward sustainable land use both from the agronomic and environmental viewpoints. In Lal R., et al., (ed.) Climate Change and Terrestrial Carbon Sequestration in Central Asia. The Netherlands. Taylor&Francis, pp. 279-331.

110. Six, J. and Jastrow J.D. 2002. Organic Matter Turnover. In Lal R. (ed.) Encyclopedia of Soil Science. Dekker, NY, pp.936-942.

Chapter 2

MICROAGGREGATE STABILITY OF TROPICAL SOILS AND ITS ROLES ON SOIL EROSION HAZARD PREDICTION

C.A. Igwe and S.E. Obalum[1]

[1]Department of Soil Science, University of Nigeria, Nsukka, Nigeria

INTRODUCTION

Soil aggregate stability influences a wide range of physical and biogeochemical processes in the agricultural and natural environments, including soil erosion [3]. The relative preponderance of aggregates of various sizes in the soil and their stability to external forces are, therefore, an issue of major concern to soil scientists. By definition, an aggregate is a composite body or granule of loosely bound mineral particles within a soil, the binding of which is characteristically mediated by a relatively minor amount of organic matter [Encyclopedia of Soil Science, ESS 2008]. The mineral and organic particles involved in such a natural conglomeration, otherwise known as aggregation, cohere to each other more than to the neighbouring particles and/or aggregates [ESS 2008]. Soil aggregates are therefore soil structural units of which classical soil research recognizes two major size-based categories, macroaggregates and microaggregates. Collapse of macroaggregates yields microaggregates. Thus, macroaggregates may be viewed as having microaggregates as their building blocks. Sometimes, external forces acting on a soil can also foster formation of aggregates from dispersed materials. It is the interplay of aggregate formation and breakdown that results in soil structure [54]. Although extremities in either of these structure-promoting processes are undesirable, they are considered an agronomic and environmental problem only in the case of breakdown. This is because it is much easier to break down over-sized aggregates into favourably sized ones than to achieve aggregation in structurally dilapidated soils. Consequently, studies on the responses of soil aggregates to natural and anthropogenic forces appear to tilt more towards stability or otherwise of soil aggregates than to their coalescence by these forces. Soil aggregation includes the processes of formation and stabilization, both of which occur continuously and concurrently [3]. Soil aggregate/structural stability may be defined as a measure of the ability of the soil structural units to resist change or the extent to which they remain intact when mechanically stressed by environmental factors [ESS 2008]. The environmental factors that become important in this regard generally depend on climate and soil characteristics related to the nature of the parent materials and age of the soil. Another important factor is the intensity of disturbance related to land use and management [1]. To understand the importance of climate, it will be good to first state that water is such an indispensable entity in the discussion of soil aggregate stability that the subject is sometimes referred to aggregate stability to water. Climate sets the limit of change in the state of water in the soil, whether the soil's response to the major climatic variables (rainfall and temperature) would be limited to mere wetting and drying or would also include freezing and thawing. In the tropical climate, soils are subject to frequent wetting and drying cycles in the short term during the rainy season and in the long term between the distinct rainy and dry seasons. Although freezing and thawing constitutes a greater stressor to soil aggregates, it rarely occurs in the tropics and therefore should be de-emphasized here for the sake of the scope of this chapter. In terms of inherent soil characteristics, tropicals soils generally show a higher aggregate stability compared to temperate soils [55], and this is due mainly to mineralogy of the former being characterized by dominance of oxides and kaolin clays [54]. Nevertheless, aggregate stability remains a valid topic in the tropics, especially in the broad area of environmental management, because many soils in the region are regarded as structurally fragile and unstable. This is due to other soil-related and certain climatic peculiarities of the region.The soil-related factors militating against the aggregate stability of the majority of tropical soils include the sandy nature of their parent materials, which often reflects in the texture of the soils. It has been reported that the resistance of soil aggregates to raindrop impact decreases with a decrease in clay content of the soil [40]. Most other soils occurring in areas with heavy rainfall, even when not originally sandy, have been so intensively washed by runoff and leaching that their texture tends toward coarseness [33]. It is perhaps because of the vast area occupied by soils in these categories, commonly referred to as 'tropical sandy soils' in the literature, and the effect of the sandy texture on their aggregate stability that considerable research goes into their management [FAO 2007]. The coarse texture of the soils, coupled with the low concentration and high mineralization rates of organic matter (the typical aggregating agent) in them, implies impaired aggregation. Again, most tropical soils have long weathering history as is often evident in their low silt content [33], and this also contributes to frustrating aggregation processes in the soils. Indiscriminate deforestation, inappropriate land use and non-sustainable soil management options are also a common feature of agriculture in the region. In terms of climate, the aforementioned long-term wetting and drying cycles in most of the tropical region can have important implications for the aggregate stability of the soils. Also, the characteristic rainstorms and the associated heavy raindrops in especially the humid tropics can have considerable splash effect [38] and, therefore, are a force to reckon with in soil aggregate destabilization. It is thus clear that most tropical soils are structurally fragile and susceptible to many forms of erosion including accelerated and catastrophic erosion. [3] noted that good soil structure, known by the presence of well formed and stable aggregates, is the most desirable of all soil attributes for sustaining agricultural productivity and for preserving environmental quality. In the above context, a good understanding of the aggregate stability of tropical soils and its relationship with their erodibility is needed to guide the management of these soils against erosive and similar degradative forces. In spite of the generally higher aggregate stability of tropicals soils compared to temperate soils [55], soil erosion remains a major threat to agricultural productivity in the tropical region. Proper management is necessary to position these soil resources for continued support of agricultural and allied activities while not compromising environmental quality. Soil aggregate stability has been shown to give some guide on the relative stability of Ultisols from sub-tropical China to externally imposed destructive forces and, hence, to be an appropriate indicator of the relative susceptibility of the soils to detachment, runoff and interrill erosion [63, 53]. Our focus here is on microaggregation and the relationship between microaggregate stability and erodibility of tropical soils.

APPROPRIATE AGGREGATE STABILITY INDICES FOR ASSESSING EROSION HAZARDS IN TROPICAL SOILS

The derivation of many aggregate stability indices involves all aggregate-size classes and, as a result, such indices provide information on the overall stability of the soil. A typical example and, perhaps, the mostly widely used of such indices is the mean-weight diameter (MWD) of aggregates. However, the MWD is often regarded as index of macroaggregate stability of soils, probably because of the preponderance of macroaggregate-size classes over microaggregate-size classes in its computation. Where authors of the papers reviewed in this chapter fail to specify which of macro- and microaggregate stability their indices represent, we regard the indices as representing macroaggregate stability rather than microaggregate stability, provided their determination did not involve dispersion. The MWD and indeed all such aggregate stability indices which integrate aggregate-size classes into one number are regarded as macroaggregate stability indices in this chapter. The use of such indices to assess erodibility may prove suitable in temperate soils, but may not in highly weathered tropical soils known for their oxyhydroxidic mineralogy and very stable microgranular structure [17]. The question remains which of macro- and microaggregate stability more closely relates to erodibility of the majority of tropical soils. To answer this question, we need to first understand the mechanisms that are generally responsible for the breakdown of macro- and microaggregates. The main mechanisms of aggregate breakdown for macro- and microaggregates are slaking and dispersion, respectively. Slaking is the initial break-up of macroaggregates into microaggregates when immersed in water, caused by pressure due to entrapped air [38] and/ or by differential swelling [ESS 2008]. Unlike slaking, dispersion liberates the soil colloidal particles that are more transportable during erosion. Hence, microaggregate stability is often referred to as colloidal stability. This suggests that microaggregate stability may be a better indicator of potential soil erosion hazards. Some studies have related potential soil loss or, more specifically, the erodibility of tropical soils to their aggregate stability at both the macro and micro levels. These studies tend to support the view that erosion in the soils is related more to microaggregate stability than to macroaggregate stability. For instance, Igwe et al. [19] compared the predictability of soil loss by selected macro- and microaggregate stability indices for some soils from southeastern Nigeria. They found that all microaggregate stability indices predicted soil loss better than their macroaggregate stability counterparts. Some other researchers reported weak correlations between soil erodibility and macroaggregate stability indices for some Nigerian soils [30, 31]. The soils in question are by virtue of their parent materials dominated by quartz and, as is the case with many tropical soils, are at an advanced stage of weathering. Hence, such other minerals as Fe-oxyhydroxides and kaolinite abound in them, and these are the minerals that are known to cause highly stable aggregation [54]. Since these predominant minerals do not expand rapidly when immersed in water, slaking proceeds rather slowly in the soils. The implication is that the soils show fairly high macroaggregate stability which is a misrepresentation of their high erodibility and erosion status [33]. Considering the widely accepted role of soil organic matter in aggregate formation and stabilization/destabilization, the choice of microaggregate stability for the prediction of potential eroson hazards in tropical soils would also be explained by the relative influence of organic matter on macro- and microaggregate stability. Macroaggregates are generally considered more sensitive to soil organic matter concentrations– and hence are less stable–than microaggregates [58]. Whereas the theory of macroaggregates being less stable than microaggregates may hold true for tropical soils, that of macroaggregates being more sensitive to soil organic matter concentrations than microaggregates remains a controversial topic. It has been shown that the relationships between aggregate stability indices and organic matter concentrations in tropical soils are generally characterized by weak correlations [55], and these are thought to be due mainly to the relatively lower organic matter status of the soils. However, inconsistencies characterize the response of macro- and microaggregation to organic matter concentrations in tropical soils. The relationship between macroaggregate stability and soil organic matter concentration has been reported to be non-significant [17, 31, 64] or postively significant [7, 18, 26] or negatively significant [23]. There are indications that these relationships may depend on method of assessment of macroaggregate stability as well as on location. Soil clay content is another factor that may dictate the nature of organic matter effect on macroaggregate stability of tropical soils [61]. Similarly, the relationship between microaggregate stability and organic matter concentration in tropical soils has been reported to be non-significant [23] or positively significant [12, 42, 64, 51] or negatively significant [30, 33, 34]. There are indications that these relationships may depend on microaggregate stability index adopted by the authors as well as on the contents of organic matter in the soil relative to other microaggregating agents.

SOIL MICROAGGREGATION AND MICROAGGREGATE STABILITY

There are a lot of inconsistencies in the literature regarding the appropriate size boundary between macro- and microaggregates. The placement of size boundary for the classification of aggregates into macro- and microaggregates and the delineation of their upper and lower limits, respectively appear to depend on the researcher's orientation and location. We adopt here the categorization scheme proposed by Oades and Waters [46], which specifies the boundary between macro- and microaggregates as 0.25 mm, and this is consistent with the use of 0.25 mm as the boundary between water-stable and water-unstable aggregates in aggregate stability studies. In the hierarchy of aggregate size order, the lower boundary of microaggregates is taken to be 0.02 mm [46]. However, these upper and lower boundaries may be exceeded in highly weathered tropical soils where the association between microaggregates and clay-sized granules often form a kind of continuum of very stable aggregates [59, 49]. The stable microgranular structure is often manifested in form of pseudo-sands composed of clay particles that are strongly cemented together by Fe oxides [31].

Microaggregates are formed in a number of ways, each influenced by a number of factors. The process of microaggregation combines break-up of aggregates due to slaking and aggregates due to subsequent attrition [14]. Factors that influence microaggregation may differ between the temperate and tropical regions. Some researchers working in a German temperate soil reported that microaggregation depended strongly on the size distribution of primary particles rather on land use [39]. Conversely, an assessment of microaggregate stability under different land use types in a Nigerian tropical soil revealed a strong dependence of the soil microaggregation on land use [51]. This implies that the agents of stabilization of microaggregates in tropical soils are sensitive to land use.

The high aggregate stability for which tropical soils are reputed is not limited to macroaggregates. As already noted, microaggregates formed in tropical soils at advanced stages of weathering are also of very high stability [59, 46, 28]. In spite of this, microaggregate stability may still be a good indicator of the erodibility of tropical soils because of its direct link with silt and clay dispersion. Mineralogy appears to have a great influence on microaggregate stability of soils [45]. In this regard, the major microaggregating agents in tropical soils are Fe and Al oxides [17, 64, 31, 2, 28, 57]. However, in hardsetting lowland soils with low organic matter concentration and which are prone to seasonal flooding, microaggregation may be achieved through practices that enhance the organic matter concentration in them, since the roles of Fe and Al oxides in such soils may be dispersive rather than microaggregating [27]. Also, the microaggregating effect of Fe_2O_3 has been reported to be masked in some soils with relatively high concentrations of organic matter (1.39-6.79%) while that of exchangeable Ca and Mg became evident due to the tie-up of these elements with organic matter and hence their minimal leaching [Opara 2009]. Closely related to the effect of Fe and Al oxyhydroxides on the microaggregate stability of tropical soils is that of the non-expanding clay types, which dominate the clay mineralogy of the soils [30, 2006, 61, 57].

In tropical soils, soil organic matter may act as a dispersing/deflocculating agent [31], as a microaggregating agent [26, 51] or as a facilitator to the microaggregating effect of Fe-Al oxides [28], depending on its relative abundance in the soils. By contrast, the effect of soil organic matter concentration on microaggregate stability of temperate soils appears not to be prounounced [39]. Apart from protecting the surface against raindrop impact, organic matter may impart hydrophobic characteristics to the soil, thereby reducing the slaking that usually precedes dispersion [38]. In some Fe-Al oxidic tropical soils from Malaysia, it was polysaccharide constituent of soil organic matter rather than total organic matter that influenced microaggregate stability [57]. Notably, the soil content of Fe and Al oxyhydroxides is not easy to manipulate by regular soil management practices [5]. The inference that can be drawn here is that the view that organic matter is not the main aggregating agent in tropical soils rich in Fe-Al oxyhydroxides [5] may not always apply to microaggregation, but the exact role of organic matter may depend on its concentration in the soil and on its chemical composition as may be determined by the prevailing land use and soil management.

AGGREGATE BREAKDOWN MECHANISMS AND EROSION PROCESSES

Some tropical soils occuring in high-intensity rainfall zones have the tendency to slake and form seals, thereby resulting in considerable runoff and soil erosion [50, 13]. Although rainfall impact and slaking cause much greater breakdown of macroaggregates than microaggregates, these two factors can also be important for microaggregate stability and soil erodibility in at least two ways. First, slaking precedes dispersion. And this is the reason why, even though slakability is different from dispersibility, soils with high slaking potential are at high risk of interill erosion [41]. Second, sealing and crusting often accompany slaking. Seal is defined as the orientation and packing, at the very surface of the soil, particles dispersed from soil aggregates due to the impact of rain drops, thereby rendering the soil relatively impermeable to water [44]. This is the first stage of seal formation. As the ponded water infiltrates or evaporates, the soil particles suspended in it get deposited on the soil surface, thereby increasing the thickness of the seal. This is the second stage of seal formation. The entire seal eventually dries out to become crust, a thin but much more compact and hard layer than the material directly beneath [44, 60]. Both seals and crusts are therefore formed in the same way and occur commonly in the semi-arid regions [44, 60]. Crusts formed due to the first and second stages of seal formation are called structural crusts and depositional or sedimentary crusts, respectively [44, 38].

Most tropical soils are highly weathered and lacking in expanding clay types. Where they occur, the associated shrink-swell hazard is mostly concentrated in the subsoil where there is increased content of clay particles due to translocation and illuivation or residual accumulation of clay [33]. Because of this, slaking is due more to compression of air entrapped inside aggregates during wetting than to swelling. In the absence of swelling, the intense rainstorms experienced in the tropical region may result in sedimentary sealing and crusting especially in soils with reasonably high clay content but with disproportionately low concentration of organic matter [62]. Surface sealing and crust formation are an important factor in erosion processes, for they influence detachability of soil particles from aggregates, as well as infiltration rate and surface roughness which determine runoff volume and speed, respectively [38].

For soil erosion in interrill areas, three generally recognized sub-processes completely define soil erosion; and they include detachment, transport and sedimentation [38]. Some researchers working with sandy-loam soils in the semi-arid tropics have obtained results which suggest that the erodibility of a soil depends on the relative proportion of aggregates in the soil, being higher when the aggregate size distribution shows a greater proportion of large-sized aggregates [35]. Others working with low-activity-clay tropical soils reported that the saturated hydraulic conductivity increased with an increase in structural stability of the soils [61, 48]. Increased saturated hydraulic conductivity implies reduced weakening and dispersion of the soil aggregates following rainfall and/or irrigation and, hence, less susceptibility to erosion. It appears therefore that, with respect to erosion, the predominance of large-sized aggregates in soils is not always an indicator of good soil structure, but the stability of the soil pore system is.

It has been shown that, in tropical soils, disruption of macroaggregates leaves them as microaggregates rather than as primary particles [17]. Disintegration of soil macroaggregates into microaggregates following rainfall, slaking, dispersion and sealing can decrease infiltration and saturated hydraulic conductivity of the soil [36, 37]. These effects which ultimately increase soil loss can be more severe in soils of low organic matter concentration [37], as those occuring in the tropical region. The main mechanism of microaggregate breakdown is dispersion into primary particles, and this is influenced by the electrolyte concentration of the soil solution and the applied water, exchangeable sodium percentage and mechanical disturbance [38]. Electrolyte concentration and the dispersion it induces can lead to a situation whereby re-deposition of the dispersed particles cause clogging of water-conducting pores in the soil, in which case the hydraulic conductivity becomes drastically reduced [10]. The roles of exchangeable sodium percentage and electrolyte concentration in microaggregate stability are also evident in tropical soils [31, 32], probably due to the effect of ions on the amount of aggregates cemented by Fe-Al oxyhydroxides.

Generally, polyvalent cations cause flocculation whereas the monovalent cations cause dispersion [38]. It appears, however, that in hardsetting tropical soils with low organic matter concentration and that are prone to seasonal flooding, the flocculating role of polyvalent cations and the dispersive role of monovalent cations are usually not evident [27]. On the other hand, polyvalent cations (Ca^{2+} and Mg^{2+}) are good microaggregating agents under upland soil conditions, provided there is sufficient organic matter in the soil to retain these cations against leaching [51]. For a range of tropical soils all from Nigeria, factors that have been identified to influence soil dispersion include presence and concentration of monovalent cations (mostly K^+) in prospective irrigation water [27], soil reaction (pH), sodium adsorption ratio, and soil properties related to cation exchange [32, 23, 26] In the same region, elemental contents in silt fraction were reported to influence microaggregate stability [24].

ASSESSMENT OF SOIL MICROAGGREGATE STABILITY

Microaggregate stability is normally assessed by the extent of dispersion of microaggregates into granules and/or primary particles. This is difficult to do under field conditions where the dynamic nature of this soil property may not permit attainment of reliable data. Consequently, most methods of assessment of microaggregate stability are based either on conceptual model of microaggregation involving the finer and colloidal particles or on the response of isolated microaggregates to simulated dispersive force in the laboratory. Although the disintegration forces applied in the laboratory may attempt to simulate those found in the field, they do not fully duplicate field conditions [3]. Forces applied to achieve dispersion during microaggregate stability tests may even be bigger and too sudden compared to the ones that cause dispersion under field conditions. Results of such tests are, however, still useful for they allow for a discrimination between soils in accordance with field observations [3], thereby providing information that can guide management decisions. Some of the methods that have been applied to tropical soils are summarized (Table 1). The information presented in this table shows that all the indices have to do with clay and/or silt dispersion in water. Although either of the water-dispersible clay (WDC) and water-dispersible silt (WDS) can be used to do the assessment of microaggregate stability, most researchers prefer using indices that include both.

One observation that is noteworthy is the seemingly lack of agreement among the soil microaggregate stability indices included in this review. This lack of agreement is evident in the inconsistent pattern in which these indices relate to other soil properties, including soil contents of oxides and organic matter, both of which have been shown to be very important in microaggregation. For instance, WDC and clay dispersion ratio (CDR), both of which are indices of colloidal stability, have been reported to correlate with soil organic matter concentration in a contrasting manner [30]. It appears thus that, under certain conditions, some colloidal stability indices serve better, but under some other conditions, the same colloidal stability indices may not be suitable.

SOIL EROSION HAZARDS IN TROPICAL SOILS AND THE NEED FOR PREDICTION

The more widespread forms of erosion are rill and interrill erosion. Soil erosion can have both on-site and off-site effects which are the lowering of soil productivity and deposition of sediments, respectively. Crop yields are usually used as a proxy measure of soil productivity loss to erosion. Deposition of sediments, mostly colloidal particles detached from the soil by agents of erosion, occurs after they are transported by surface runoffs generated during rainfall (in the case of water erosion) and turbulent winds (in the case of wind erosion). Water erosion also results in the transport of runoff-laden solutes and dissolved contaminants and is thus a major source of land and water pollution. The problem is experienced more in the humid and sub-humid tropics where the rains often come as rainstorms. Here, soil loss to water erosion can be over 50 tons $ha^{-1}yr^{-1}$ [50, 15]. In contrast, the impact of wind erosion is felt more in the semi-arid and arid tropical climates, with soil loss rate that often surpasses that due to water erosion. In the West African Sahel, for instance, soil loss to wind erosion can be in the range of 58-80 tons $ha^{-1}yr^{-1}$ [34].

In those areas where water erosion is the bigger problem, taxonomically different soils can respond differently to erosion under similar conditions. For instance, Inceptisols and Entisols have been reported to be more erodible than Ultisols, due to higher Fe and Al contents of the latter [23]. With respect to crop yields, the productivity of adversely eroded soils can be restored through careful selection of appropriate soil management practices. However, except in a few cases where materials deposited by runoff are properly harnessed, the off-site effect of soil erosion always constitutes environmental problems. In contemporary agriculture where the emphasis is on not just achieving high yields but also on making agricultural enterprise environment-friendly, such environmental problems arising from soil erosion should be viewed as undermining agricultural productivity.

The problem of soil erosion and the associated negative impacts on agriculture and the environment is particularly severe in tropical sub-Saharan Africa, where it is a major cause of declining and stagnating soil productivity [48]. When considering appropriate soil conservation as an option, the first step is to try to understand the roles of microaggregate stability in checkmating soil erosion and in predicting soil erosion hazards. Prediction of soil erosion hazards involves a quantitative assessment of potential soil erosion in a land resource of an area. Such quantitative information is used in soil erosion hazard mapping for both short-term and long-term planning against erosion and associated deleterious effects, and this can have many agricultural and environmental benefits. Many atimes, erosion hazard maps are viewed as a tool for detailed farm planning and management [30]. Information on potential erosion hazards can also be used to embark on precautionary soil and water conservation measures. For instance, conservation specialists can use such information to select appropriate engineering designs and structures aimed at forestalling the occurrence of erosion in the first place, or controlling erosion in already eroded areas. Once started, rill and interill erosion need to be timely arrested, otherwise they may escalate into gully erosion, which is the more spectacular form of erosion that often threatens the integrity of the environment.

MICROAGGREGATE STABILITY AND EROSION HAZARD PREDICTION FOR TROPICAL SOILS

Virtually all known methods of assessing microaggregate stability, discussed earlier, employ the extent of dispersion into primary particles. The relevance of microaggregate stability for assessing potential erosion hazards lies, therefore, on the effects of dispersion on soil hydrophysical processes. Dispersion generally induces processes that are related to soil erodibility such as very fast crusting, slow infiltrability, and great mobility of particles in water [38]. Soil erodibility may be defined as the degree or intensity of a soil's state or condition of being susceptible to erosion [56]. It is just one of the main parameters needed for erosion hazard prediction. The most commonly used index of erodibility is the erodibility factor (K-factor) of the revised universal soil loss equation (RUSLE). Although fragments/sediments detached by raindrops can be finer than the original soil, the detachment is often accompanied by mere displacement (splash effect); the actual transport and sedimentation involve silt- and clay-sized particles [38]. Therefore, microaggregates dispersion is a pre-condition for soil erosion to be complete. There is evidence from the United States that WDC and CDR can be good estimators of erodibility of some soils in Ohio [4].

Microaggregate stability, when used as a tool for predicting soil erosion hazards, takes into account only the aspect of such hazards that are due to the soil inherent erodibility. One would therefore expect researchers to relate microaggregate stability to only soil erodibility when assessing potential erosion hazards. However, because soil erodibility is a dynamic soil property, its accurate determination can sometimes be difficult. Acquisition of data for soil erodibility is particularly difficult in the case of the K-factor of the RUSLE, as this requires some basic land-use information as well as pre-measurement soil management specifications, actual practice of which is often tedious and time-consuming. Consequently, not all researchers relate microaggregate stability to soil erodibility; some often relate it directly to soil loss to natural or simulated erosion, while keeping constant such other factors that affect erosion as rainfall, topography, vegetation, and soil management and conservation practices. We reason that, unless the relationship between microaggregate stability indices and erodibility/soil loss are not established by statistical correlations, the effects of such methodological differences may be negligible. Although a good number of studies have been conducted on aggregate stability of tropical soils, our survey of the literature reveals that not many of these studies related the erodibility of the soils or potential soil loss to aggregate stability. Soil aggregate stability or instability is such a critical factor in erosion processes that erosion is often the first thing that comes to the mind when pondering over usefulness of information on aggregate stability. It is thus surprising that the majority of studies on aggregate stability of tropical soils failed to describe its relationship with soil erosion. Again, among the few studies that did otherwise, only a small proportion used microaggregate aggregate stability indices in spite of the fact that, as we have been able to show earlier in this review, microaggregate stability more than macroaggregate stability corresponds to the dispersion and erodibility of tropical soils. We review in the preceding paragraph only those studies that related soil erodibility or potential soil loss to microaggregate stability in the tropical region.

Table 1: Indices of microaggregate stability commonly applied to tropical soils

Index	Derivation	Interpretation	References
Clay ratio, CR	$\frac{\% \text{ sand}}{\% \text{ silt} + \% \text{ clay}}$	A	Mbagwu (1986)
Degree of aggregation, DOA†	$\frac{w_a \cdot w_b}{w_a}$	B	Zhang and Horn (2001)
Water dispersible clay, WDC	Clay after particle-size analysis with deionized water only	A	Mbagwu and Auerswald (1999); Igwe (2005)
Water dispersible silt, WDS	Silt after particle-size analysis with deionized water only	A	Igwe and Nkemakosi (2007)
Aggregated clay, AC or Clay aggregation, CA	Total clay – WDC	B	Mbagwu and Auerswald (1999); Igwe (2003)
Aggregated silt + clay, ASC	Total silt and clay – WDS and WDC	B	Igwe et al. (1999a)
Clay dispersion ratio, CDR or Clay dispersion index, CDI	$\frac{\% \text{ WDC}}{\% \text{ total clay}}$	A	Igwe and Nkemakosi (2007); Opara (2009)
Clay flocculation index, CFI	Total clay – WDC/total clay	B	Igwe and Nkemakosi (2007)
Dispersion ratio, DR or Water dispersible clay and silt, WDCS	$\frac{\% \text{ WDS} + \% \text{WDC}}{\% \text{ total silt and clay}}$	A	Mbagwu (1986); Igwe (2005); Igwe and Nkemakosi (2007); Sung (2012)

†w_a and w_b stand for the proportion of particles between 0.25 and 0.05 mm obtained by microaggregate size analysis and by particle size analysis, respectively.
A – The smaller the value, the more stable the microaggregates are.
B – The bigger the value, the more stable the microaggregates are.

In southeastern Nigeria, clay ratio (CR) and dispersion ratio (DR) were reported as being close substitutes to the K-factor in the prediction of soil loss [40]. Also in this region, Igwe et al. [29] related the K-factor to selected indices of microaggregate stability for soils from diverse geological formations. Their results showed a good correlation ($r = 0.53$) between K-factor and clay flocculation index (CFI), and they recommended that the CFI alone could be used to predict soil erosion hazard in the area. The stability and soil-loss response of a stony Nigerian tropical soil undergoing intensive cultivation to simulated tillage and stone removal was investigated [30]. This laboratory study revealed that tillage and stone removal led to increases in WDC, DR and CDR; and that this failure in microaggregate stability of the soil increased erosion of the soil. Still working with soils from southeastern Nigeria, Igwe [31] reported that any of DR, CDR and WDC could be used in predicting erodibility of some the soils. The CDR and DR were also found, in a separate study, to have significantly ($r = 0.44$ and 0.39, respectively) correlated with K-factor of the RUSLE and were therefore deemed good indices for predicting erodibility of soils of eastern Nigeria [32]. All these studies demonstrate the suitability of some microaggregate stability indices for assessing soil erodibility and potential soil loss in the tropical region.

AREAS OF FURTHER RESEARCH

All the indices of microaggregate stability included in this review were developed based on silt and/or clay dispersion which occurs only in wet or submerged soils, and this limits their applicability to erodibility assessment to the case of water erosion [9]. In the semi-arid and arid tropics, wind erosion is a major source of soil and nutrient loss in agricultural soils. An index of microaggregate stability is therefore needed for such soils to also enable the assessment of potential erosion hazards in them. Similarly, there are indications that removal of gravels and stones from tropical soils characterized by high gravel content can confer higher erodibility to such soils [43, 25]. This implies that, for this category of soils, the use microaggregate stability indices determined from routine laboratory measurements as indicators of soil erodibility may be misleading. It may therefore be necessary to correct microaggregate stability results for gravel content, especially when they are intended for use in the assessment of soil erodibility. Research is needed on the best method of doing such a correction as may be confirmed by a good agreement between the ensuing results and field-measured erodibility of the soil. Also, some researchers have reported good correlations between their microaggregate stability indices and soil contents of silt [57] or clay [26, 51], just as others have reported that elemental contents in silt fraction affected microaggregation [24]. Silt is known to be the soil particle that is most suspectible to loss during erosion [52], and the data presented by Igwe and Ejiofor [2005] for a severely gullied tropical soil support this assertion. This suggests that paying attention to soil texture, especially variations in silt content, may benefit microaggregate stability studies in relation to erodibility.

It is known that oxides which abound in many tropical soils are a major promoter of their colloidal stability. The role of particularly Fe oxides in microaggregate stability may not be limited to the promotion of microaggregate formation. A study conducted in a mediterranean environment revealed that Fe oxides also acted to decrease dispersion of clay [6]. The possibility of this phenomenon and the factors promoting it in Fe-oxide-rich soils in the core tropics need to be explored. This review reveals that there are conflicting reports on the effect of organic matter concentration on soil microaggregation and microaggregate stability of tropical soils. Forms of oxides in the soil can influence not only their aggregating potential but also that of organic matter [11], and this has been demonstrated specifically for microaggregation of tropical soils [28]. On the other hand, the chemical composition of organic matter and its distribution in the aggregate-size classes (whether it is physically protected within microaggregates or not) may also contribute to determining how it influences microaggregation in the soil. More studies are therefore suggested on the role of organic matter in microaggregate stability of tropical soils, with emphasis on soils differing in both contents and forms of oxides.

In erosion processes, field capacity is expected to be an important factor because of its direct link with infiltration and runoff. It has been shown that slaking potential of a soil decreases with an increase in its field capacity [8], suggesting that the tendency for dispersion may also decrease with an increase in field capacity. However, in severely gullied soils in eastern Nigeria showing silt content of not more than 1% and mean organic matter concentration of 0.18% (both on weight basis), CDR was shown to increase (i.e. decrease in colloidal stability) with an increase in field capacity [23]. Recently, Abrishamkesh et al. [1] reported higher field capacity under a condition of higher structural stability than lower structural stability in a temperate environment. Similarly, Obalum et al. [48] reported that the lower the structural stability of some coarse-textured tropical soils, the higher the pressure at which they attain field capacity. They attributed the observation to reduced dispersion and hence increased internal drainage of the soils as their stability increased. It appears therefore that the field capacity represents a structural index related to both dispersability and stability of soil aggregates. Research is needed to fully explore the relationships among field capacity, microaggregate stability and erodibility of tropical soils differing in degree of past erosion.

CONCLUSION

The majority of tropical soils show high microaggregate stability irrespective of their low organic matter concentration. This is due mainly to their high contents of Fe and Al oxides which are known to promote microaggregation in soils of low organic matter concentration. However, there are some conflicting reports on the effects of the various players, especially oxides and organic matter, on microaggregation in tropical soils. So many natural and anthropogenic factors can lead to dispersion of the soils, but the factors tend to vary from study to study. A number of agricultural and environmental problems can arise from the dispersion of clay especially in sandy soils characterized by low concentration of organic matter [9], like the ones predominants in the tropics. The most important of these problems is soil erosion. Although only few studies have related soil erosion hazard (assessed either by soil erodibility or by soil loss) to selected indices microaggregate stability, these studies show that microaggregate stability is a useful tool for predicting erosion hazards in tropical soils. However, comparisons of results of erosion hazard prediction would be meaningful only when the same index of microaggregate stability is used. We suggest some areas for further research on microaggregation in tropical soils and the relationship between colloidal stability and soil erodibility.

REFERENCES

1. Abrishamkesh S, Gorji M, Asadi H. Long-term effects of land use on soil aggregate stabilty. International Agrophysics 2011; 25 103-108.
2. Alekseeva TV, Sokolowska Z, Hajnos M, Alekseev AO, Kalinin PI. Water stability of aggregates in subtropical and tropical soils (Georgia and China) and its relationships with the mineralogy and chemical properties. Eurasian Soil Science 2009; 42 415-425.
3. Amezketa E. Soil aggregate stability: a review. Journal of Sustainable Agriculture 1999; 14 83-151.
4. Bajracharya RM, Elliot WJ, Lal R. Interrill erodibility of some Ohio soils based on field rainfall simulation. Soil Science Society of America Journal 1992; 56 267-272.
5. Barthes BG, Kouakoua E, Larre-Larrouy M, Razafimbelo TM, de Luca EF, Azontonde A, Neves CSVJ, de Freitas PL, Feller CL. Texture and sesquioxide effects on water-stable aggregates and organic matter in some tropical soils. Geoderma 2008; 143 14-25.
6. Calero N, Barron V, Torrent J. Water dispersible clay in calcareous soils of southwestern Spain. Catena 2008; 74 22-30.

7. Chappell NA, Ternan JL, Bidin K. Correlation of physicochemical properties and sub-erosional landforms with aggregate stability variations in tropical Ultisol disturbed by forestry operations. Soil and Tillage Research 1999; 50 55-71.

8. De Boodt M. West European Methods for Soil Structure Determinations. International Society of Soil Science Communique 1, West European Group 2005; Part VI 17-18, 36-38.

9. Dexter AR, Cyzz EA. Effects of soil management on the dispersibility of clay in a sandy soil. International Agrophysics 2000; 14 269-272.

10. Dikinya O, Hinz C, Aylmore G. Dispersion and re-deposition of fine particles and their effects on saturated hydraulic conductivity. Australian Journal of Soil Research 2006; 44 47-56.

11. Duiker SW, Rhoton FE, Torrent J, Smeck NE, Lal, R. Iron (hydr)oxide crystallinity effects on soil aggregation. Soil Science Society of America Journal 2003; 67 606-611.

12. Dutartre Ph, Bartoli F, Andreux F, Portal JM, Ange A. Influence of content and nature of organic matter on the structure of some sandy soils from West Africa. Geoderma 1993; 56 459-478.

13. Ekwue EI. A simple technique for measuring soil infiltration rates during simulated rainfall. Journal of Arid Agriculture 1994; 3-7 95-104.

14. Encyclopedia of Soil Science, ESS. Aggregate stability to drying and wetting. *In* Chesworth C (Ed). *Encyclopedia of Soil Science*, Springer 2008; pp. 28-33.

15. FAO. Land and environmental degradation and desertification in Africa. FAO Corporate Document Repository, 1995. http://www.fao.org/docrep/X5318E/X5318E00.htm

16. FAO. Management of tropical sandy soils for sustainable agriculture. FAO Corporate Document Repository, 2007. 536 pp. http://www.fao.org/docrep/010/ag125e/ag125e00.htm

17. Folly A. Estimation of erodibility in the savanna ecosystem, northern Ghana. Communications in Soil Science and Plant Analysis 1995; 26 799-812.

18. Idowu OJ. Relationships between aggregate stability and selected soil properties in humid tropical environment. Communications in Soil Science and Plant Analysis 2003; 34 695-708.

19. Igwe CA, Akamigbo FOR, Mbagwu JSC. Application of SLEMSA and USLE erosion models for potential erosion hazard mapping in southeastern Nigeria. International Agrophysics 1999b; 13 41-48.

20. Igwe CA, Akamigbo FOR, Mbagwu JSC. Chemical and mineralogical properties of soils in southeastern Nigeria in relation to aggregate stability. Geoderma 1999a; 92 111-123.
21. Igwe CA, Akamigbo FOR, Mbagwu JSC. Physical properties of soils of southeastern Nigeria and the roles of some aggregating agents in their stability. Soil Science 1995b; 160 431-441.
22. Igwe CA, Akamigbo FOR, Mbagwu JSC. The use of some soil aggregate indices to assess potential soil loss in soils of southeastern Nigeria. International Agrophysics 1995a; 9 95-100.
23. Igwe CA, Ejiofor N. Structural stability of exposed gully wall in central eastern Nigeria as affected by soil properties. International Agrophysics 2005; 19 215-222.
24. Igwe CA, Nkemakosi JT. Nutrient element contents and cation exchange capacity in fine fractions of southeastern Nigerian soils in relation to their stability. Communications in Soil Science and Plant Analysis 2007; 38 1221-1242.
25. Igwe CA, Okebalama CB. Soil strength of some central eastern Nigeria soils and effect of potassium and sodium on their dispersion. International Agrophysics 2006; 20 107-112.
26. Igwe CA, Udegbunam ON. Soil properties influencing water-dispersible clay and silt in an Ultisol in southern Nigeria. International Agrophysics 2008; 22 319-325.
27. Igwe CA, Zarei M, Stahr K. Clay dispersion of hardsetting Inceptisols in southeastern Nigeria as influenced soil components. Communications in Soil Science and Plant Analysis 2006; 37 751-766.
28. Igwe CA, Zarei M, Stahr K. Colloidal stability in some tropical soils of southeastern Nigeria as affected by iron and aluminium oxides. Catena 2009; 77 232-237.
29. Igwe CA. Changes in water-dispersible clay and erosion of stony slope soil under tillage in Nigeria. Polish Journal of Soil Science 2002; 35 31-36.
30. Igwe CA. Clay dispersion of selected aeolian soils of northern Nigeria in relation to sodicity and organic carbon content. Arid Land Research and Management 2001; 15 147-155.
31. Igwe CA. Erodibility in relation to water-dispersible clay for some soils of eastern Nigeria. Land Degradation and Development 2005; 16 87-96.
32. Igwe CA. Erodibility of soils of the upper rainforest zone, southeastern Nigeria. Land Degradation and Development 2003; 14 323-334.

33. Igwe CA. Tropical soils, physical properties. *In* J Glinski, J Horabik, J Lipiec (Eds.), *Encyclopedia of Agrophysics* (1st ed.), Springer 2011; 934-937.

34. Ikazaki K, Shinjo H, Tanaka U, Tobita S, Funakawa S, Kosaki T. Field-scale aeolian sediment transport in the Sahel, West Africa. Soil Science Society of America Journal 2011; 75 1885-1897.

35. Kukal SS, Manmeet-Kaur, Bawa SS. Erodibility of sandy loam aggregates in relation to their size and initial moisture content under different land uses in semi-arid tropics of India. Arid Land Research and Management 2008; 22 216-227.

36. Lado M, Paz A, Ben-Hur M. Organic matter and aggregate-size interactions in saturated hydraulic conductivity. Soil Science Society of America Journal 2004a; 68 234-242.

37. Lado M, Paz A, Ben-Hur M. Organic matter and aggregate-size interactions in infiltration, seal formation, and soil loss. Soil Science Society of America Journal 2004b; 68 935-942.

38. Le Bissonnais Y. Aggregate breakdown mechanisms and erodibility. *In* Lal R (Ed.), *Encyclopedia of Soil Science* (2nd ed.), Taylor and Francis 2006; 40-44.

39. Leifeld J, Kogel-Knabner I. Microaggregates in agricultural soils and their size distribution determined by X-ray attenuation. European Journal of Soil Science 2003; 54 167-174.

40. Mbagwu JSC, Bazzoffi P. Soil characteristics related to resistance of breakdown of dry soil aggregates by water-drops. Soil and Tillage Research 1998; 45 133-145.

41. Mbagwu JSC, Auerswald K. Relationship of percolation stability of soil aggregates to land use, selected properties, structural indices and simulated rainfall erosion. Soil and Tillage Research 1999; 50 197-206.

42. Mbagwu JSC, Piccolo A, Mbila MO. Water stability of aggregates of some tropical soils treated with humic substances. Pedologie 1993; XLIII-2 269-284.

43. Mbagwu JSC. Erodibility of soils formed on a catenary toposequence in southeastern Nigeria as evaluated by different indexes. East African Agricultural and Forestry Journal 1986; 52 74-80.

44. Morin J. Soil crusting and sealing. In: Soil Tillage in Africa: Needs and Challenges. FAO Soils Bulletin; 1993, FAO Corporate Document Repository.

45. http://fao.org/docrep/T1696E/T1696E00.htm

46. Nwadialo BE, Mbagwu JSC. An analysis of soil components active in microaggregate stability. Soil Technology 1991; 343-350.

47. Oades JM, Waters AG. Aggregate hierarchy in soils. Australian Journal of Soil Research 1991; 29 815-828.

48. Obalum SE, Buri MM, Nwite JC, Hermansah, Watanabe Y, Igwe CA, Wakatsuki T. Soil degradation-induced decline in productivity of sub-Saharan African soils: the prospects of looking downwards the lowlands with the *sawah* eco-technology. Applied and Environmental Soil Science; Volume 2012, Article ID 673926, 10 pages. doi:10.1155/2012/673926

49. Obalum SE, Igwe CA, Hermansah, Obi ME, Wakatsuki T. Using selected structural indices to pinpoint the field moisture capacity of some coarse-textured agricultural soils in southeastern Nigeria. Journal of Tropical Soils 2011; 16 151-159.

50. Obi ME, Salako FK, Lal R. Relative susceptibility of some southeastern Nigeria soils to erosion. Catena 1989; 16 215-225.

51. Obi ME. Runoff and soil loss from an Oxisol in southeastern Nigeria under various management practices. Agricultural Water Management 1982; 5 193-203.

52. Opara CC. Soil microaggregates stability under different land use types in southeastern Nigeria. Catena 2009; 79 103-112.

53. Richter G, Negedank JFW. Soil erosion processes and their management in the German area of the Moselle river. Earth Surface Proceedings 1977; 2 261-278.

54. Shi Z, Yan F, Li L, Li Z, Cai C. Interrill erosion from disturbed and undisturbed samples in relation to topsoil aggregate stability in red soils from subtropical China. Catena 2010; 81 240-248.

55. Six J, Bossuyt H, Degryze S, Denef K. A history of research on the link between (micro)aggregates, soil biota and soil organic matter dynamics. Soil and Tillage Research 2004; 79 7-31.

56. Six J, Feller C, Denef K, Ogle SM, de Moraes-SA JC, Albrecht A. Soil organic matter, biota and aggregation in temperate and tropical soils – Effects of no-tillage. Agronomie 2002; 22 755-775.

57. Soil Science Society of America. *Glossary of Soil Science Terms*, American Society of Agronomy, Crop Science Society of America, Soil Science Society of America; 2001, Wisconsin, USA.

58. Sung CTB. Aggregate stability of tropical soils in relation to their organic matter constituents and other soil properties. Pertanika Journal of Tropical Agricultural Science 2012; 35 135-148.

59. Tisdall JM. Formation of soil aggregates and accumulation of soil organic matter. In: Structure and Organic Matter Storage in Agricultural Soils (eds. Carter MR, Stewart BA). Advances in Soil Science 1996. CRC Press, Boca Raton, FL. pp. 57-96.
60. Trapnell CG, Webster R. Microaggregates in red earths and related soils in east and central Africa, their classification and occurrence. Journal of Soil Science 1986; 37 109-123.
61. Valentin C. Soil crusting and sealing in West Africa and possible approaches to improved management. In: Soil Tillage in Africa: Needs and Challenges. FAO Soils Bulletin; 1993, FAO Corporate Document Repository. http://fao.org/docrep/T1696E/T1696E00.htm
62. Wuddivira MN, Camps-Roach G. Effects of organic matter and calcium on soil structural stability. European Journal of Soil Science 2007; 58 722-727.
63. Wuddivira MN, Stone RJ, Ekwue EI. Clay, organic matter, and wetting effects on splash detachment and aggregate breakdown under intense rainfall. Soil Science Society of America Journal 2009; 73 226-232.
64. Yan F, Shi Z, Li Z, Cai C. Estimating interrill soil erosion from aggregate stability of Ultisols in subtropical China. Soil and Tillage Research 2008; 100 34-41.
65. Zhang B, Horn R. Mechanisms of aggregate stabilization in Ultisols from subtropical China. Geoderma 2001; 99 123-145.

Chapter 3

STABILITY OF ORGANIC MATTER IN ANTHROPIC SOILS: A SPECTROSCOPIC APPROACH

M.C. Hernandez-Soriano[1], A. Sevilla-Perea[2], B. Kerré[1], and M.D. Mingorance[2]

[1]Division of Soil and Water Management, KU Leuven, Belgium

[2]Instituto Andaluz de Ciencias de la Tierra, University of Granada - CSIC, Spain, Spain

INTRODUCTION

Stability of Soil Organic Matter

The soil organic matter (SOM) plays an essential role in soil biogeochemical processes (Bot and Benites, 2005). Thus, a productive and healthy soil must present a balance among SOM protection and soil biological functioning (Wander, 2004). However, the prediction of organic matter dynamics in soil is hampered by the complexity of SOM distribution and chemical composition (Foereid et al., 2012). The integration of organic inputs in the physicochemically defined organic pools in soil (Six et al., 2002) and their effect on native organic matter has been described to vary with land use, soil physicochemical properties (Strong et al., 2004; Denef and Six, 2005); and composition of the organic inputs (Kimetu and Lehmann, 2010).

The term soil organic matter refers to all organic substances in the soil: plant and animal residues, substances synthesized through microbial and chemical reactions and biomass of soil micro-organisms. The processes responsible for the stabilization of SOM constitute an essential component of global biogeochemical cycles (Lehmann et al., 2007). Overall, the chemical composition of the organic matter (OM) and the interactions with other soil components such as the mineral phase largely drive the mechanisms for SOM stabilization (Baldock and Skjemstad, 2000), which can be summarized as: (1) biochemical stabilization, (2) physical stabilization and (3) chemical stabilization (Six et al., 2002; von Lützow et al., 2006). The extent of protection offered by each mechanism (Fig. 1) depends on the chemical and physical properties of the mineral matrix and the morphology and chemical structure of the organic matter (Six et al., 2002). Thus, each mineral matrix presents a unique and finite capacity to stabilize organic matter (Baldock and Skjemstad, 2000).

The *physical stabilization* is the preferential location of OM in the soil structure which results in lower access to OM by soil micro-organisms. Thus, integration of OM in soil aggregates reduces the availability of OM for microbial transformation (Six et al., 2002).

The *biochemical stabilization* is a selective enrichment of organic compounds, and refers to the inherent recalcitrance of specific organic molecules against degradation by microorganisms and enzymes. Thus, compounds like lignin, lipids and polyphenols will remain more stable in the soil matrix compared to more labile compounds like polysaccharides and proteins (Six et al., 2002; Kögel-Knabner et al., 2008).

The *chemical stabilization* involves all intermolecular interactions between organic and inorganic substances leading to a decrease in availability of the organic substrate due to surface condensation and changes in conformation, i.e., sorption to soil minerals and precipitation. The chemical stabilization of SOM results mainly from the interaction of SOM with minerals and metal ions (Fig. 1). These interactions include organo-mineral associations such as complexation of organic substances with polyvalent cation bridges, weak hydrophobic interactions (Van der Waals and H-binding) and sorption of SOM to soil minerals (von Lützow et al., 2006; Jastrow et al., 2007). Therefore, some authors have pointed clay fraction as an inhibitor of SOM decomposition (Kleber et al., 2007). For instance, Merckx et al. (1985) described that the stabilization of C and N in soils is positively correlated to the content of clay and silt. Moreover, other authors have indicated that the specific type of clay present in the soil, i.e. clay mineralogy, is most relevant for the capability of a particular soil to stabilize OM (Sollins et al., 1996; Denef and Six, 2005). Consequently, it might be adequate to evaluate specific surface and surface reactivity of soil minerals as predictors of OM stabilization rather than clay content (Baldock and Skjemstad, 2000).

According to Kogel-Knabner et al. (2008), the protection of OM against decomposition by the described mechanisms decreases in the order: chemically protected > physically protected > biochemically protected > non-protected (Fig. 1).

The complexation of OM with mineral surfaces occurs mostly through ligand exchange, which is an organo-mineral association between OH groups on mineral surfaces and ionized phenolic OH and carboxylic groups of the OM (Korshin et al., 1997). This interaction is particularly relevant in acidic soils with minerals presenting protonated OH groups and is reverted with increasing pH values in soils (von Lützow et al., 2006).

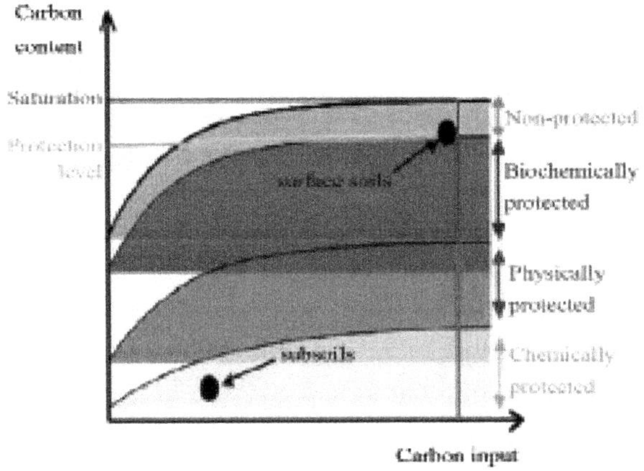

Figure 1: Protection of organic matter (OM) in soil. Chemical protection refers to the interactions of OM with minerals; physical protection renders OM poorly accessible to microbes and enzymes; and biochemical protection results from the differential degradability of organic structures. Saturation is thus defined as the theoretical C storage capacity of a soil (adapted from Six et al., 2002).

Another important mechanism for organo-mineral association is the formation of polyvalent cation bridges. The presence of negatively charged acidic functional groups (COO-) in organic molecules results in repulsion from negatively charged surfaces of clay minerals. However, polyvalent cations function as a cation bridge between those two negatively charged sites (von Lützow et al., 2006). In alkaline and neutral soils, the most abundant multivalent cations are Ca2+ an Mg2+ while Fe3+ and Al3+ predominate in acid soils, and present higher binding strength. Besides, the role of cation bridges in the stabilization of SOM is particularly relevant for soils with a predominance of 2:1 clays such as smectite and illite (Jastrow et al., 2007).

Adsorption processes also contribute to protect OM against biological degradation (Balesdent et al., 2000). Although feldspars and quartz are the most common minerals in soils, their specific surfaces are rather low (approximately 0.1 m^2/g), while clay particles provide a significant surface area (specific surface >10 m^2/g) for the adsorption of OM (Jastrow et al., 2007). Therefore, soils with a high content of clays may provide higher SOM protection than sandy soils, resulting in higher total contents of organic matter (Blanco-Canqui and Lal, 2004). Multivalent cations also contribute to the stabilization of OM by inducing flocculation. Clay particles saturated with multivalent cations remain in a flocculated state, which reduces the exposure of adsorbed organic materials on the clay surface. Thus, flocculation and condensation of organo-mineral complexes can effectively isolate and protect OM from decomposition (Baldock and Skjemstad, 2000). This mechanism has been mainly described for soils with low-charge clays (Jastrow et al., 2007).

Kleber et al. (2007) proposed a conceptual model for organo-mineral interactions in soils, partly based on the self-organizing molecular structure of SOM. These authors suggest that SOM is adsorbed on the mineral surfaces in three discrete zones. Thus, in a contact zone, amphiphilic+ organic fragments accumulate mainly on charged surfaces through electrostatic interactions, thereby directing hydrophobic parts outwards toward the aqueous solutions. This organization results in a membrane-like bilayer with a hydrophobic core. In the outer region, denominated as kinetic zone, further accumulation of organic fragments likely occurs, and the process is assumed to be mediated by the presence of multivalent cations. However, further research is still necessary to advance our understanding on the dynamics and structure of organo-mineral interactions to validate this zonal concept.

Although abundant research has been previously conducted, the described mechanisms responsible for the reduction of OM decomposition rates, such as sorption of OM to minerals, are not yet well understood. Different stabilization mechanisms may act simultaneously and those proposed in the literature remain speculative and poorly supported by data, mainly due to methodological constraints (von Lützow et al., 2006). Therefore, novel research strategies attempt a better understanding of the mechanisms of OM stabilization in soil by studying the molecular composition of SOM in specific soil fractions. The implementation of spectroscopic techniques (Cory and McKnight, 2005; Lehmann et al., 2010) provides novel and promising methodological strategies to undertake the challenge of characterizing SOM composition and spatial distribution in soil.

Stability of Organic Matter in Artificial Soils

The production of artificial soils or technosols (WRB, 2006) aims to recover landscapes or increase soil productivity. Overall, the addition of soils with organic materials aims to benefit soil quality or increase crop yields by regulating nutrient supply and improving soil structure (Wagner et al., 2007). Addition of single or composite organic wastes is expected to have a positive effect in initiating soil aggregation of structurally degraded topsoils (Wagner et al., 2007 and references therein). For instance, composted or stabilized municipal sewage sludge is frequently applied to soil as organic amendment for restoration purposes. Otherwise, abandoned Fe mine tailings provide a source of Fe-rich mud (FeM), that constitutes an environmental challenge for its adequate disposal, being currently stored in large open-air ponds. A suitable approach might be to use such FeM as a substrate to obtain artificial soils. The Fe oxides and hydroxides present in the FeM might provide a suitable surface for adsorption of organic compounds and may favor the formation of organo-mineral associations. which may result in a pool of chemically stabilized OM. Chemical stabilization by organo-mineral associations is a main mechanisms leading to soil aggregation (Kögel-Knabner et al., 2008) and such strategy might result in pools of carbon with long residence time in soil (Macias and Camps Arbestain, 2010).

The objective of this study was to evaluate the chemical composition of OM in artificial soils obtained from organic wastes combined with the FeM at different ratios. The analysis of OM composition at a molecular level and the characterization of the spatial distribution among different pools by Fourier transform infrared spectroscopy (FTIR) can be directly related to SOM stability, soil respiration and OM decomposition rates. Previous research has demonstrated that intensities of distinct peaks obtained by FTIR analysis can be a measure of decomposition of organic carbon in soil (Haberhauer et al., 1998).

Otherwise, dissolved organic matter (DOM) constitutes a highly available carbon source for microorganisms while playing a fundamental role in the mobilisation of organic compounds (Kalbitz et al., 2003). Variations in the composition of the OM present in this pool are an essential component to the knowledge on SOM dynamics (Kalbitz et al., 2000). Therefore, this pool of OM deserves particular attention in our attempt to characterize the evolution of SOM pools. The spectrofluorometric analysis of the soil solution extracted from the different scenarios assayed provides a fingerprinting of the composition of DOM (Cory and McKnight, 2005). Thus, excitation-emission matrix spectroscopy provides the sensitivity to examine subtle changes in DOM fluorescence and provide a valuable insight into variations on the DOM pool composition.

MATERIALS AND METHODS

Studied Area

The studied area is an Fe mine dump in Southeast Spain (Alquife, Granada) planned to be used for residential, leisure and agro-industrial activities. The soil is a degraded technosol with high infiltration rate under a continental

Mediterranean climate.

Technosols

A collection of five technosols (Table 1) was obtained by combining a composted mixture (19.4% organic carbon) of sewage sludge from wastewater treatment and olive pruning (SVC) with FeM (44% Fe oxides) and/or a biodiesel byproduct (DRS) with a high concentration of glycerol, following saturation and incubation at 28° C for 30 d. The mineral waste was originated in milling activities carried out in the mine site. A second collection of five technosols was obtained by controlled acidification (H_2SO_4 5M) of mixtures as such described for the first set of technosols (Table 1).

Table 1: Composition of technosols

Technosol	**SVC**	**FeM**	**DRS**	**%OC**
T1	100%	-	-	19.4
T2	90%	-	10%	21.0
T3	90%	10%	-	17.5
T4	85%	5%	10%	19
T5	70%	30%	-	13.6

A second batch of technosols (TS) was prepared with equivalent composition but SVC and FeM were saturated and incubated for 1 week before mixing. For TS3 and TS4, FeM was acidified with $H_2SO_4$5N during the preconditioning step.

Samples of the different technosols were collected after 2, 9, 20 and 30 d of incubation and stored at -20° C for posterior analysis.

Carbon Mineralization Rates

To determine carbon mineralization upon application of the technosols in the mine dump, subsamples of soil from two different plots in the dump (AL7 and AL14) were amended at 2% with the different technosols and placed in air tight incubation jars with a volume of 300 mL and moisture content adjusted to field capacity. The lids of the incubation jars were fitted with three-way valves to allow sampling the air from the headspace. The jars were stored in an incubation room at 25° C for circa 120 d. Headspace in the jars was periodically sampled with 60 mL syringes and the CO_2 concentration measured with an infrared gas analyzer (LI-COR; Li-820). The amount of carbon respired was calculated using the ideal gas equation and expressed as percentage of carbon respired relative to the total carbon content in the amended soil.

FTIR-microscopy

Microaggregates-like structures (100-200 μm) were isolated and collected from the different technosols after 2 and 30 d of incubation and analyzed with a Fourier transform infrared spectrophotometer (Varian 620-IR IR microscope) coupled to a microscope (FTIR-microscope) using a KBr splitter and a liquid nitrogen cooled Focal Plane Array detector for spectrochemical imaging and a CCD camera. The spectra were recorded for the microaggregates-like structures in the mid-infrared range (4000–800 cm^{-1}) by combining 32 scans at a resolution of 1 cm^{-1}. The spectra were recorded in absorbance units. Peak area integration and analysis of the spectral features distribution in the microaggregates were performed using the software Agilent Resolutions Pro. Spectra and image analysis presented were obtained as the average of 5 spectra.

Water Soluble Organic Matter

The fraction of water soluble organic matter (WSOM) was obtained from the technosols sampled at different incubation times, through centrifugation (10 min at 3000 g) using the 'double chamber' method (Bufflap and Allen, 1995). After centrifugation, the soil solution samples were immediately filtered through a 0.45-μm filter. The solutions were analyzed for dissolved organic carbon (DOC) using a TOC-analyser (Analytical Sciences Thermalox). The UV-absorbance was measured with a UV-VIS spectrophotometer (Perkin-Elmer, Lambda 20, quartz cells).

Variation in the ratio of absorbance to DOC was used to characterize the quality of DOM, through the specific UV absorbance at 254 and 340 nm (Tipping et al., 2009).

Spectrofluorometry

The soil solution samples were diluted such that the absorbance at 254 nm was less than 0.2 prior to the collection of fluorescence spectra (Miller et al., 2010). Fluorescence excitation-emission spectra were obtained for the pore water solutions using a JY HORIBA Fluorolog-3 spectrofluorometer with an excitation range set from 240 to 400 nm and an emission range set from 300 to 500 nm in 2 nm increments. Instrumental parameters were excitation and emission slits, 5 nm; response time, 8 s; and scan speed, 1200 nm min^{-1}. Spectra were analyzed using the software FluorEssence.

Adsorption of Gallic Acid on Fe Mud

Sorption isotherms were carried out using a batch equilibration method, with 5 g of FeM and 20 mL of an aqueous solution of gallic acid (GA, Sigma Aldrich) at concentrations ranging 5–50 mM. The samples were mechanically shaken end-over-end in a thermostatic chamber at 20 ± 1 °C for 24 h. The samples were centrifuged at 3500 rpm and 15 °C for 15 min. The isotherms were run in duplicate. A GA solution without addition of FeM was used as control, to account for possible degradation during the batch process.

The difference between initial concentration of GA and the concentration of GA remaining in solution after reaching equilibrium was attributed to sorption of GA on FeM. The sorption equilibrium partition coefficient Kd (L kg^{-1}) was calculated as $K_d = X / C_e$, where X is the concentration of GA in the FeM (mg kg^{-1}) and Ce is the concentration of GA in the solution at equilibrium (mg L^{-1}). The adsorption experiment was described by the empirical Freundlich equation $(X = K_f C_e^n)$, where K_f is the Freundlich adsorption coefficient (L kg^{-1}) and n a constant which depends on the adsorbate, the adsorbent and the temperature.

RESULTS

Carbon Mineralization Rates

Results obtained from the carbon mineralization assays are summarized in Figures 2-4, which describe cumulative respiration determined for the application of the different technosols to samples of soil collected from the mine dump.

Soil addition with the first batch of technosols (Table 1) increased carbon mineralization rates for all the technosols applied (Fig. 2) compared to control soil (C). For technosols produced solely from SVC and Fe mud, a higher ratio SVC:FeM (T3: SVC:FeM, 90:10) resulted in lower CO_2 production than T5 (SVC:FeM 70:30), regardless the OC content.

The preconditioning step significantly affected carbon mineralization rate. Addition of technosols TS2 and TS4 (saturated) to the dump soil resulted in lower mineralization rates than application of non-preconditioned technosols

(Fig. 3). Thus, for technosols obtained solely from SVC/FeM, saturation of wastes before mixing (TS5) significantly decreased the mineralization of OC added to the soil when FeM was acidified during the preconditioning (TS3, Fig. 4).

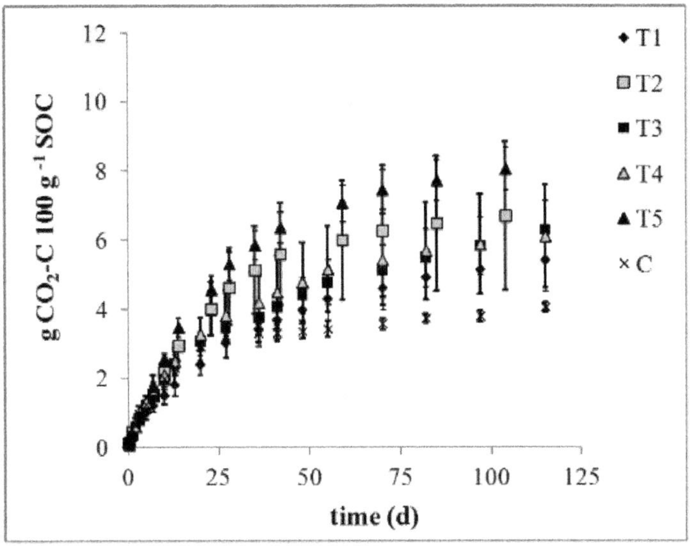

Figure 2: Cumulative CO_2-C respired for T1-T5 and control soil.

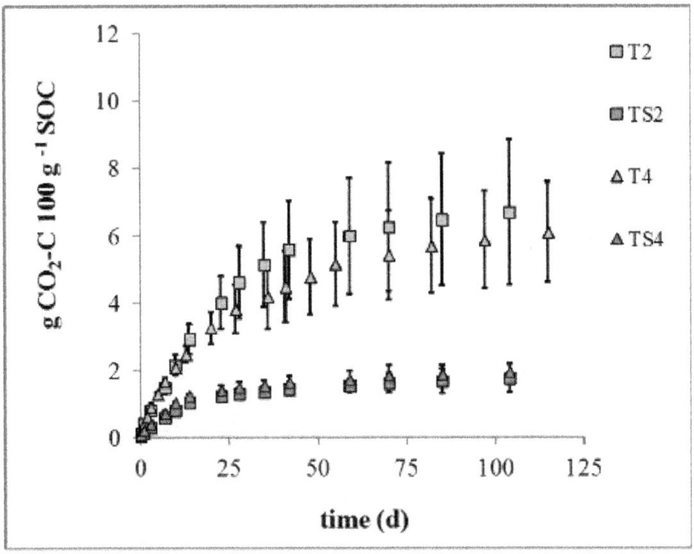

Figure 3: Cumulative CO_2-C respired for T2, T4, TS2 and TS4.

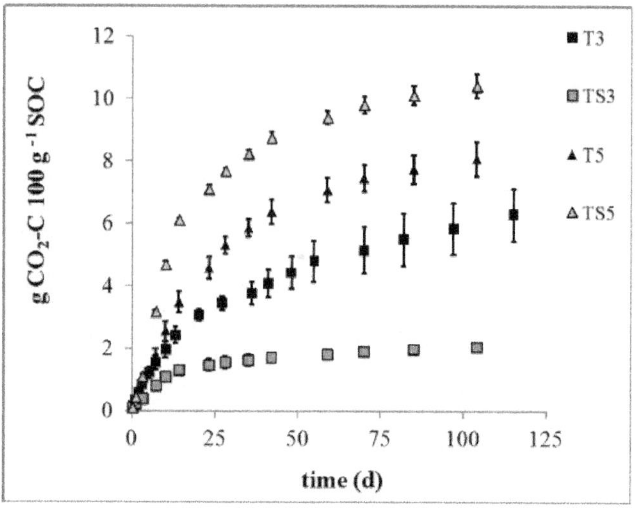

Figure 4: Cumulative CO_2-C respired for T3, T5, TS3 and TS5.

Dissolved Organic Carbon: Concentration and Composition

Total organic carbon content and UV-absorbance analysis performed for the water soluble organic carbon collected from the technosols indicated that DOC from technosols prepared with SVC+FeM (90/10) was highly humified, as suggest by SUVA and extinction coefficients values at 254 and 340 nm (Fig. 5 and 6).

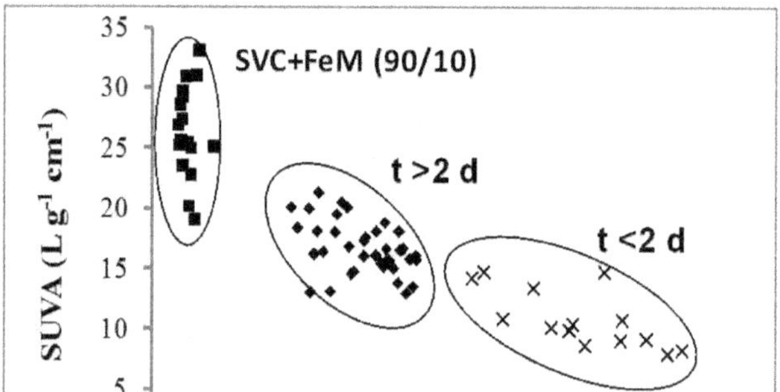

Figure 5: Relationship SUVA-DOC. Squares, technosol prepared with SVC+FeM (90/10); diamonds, rest of technosols after incubation time > 2 d; X, rest of technosols after 2 incubation days.

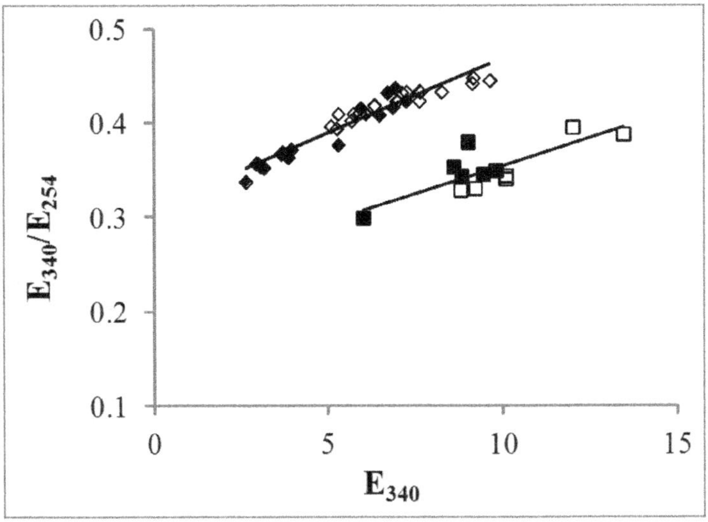

Figure 6: Relationship ratio of extinction coefficients at 340 and 254 nm and the extinction coefficient at 340. Full squares, T3, T9 (t=2d); empty squares, T3, T9 (t=30 d); full diamond, T1-T2, T4-T8, T10-T12 (t=2 d); empty diamond, T1-T2, T4-T8, T10-T12 (t=30 d).

Table 2 summarizes the values determined for the different fluorescence indexes derived from the spectrofluorometric analysis. Overall, higher humification index values (HIX) were derived for increasing concentration of FeM in a particular technosol, which confirms higher humification of the water soluble organic carbon. The freshness index values (β:α) confirmed the expected predominance of recently derived DOC. Besides, the high fluorescence index values determined (FI) indicated that DOC was originated from intense microbial activity.

Overall, preconditioning the materials prior to the production of the technosols resulted in higher values of the HIX, likely due to an increase in organo-metal interactions, while changes on the β:α and FI values were neglectable. Variation of the fluorescence indexes over time was not conclusive.

Otherwise, results from the excitation emission matrixes (Ex/Em) collected for T1 and T3 (Fig. 7)indicated a substantial increase in the fraction of UV (Ex/Em 260/400-460) and visible (Ex/Em 320-360/400-460) humic-like organic matter for T3 compared to T1, which suggest the presence of a pool of highly stable, low degradation rate OM. Moreover, the strong increase in fluorescence intensity suggests that the added OM might complex metal ions in solution, which can result in a protective effect for DOM against rapid mineralization. The attenuation of the signal at Ex/Em 320-360/400-460 over time suggest the precipitation of organo-metal complexes in the solid phase.

Table 2: Spectroscopic analysis

Technosol	Incubation (d)	HIX	β:α	FI
T1	2	5.25	0.68	2.17
T1	30	5.72	0.66	2.09
T3	2	7.25	0.60	1.99
T3	30	7.16	0.58	1.88
T4	2	4.62	0.7	2.18
T4	30	6.14	0.67	2.12
TS1	2	4.91	0.74	2.19
TS1	30	5.51	0.67	2.25
TS3	2	6.85	0.60	1.96
TS3	30	7.13	0.62	1.91
TS4	2	6.57	0.58	2.21
TS4	30	7.46	0.58	1.82

HIX: Humification index. β:α: freshness index. FI: Fluorescence index.

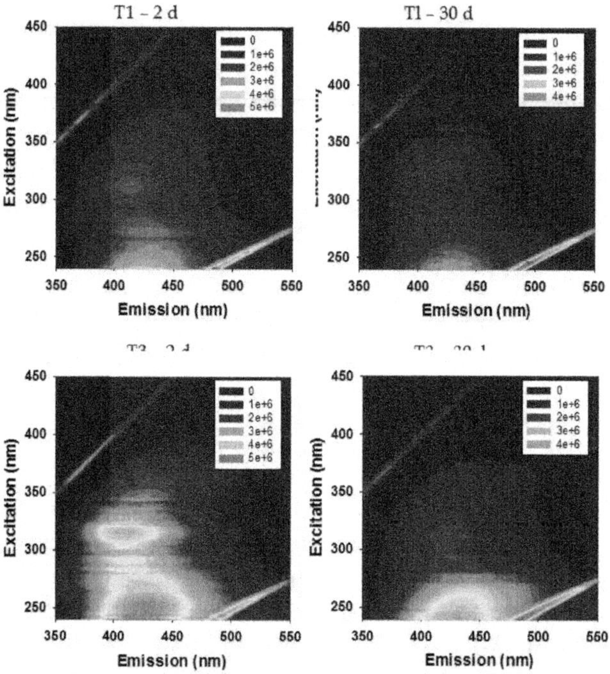

Figure 7: Excitation-Emission matrixes for technosols T1 and T3 after 2 and 30 d of incubation.

FTIR-microscopy

Spectra obtained for microaggregate-like structures (Fig. 8) showed a consistent absence of aliphatic-C (2900 cm^{-1}); the presence of aromatic compounds-C assigned to signals at 1400-1500 cm^{-1} and at 1600 cm^{-1}; and aromatic overtones at 1790, 1865 and 1998 (T3, Fig. 8), according to previous literature (Demyan et al., 2012). Polysaccharide-C were identified in the fingerprint region (between 800 and 1200 cm^{-1}) while peak at 3620 cm^{-1} are related to the presence of clay like compounds (Lehmann et al., 2007). Additionally, a peak at 3700 cm^{-1} was obtained for the analysis of T5, which might also be related to clay-like compounds.

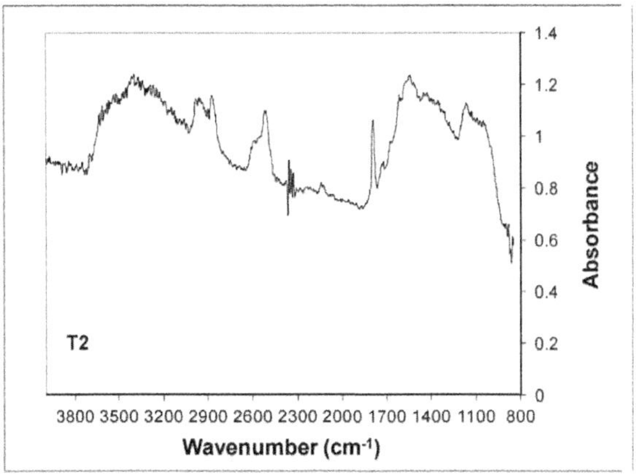

(a)

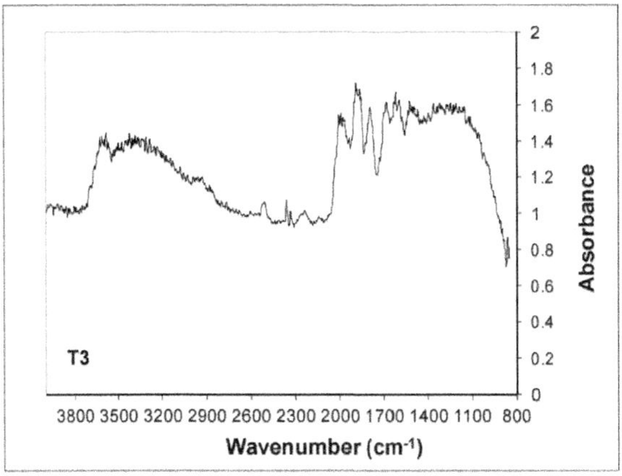

(b)

Figure 8: FTIR spectra collected for technosols T2 and T3.

Analyzing the distribution of such spectral features in soil microaggregates revealed polysaccharides homogeneously dispersed on the surface of the microaggregates, as depicted for T1 (Fig. 9) and T3 (Fig. 10). Otherwise, distribution analysis for T1 and T3 (Fig. 9 and 10) suggested the presence of cores of aromatic compounds (Wan et al., 2007).

800-1200 cm^{-1}

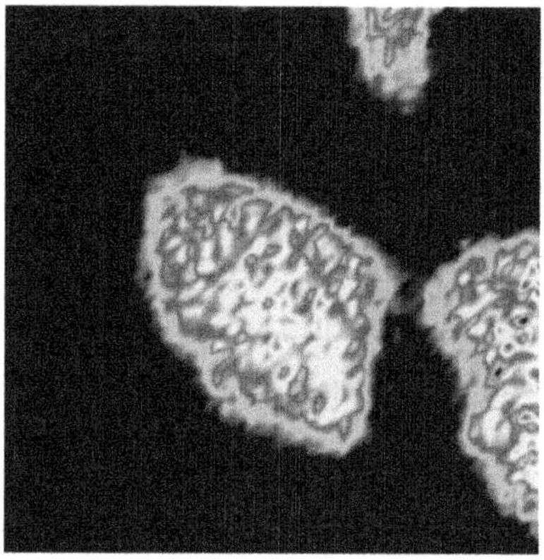

(a)

1400-1500 cm^{-1}

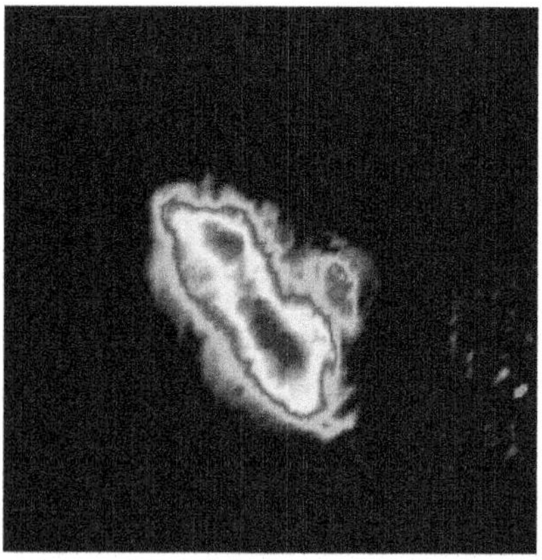

(b)

Figure 9: Distribution of polysaccharides (800-1200 cm^{-1}) and aromatic compounds (1400-1500 cm^{-1}) in a microaggregate from T1 (without FeM).

800-1200 cm^{-1}

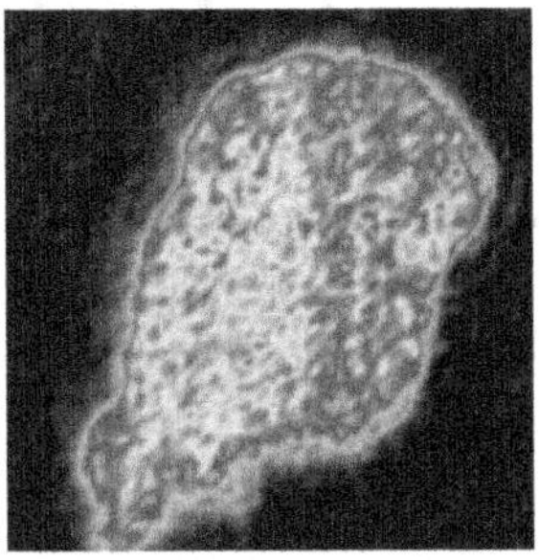

(a)

1400-1500 cm^{-1}

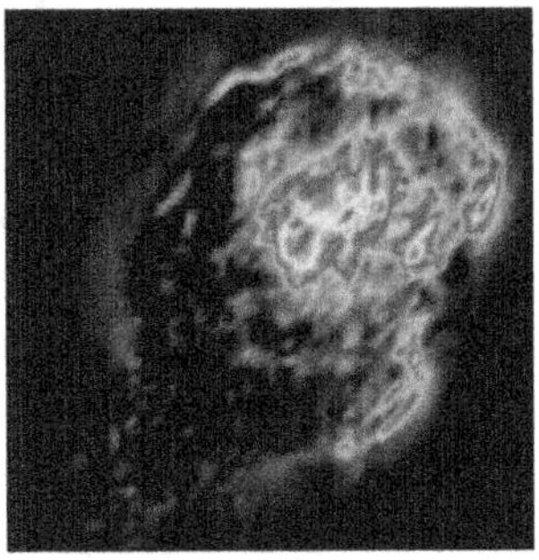

(b)

1410 cm^{-1}

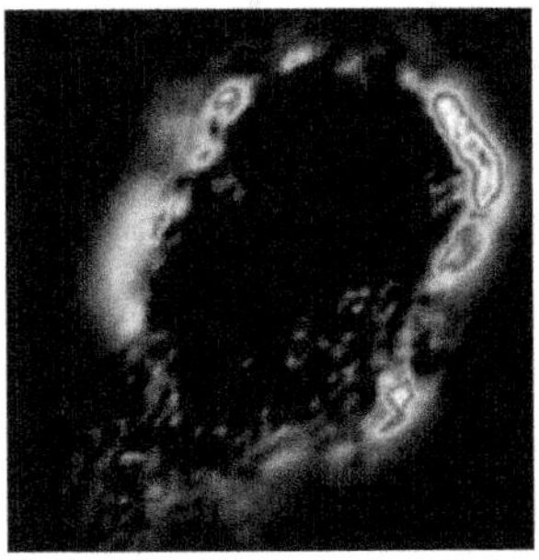

(c)

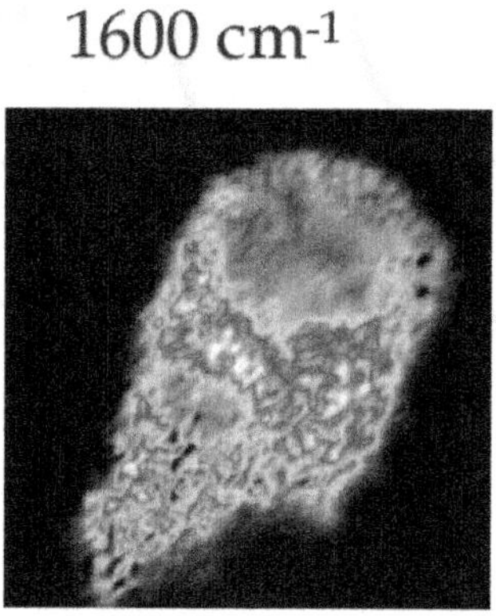

(d)

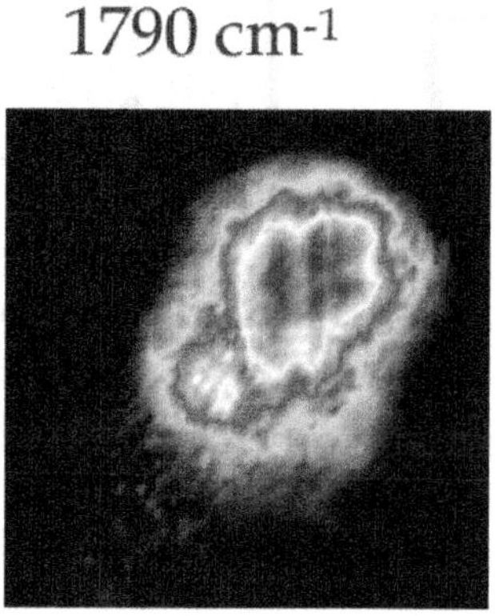

(e)

Figure 10: Distribution of chemical compounds in a microaggregate from T3 (SVC plus 10% FeM) obtained from FTIR spectra. Polysacharides at 800-1200 cm^{-1}, aromatic compounds at 1400-1500, 1600, 1410 and 1790 $cm^{-1}$and organo-mineral associations at 1410 cm^{-1}.

The distribution of clay-like compounds in the microaggregates isolated from T1, T3 and T5 indicates an increasing presence of such compounds for higher concentrations of FeM in the technosols (Fig. 11).

T1

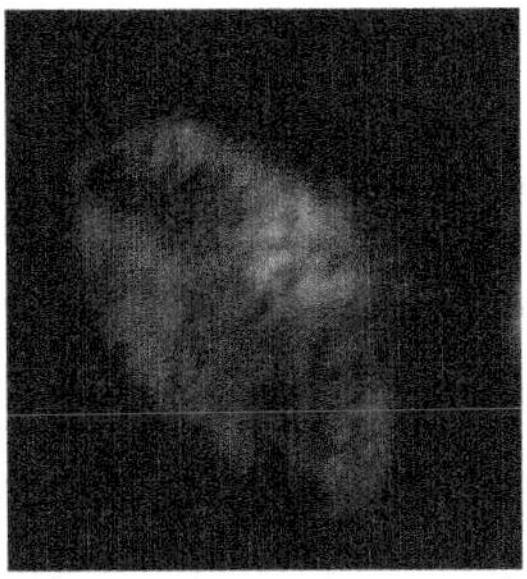

(a)

T3

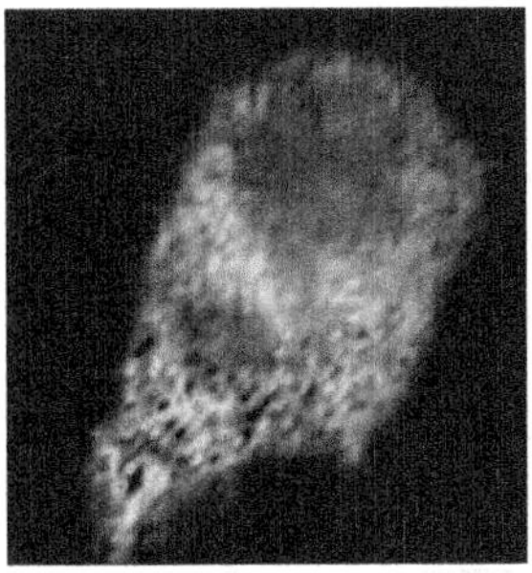

(b)

T5

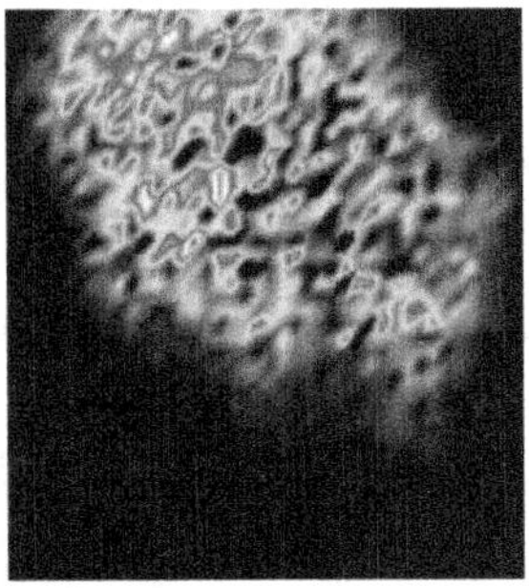

(c)

Figure 11: Distribution analysis for the signal recorded at 3620 cm^{-1} in micro-aggregates from the three technosols.

Thus, a weak signal was detected for T1 that might be related to oxidized compounds that overlap with the signal corresponding to clay-like compounds, while a strong signal was obtained for T3 and T5, consistently with the different ratios of FeM in their composition. Moreover, the distribution of clay-like compounds obtained for T3 (Fig. 11) overlaps with the distribution depicted for aromatic compounds (1600 cm^{-1}, Fig. 10), which confirms the presence of organo-mineral associations.

Adsorption of Gallic Acid on Fe Mud

Results from batch adsorption assays confirmed the capability of FeM to adsorb gallic acid (300 mmol kg^{-1}), probably through interaction of the carboxyl and phenolic groups with the Fe oxide surface, as determimed by the decrease in the signals at 220 and 270 nm for increasing concentrations of FeM in solution (Fig. 12). The adsorption constant derived, K_d=231.5 L kg^{-1}, indicates high adsorption of the acid in the FeM n=0.1, weak sorption of the second layer.

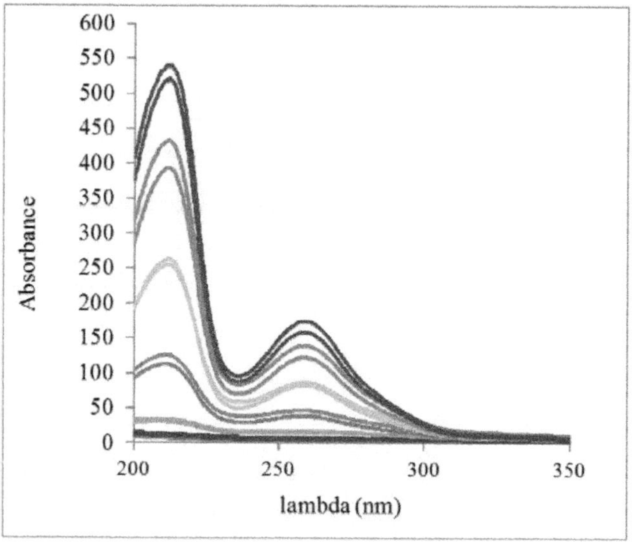

Figure 12: Gallic acid solution UV spectra. K_d=231.5 L kg^{-1}, high adsorption of the acid in the FeM n=0.1, weak sorption of the second layer.

DISCUSSION

Fresh organic inputs applied to a harsh environment such as a mine dump in the arid Mediterranean climate can be expected to be rapidly mineralized, as previously described by Novara et al. (2012). The evolution of DOM in a particular soil largely determines the SOM stability and protective capacity of the soil. Hence, the aromaticity and fluorescent properties of DOM provide an adequate characterization of this pool of OM, which allow adequately predicting carbon mineralization rates and protective capacity for a given soil scenario (Murphy et al., 2010).

The extent and rate of DOM biodegradation and humification were in agreement with previous studies (Kalbitz et al., 2003). The increase in DOM with low aromaticity upon addition of such amendments might enhance the microbial activity in the soil, but poor beneficial effects can be expected in the long term due to the short residence time of the OM added. The application of composite amendments including a source of metal ions might contribute to a longer permanence of OM in the soil (Kaiser and Kalbitz, 2012), which will largely benefit soil quality in the long term. Thus, application of composites with a low percentage of FeM resulted in higher humification indexes. Moreover, soil preconditioning by acidification and incubation under saturated conditions promoted the formation of organo-metal complexes, which resulted in lower mineralization rates (Kaiser and Kalbitz, 2012).

Otherwise, the presence of FeM in the artificial soils provides a pool of Fe oxides and Fe and Al hydroxides that presents a clay-like behaviour (Lehmann et al., 2007). This was confirmed by peak position at 3620 cm^{-1} in spectra collected for T3 (Fig. 10-11). The slight signal recorded for T1 can be attributed to hydroxyl groups in the SVC. Therefore, the incorporation of FeM in the production of technosols results in organo-mineral associations, due to complexation of Fe with phenol/carboxyl groups, which contributes to the protection and stabilization of fresh inputs of organic carbon. Complexation of fulvic acids with Fe oxides surfaces has been linked to the occurrence of a band around 1410 cm^{-1} (Gu et al., 1994). For technosols from SVC+FeM, distribution analysis suggests that such complexes locate in the edges of the microaggregate as depicted for T3 in Fig. 10. However, the protection is limited by the low adsorption capacity of this FeM, which can be attributed to a negatively charged surface (pH=8.5). Organic compounds present in the fresh inputs such as gallic acid (pH=3.5) might induce short-term acidification on the Fe oxide surface, which could allow the adsorption of carboxyl and phenolic groups (Ni et al., 2011). However, increasing the ratio of FeM in the technosol might counteract such effect, resulting in lower protection effect, which explains the higher mineralization rate determined for such technosols.

Polysaccharides were ubiquitous on the microaggregates analyzed and homogeneously dispersed on the surface of the microaggregates, which is consistent with an increase in the microbial activity due to the addition of fresh inputs of organic carbon (Six et al., 2004). The presence of cores of aromatic compounds in the microaggregates-like structures analyzed was consistent with hypothesis previously established in the literature for the formation of aggregates (Six et al., 2002; Six et al., 2004; Wan et al., 2007).

Overall, the results obtained with this study have demonstrated that succesful production of technosols as organic amendments to ameliorate soil quality might highly benefit of the incorporation of a mineral substrate at an optimized ratio. The spectroscopic characterization of DOM and soil aggregates provides a low cost, effective analysis to determine the effect of a particular amendment in soil structure and OM stability.

CONCLUSIONS

The production of technosols from low-cost wastes provides a suitable strategy for wastes disposal while providing a valuable resource for plant sustaining in soil. Application of composite amendments to degraded soils might constitute a highly effective approach for increasing soil health and productivity and a suitable alternative to conventional strategies based on single organic amendments. Moreover, such composites constitute a balanced soil amendment, which involves a compromise between enhancement of soil biological activity and the establishment of pools of stabilized organic matter.

The approach presented provides general guidances for designing optimized mixtures of C-rich organic materials through characterization of the DOM pool, soil aggregates, and the potential of the composites for chemical stabilization of OM. Thus, spectrophotometric fingerprinting of DOM and molecular characterization of OM in soil aggregates have been demonstrated to provide soil quality benchmarks to develop technosols tailored for an specific environmental scenario.

Rehabilitation plans can be designed according to soil-plant requirements as well as safe and effectively cost means for disposal wastes. Fe-enriched amendments might constitute an essential component for technosols, playing a key role in the chemical stabilization of organic matter in soil.

Overall, an optimized combination of mineral and organic wastes may result in a pool of chemically stabilized organic matter. The proposed technosols present a significant potential to create a sink of C while providing an inexpensive in-situ strategy for wastes disposal.

ACKNOWLEDGEMENTS

MCHS thanks the KU Leuven (Belgium) for a postdoctoral fellowship (PDMK/10/080). The research reported was partially supported by a Hercules project (2011-2012, KU Leuven, Belgium) and Junta de Andalucía-P08-RNM3526 (Spain).

REFERENCES

1. J. A Baldock, J. O Skjemstad, 2000Role of the soil matrix and minerals in protecting natural organic materials against biological attack. Organic Geochemistry 31697710
2. J Balesdent, C Chenu, M Balabane, 2000Relationship of soil organic matter dynamics to physical protection and tillage. Soil and Tillage Research 53215230
3. H Blanco-canqui, R Lal, 2004Mechanisms of Carbon Sequestration in Soil Aggregates. Critical Reviews in Plant Sciences 23481504
4. A Bot, J Benites, 2005The importance of soil organic matter: the key to drought resistant soil, sustained food and production.. FAO, Rome.
5. S. E Bufflap, H. E Allen, 1995Comparison of pore-water sampling techniques for trace metals. Water Res. 2920512054
6. R. M Cory, D. M Mcknight, 2005Fluorescence Spectroscopy Reveals Ubiquitous Presence of Oxidized and Reduced Quinones in Dissolved Organic Matter. Environmental Science & Technology 3981428149
7. M. S Demyan, F Rasche, E Schulz, M Breulmann, T Müller, G Cadisch, 2012Use of specific peaks obtained by diffuse reflectance Fourier transform mid-infrared spectroscopy to study the composition of organic matter in a Haplic Chernozem. European Journal of Soil Science 63189199
8. K Denef, J Six, 2005Clay mineralogy determines the importance of biological versus abiotic processes for macroaggregate formation and stabilization. European Journal of Soil Science 56469479
9. B Foereid, P. H Bellamy, A Holden, G. J. D Kirk, 2012On the initialization of soil carbon models and its effects on model predictions for England and Wales. European Journal of Soil Science 633241
10. B Gu, J Schmitt, Z Chen, L Liang, J. F Mccarthy, 1994Adsorption and desorption of natural organic matter on iron oxide: mechanisms and models. Environmental Science & Technology 283846
11. G Haberhauer, B Rafferty, F Strebl, M. H Gerzabek, 1998Comparison of the composition of forest soil litter derived from three different sites at various decompositional stages using FTIR spectroscopy. Geoderma 83331342
12. J Jastrow, J Amonette, V Bailey, 2007Mechanisms controlling soil carbon turnover and their potential application for enhancing carbon sequestration. Climatic Change 80523
13. K Kaiser, K Kalbitz, 2012Cycling downwards- dissolved organic matter

in soils. Soil Biology & Biochemistry 522932

14. K Kalbitz, J Schmerwitz, D Schwesig, E Matzner, 2003Biodegradation of soil-derived dissolved organic matter as related to its properties. Geoderma 113273291
15. K Kalbitz, S Solinger, J Park, H Michalzik, B Matzner, E., 2000Controls on the dynamics of dissolved organic matter in soils: a review. Soil Science 165277304
16. J Kimetu, J Lehmann, 2010Stability and stabilisation of biochar and green manure in soil with different organic carbon contents. Aust. J. Soil Res. 48577585
17. M Kleber, P Sollins, R Sutton, 2007A conceptual model of organo-mineral interactions in soils: self-assembly of organic molecular fragments into zonal structures on mineral surfaces. Biogeochemistry 85924
18. I Kögel-knabner, G Guggenberger, M Kleber, E Kandeler, K Kalbitz, S Scheu, K Eusterhues, P Leinweber, 2008Organo-mineral associations in temperate soils: Integrating biology, mineralogy, and organic matter chemistry. Journal of Plant Nutrition and Soil Science 1716182
19. G. V Korshin, M. M Benjamin, R. S Sletten, 1997Adsorption of natural organic matter (NOM) on iron oxide: Effects on NOM composition and formation of organo-halide compounds during chlorination. Water Research 3116431650
20. J Lehmann, J Kinyangi, D Solomon, 2007Organic matter stabilization in soil microaggregates: implications from spatial heterogeneity of organic carbon contents and carbon forms. Biogeochemistry 854557
21. J Lehmann, D Solomon, S Balwant, G Markus, 2010Organic Carbon Chemistry in Soils Observed by Synchrotron-Based Spectroscopy. Developments in Soil Science. Elsevier, 289312
22. F Macias, Camps Arbestain, M., 2010Soil carbon sequestration in a changing global environment. Mitigation and Adaptation Strategies for Global Change 15511529
23. R Merckx, den Hartog, A., van Veen, J.A., 1985Turnover of root-derived material and related microbial biomass formation in soils of different texture. Soil Biology and Biochemistry 17565569
24. M. P Miller, B. E Simone, D. M Mcknight, R. M Cory, M. W Williams, E. W Boyer, 2010New light on a dark subject: comment. Aquatic Sciences 72269275
25. K. R Murphy, K. D Butler, R. G. M Spencer, C. A Stedmon, J. R Boehme, G. R Aiken, 2010Measurement of Dissolved Organic Matter

Fluorescence in Aquatic Environments: An Interlaboratory Comparison. Environmental Science & Technology 4494059412

26. J Ni, J. J Pignatello, B Xing, 2011Adsorption of aromatic carboxylate ions to black carbon (biochar) is accompanied by proton exchange with water. Environmental Science & Technology 4592409248
27. A Novara, La Mantia, T., Barbera, V., Gristina, L., 2012Paired-site approach for studying soil organic carbon dynamics in a Mediterranean semiarid environment. CATENA 8917
28. J Six, H Bossuyt, S Degryze, K Denef, 2004A history of research on the link between (micro)aggregates, soil biota, and soil organic matter dynamics. Soil & Tillage Research 79731
29. J Six, R. T Conant, E. A Paul, K Paustian, 2002Stabilization mechanisms of soil organic matter: Implications for C-saturation of soils. Plant and Soil 241155176
30. P Sollins, P Homann, B. A Caldwell, 1996Stabilization and destabilization of soil organic matter: mechanisms and controls. Geoderma 7465105
31. D. T Strong, H De Wever, R Merckx, S Recous, 2004Spatial location of carbon decomposition in the soil pore system. European Journal of Soil Science 55739750
32. E Tipping, H. T Corbishley, J. F Koprivnjak, D. J Lapworth, M. P Miller, C. D Vincent, J Hamilton-taylor, 2009Quantification of natural DOM from UV absorption at two wavelengths. Environmental Chemistry 6472476
33. M Von Lützow, I Kögel-knabner, K Ekschmitt, E Matzner, G Guggenberger, B Marschner, H Flessa, 2006Stabilization of organic matter in temperate soils: mechanisms and their relevance under different soil conditions- a review. European Journal of Soil Science 57426445
34. S Wagner, S. R Cattle, T Scholten, 2007Soil-aggregate formation as influenced by clay content and organic-matter amendment. Journal of Plant Nutrition and Soil Science 170173180
35. J Wan, T Tyliszczak, T. K Tokunaga, 2007Organic carbon distribution, speciation, and elemental correlations within soil micro aggregates: Applications of STXM and NEXAFS spectroscopy. Geochimica Et Cosmochimica Acta 7154395449
36. M Wander, 2004Soil organic matter fractions and their relevance to soil function In: Magdoff, F.a.R.R.W. (Ed.). Soil Organic Matter in Sustainable Agriculture. CRC Press LLC, Upper Saddle River, NJ.
37. I. W. G Wrb, 2006World reference base for soil resources In: FAO (Ed.). World Soil Resources Reports. FAO, Rome.

Chapter 4

SOIL FERTILITY STATUS AND ITS DETERMINING FACTORS IN TANZANIA

M.C. Hernandez-Soriano[1], A. Sevilla-Perea[2], B. Kerré[1], and M.D. Mingorance[2]

[1]Division of Soil and Water Management, KU Leuven, Belgium

[2]Instituto Andaluz de Ciencias de la Tierra, University of Granada - CSIC, Spain, Spain

INTRODUCTION

The pedogenetic conditions in Tanzania vary widely. In particular, the country has a wide variety of parent materials of soils because of the presence of volcanic mountains, the Great Rift Valley, and several plains and mountains with different elevations (hence, different temperatures). In addition, the amount and seasonal distribution pattern of the annual precipitation vary, from less than 500 mm to more than 2500 mm. The potential land use and agricultural production differ greatly among regions, due to the presence of different soils. There have been several reports on the distribution patterns of soils and their physicochemical and mineralogical properties. According to a review of the history of soil surveys in Tanzania by Msanya et al. (2002), the major soil types described in the country are Ferric, Chromic, and Eutric Cambisols (39.7%); followed by Rhodic and Haplic Ferralsols (13.4%) and Humic and Ferric Acrisols (9.6%). To obtain basic information on soil mineralogy, Araki et al. (1998) investigated soil samples collected from regions at different altitudes in the Southern Highland and reported that the cation exchange capacity (CEC) per unit amount of clay content showed a negative correlation with elevation, which was accompanied by clay mineralogical transformation from mica to kaolinite. The authors suggested that soil formation on different planation surfaces is mainly controlled by the geological time factor whereby the lower surfaces are formed at the expense of the higher surfaces. Szilas et al. (2005) analyzed the mineralogy of well-drained upland soil samples collected from important agricultural areas in different ecological zones in the sub-humid and humid areas of Tanzania. They concluded that all soils were severely weathered and had limited but variable capacities to hold and release nutrients in plant-available form and to sustain low-input subsistence agriculture. Generally, there seems to be a consensus that the soils in Tanzania and the neighboring countries are not very fertile. The relevance of soil organic carbon management and appropriate fallowing systems such as agroforestry have been pointed out since as critical for sustaining agricultural production (Kimaro et al., 2008; Nandwa, 2001).

In the present study, the regional trend in soil fertility with respect to the soil mineralogical and chemical properties was investigated. Soil properties were correlated with different pedogenetic factors such as geology and climate. A comprehensive understanding of the distribution of some soil properties as influenced by soil-forming factors is essential for planning an appropriate land-use strategy. Besides, this knowledge will allow developing and sustaining agricultural production, while preserving natural resources such as forest and woodland ecosystems.

MATERIALS AND METHODS

Soil Samples

Ninety-five topsoil samples were collected from different regions of Tanzania. All the sampling points were located on slopes or plains, covering regions with different parent materials and with a wide variety of annual precipitation (less than 250 to more than 1500 mm) (Fig. 1; prepared based on Atlas of Tanzania [1967]). Apparent lowland soils were excluded from the analysis. The parent materials of the soils were broadly classified according to the following categories: (1) volcanic rocks (mostly basic), (2) granite and other plutonic rocks, (3) sedimentary and metamorphic rocks, and (4) Cenozoic rocks and recent deposits. The sampling plots corresponded to croplands or areas covered by either seminatural vegetation (forest or woodland) or secondary vegetation that had grown after human disturbance.

Analytical methods The soil samples collected were air-dried and passed through a 2-mm mesh sieve. Soil pH in water or 1 mol L^{-1} KCl solution was measured with a glass electrode with a 1:5 soil:solution ratio. The pH(NaF) was measured with a glass electrode in 1 mol L^{-1} NaF solution after stirring for 2 min; the soil to solution ratio was 1:50. The CEC and the amount of exchangeable bases were measured after extracting with 1 mol L^{-1} NH_4OAc at pH 7.0 and then with a 10% NaCl solution (Thomas, 1982). The NH4 + extracted with 10% NaCl solution was distilled after the addition of concentrated NaOH solution, and collected into a 2% H_3BO_4 solution. Subsequently, the NH4 content was determined by HCl titration (0.01 mol L^{-1}). The exchangeable base (Na, K, Mg, and Ca) content in the NH_4OAc solution was determined by atomic absorption spectrophotometry (AAS) (Shimadzu, AA-840-01). The exchangeable Al and H were extracted using 1 mol L^{-1} KCl. The exchange acidity (Al + H) was determined to pH 8.3 by titration with 0.01 mol L^{-1} NaOH wherein phenolphthalein was used as an indicator. Then, after the addition of 4% NaF solution to liberate OH– from the $Al(OH)_3$ precipitates, the exchangeable Al was determined by back titration to obtain the same pH (8.3) using 0.01 mol L^{-1} HCl. The exchangeable H content was determined as the difference between the exchange acidity and the exchangeable Al. The total C and total N content were measured with an NC analyzer (Sumigraph NC-800; Sumika Chem. Anal. Service, Ltd., Tokyo, Japan). The available phosphate was determined by the modified BrayII method (soil:solution = 1:20; shaking time 60s; Bray & Kurz, 1945; Olsen & Sommers, 1982). The particle size distribution was determined using a combination of sieving and pipette methods, in which a complete dispersion of silt and clay particles was achieved by adjusting the pH to 9–10 and supersonication, after pretreatment with H_2O_2 at 80°C to remove organic matter (Gee & Bauder, 1986). The clay mineral composition was semiquantified by the relative peak areas corresponding to mica (1.0 nm), kaolin minerals (1.0 and 0.7 nm), and expandable 2:1 minerals (1.4 nm) in the X-ray diffractograms obtained by using Cu–Kα radiation (RAD–2RS; Rigaku, Tokyo, Japan). The free oxides (Fe, Al, and Si) were extracted by the following two methods: (1) extraction in the dark with acid (pH 3) 0.2 mol L^{-1} ammonium oxalate (McKeague & Day, 1966) to obtain Feo, Alo, and Sio and (2) extraction with a citrate-bicarbonate mixed solution buffered at pH 7.3 by the addition of sodium dithionite (DCB) at 80°C (Mehra & Jackson, 1960) to obtain Fed and Ald. The Fe, Al, and Si content in each extract were determined by multi-channel inductively coupled argon plasma atomic emission spectroscopy (ICP-AES) (SPS-1500; Seiko, Chiba, Japan) after filtration of the extracts by 0.45 μm Millipore filters.

The data analysis was performed with the software SYSTAT version 8.0 (SPSS, 1998).

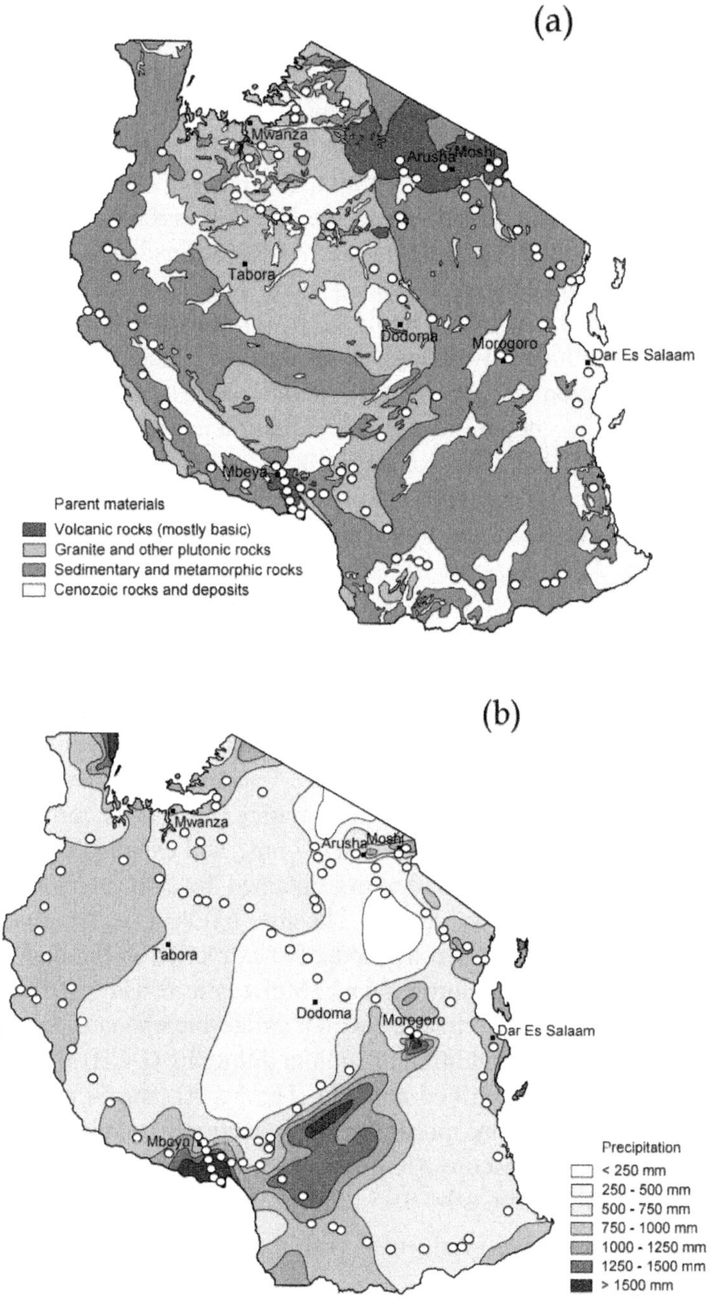

Figure 1: Geological (a) and climatic (b) conditions of the sampling plots

RESULTS AND DISCUSSION

Physicochemical and Mineralogical Properties of the Soils

Selected physicochemical and mineralogical properties for the soils studied, and the corresponding statistical analysis, are summarized in Table 1. The surface soils studied were, in general, slightly acidic, with the average values of pH(H_2O) and pH(KCl) being 6.17 and 5.37, respectively. The exchangeable Al content was low and the base saturation was high, exceeding 95% on average; hence, soil acidity was not considered a serious constraint for agricultural production. Although the average soil texture was sandy clay loam to clay loam, the particle size distribution varied widely. The average C content was 20.7 g kg–1, and the dominant clay mineral was kaolinite, followed by clay mica. However, the values obtained for most of the listed properties varied significantly over the regions under study. The coefficients of variation often exceeded 100%; which indicates a significant variability among the soil characteristics for the different regions across Tanzania.

Table 1: Physicochemical and mineralogical properties of the soils studied

Variable	Number of samples	Ave.(STD)	Min.-Max.	CV (%)
pH(H_2O)	95	6.17 (0.80)	4.36-8.66	13.0
pH(KCl)	95	5.37 (0.89)	3.71-7.96	16.5
pH(NaF)	95	8.15 (0.66)	7.12-11.01	8.0
EC (μS dm^{-1})	95	74.3 (59.4)	10.0-325	79.9
CEC ($cmol_c$ kg^{-1})	95	14.0 (11.0)	1.61-59.5	78.6
Exch. Na ($cmol_c$ kg^{-1})	95	0.18 (0.29)	0.00-1.92	161
Exch. K ($cmol_c$ kg^{-1})	95	1.10 (1.12)	0.10-5.62	102
Exch. Mg ($cmol_c$ kg^{-1})	95	2.78 (2.14)	0.18-11.4	77.0
Exch. Ca ($cmol_c$ kg^{-1})	95	6.98 (8.76)	0.00-49.5	126
Exch. Al ($cmol_c$ kg^{-1})	95	0.21 (0.51)	0.00-2.99	241
Exch. bases ($cmol_c$ kg^{-1})	95	11.0 (11.4)	0.43-60.7	103
Base satur. (%)	95	95.4 (10.5)	49.0-101	11.0
Sand (%)	95	63.6 (23.3)	3.4-96.7	36.7
Silt (%)	95	11.2 (11.3)	0.2-48.1	101
Clay (%)	95	25.2 (17.6)	1.5-81.4	69.8
Total C (g kg^{-1})	95	20.7 (24.4)	2.13-152	124
Total N (g kg^{-1})	95	1.49 (1.84)	0.21-13.7	129
Available P (gP_2O_5 kg^{-1})	95	0.15 (0.24)	0.01-1.0	161
Feo (g kg^{-1})	95	2.46 (3.28)	0.02-14.7	133
Alo (g kg^{-1})	95	3.61 (9.34)	0.08-64.3	259
Sio (g kg^{-1})	95	1.10 (3.32)	0.00-21.7	303
Fed (g kg^{-1})	95	23.7 (25.8)	0.19-159	109
Ald (g kg^{-1})	95	4.55 (7.48)	0.01-50.9	164
0.7 nm minerals (%)	90	72.5 (27.0)	5.4-100	37.3
1.0 nm minerals (%)	90	19.6 (21.4)	0.0-91.2	109
1.4 nm minerals (%)	90	7.9 (17.9)	0.0-94.6	227

Table 2 summarizes the data obtained, categorized according to the parent materials and land use. In terms of soil parent materials, the physicochemical and mineralogical properties of the volcanic-derived soils (n = 12) were significantly different from the other soil groups in terms of CEC, total C content, available P, and free oxide-related properties. Moreover, the proportion of 1.4-nm minerals was significantly higher for the soils originated from Cenozoic rocks or deposits. On the other hand, these soil properties generally did not significantly differ for different land uses.

Principal Component Analysis for Summarizing Soil Properties

A principal component analysis was performed to evaluate soil parameters related to soil fertility. The variables selected were pH(H_2O); pH(KCl); pH(NaF); CEC; amounts of exchangeable Na^+, K^+, Mg^{2+}, Ca^{2+}, and Al^{3+}; sand, silt, and clay content; total C and total N content; available P content; and Fe_o, Al_o, Si_o, Fe_d, and Al_d content. Table 3 summarizes the factor pattern for the first five principal components after varimax rotation. The analysis resulted in the soil parameters categorized into five principal components, which explained 85.4% of the total variance.

Highly positive coefficients were obtained for pH(NaF), total C and total N, Fe_o, Al_o, Si_o, and Al_d for the first component (Table 3). These variables correspond to the soil properties related to the presence of organic materials that are bound to amorphous compounds, which might be originated on recent volcanic activity. Hence, the first component is referred to as the "soil organic matter (SOM) and amorphous compounds" factor. The second component presents strongly negative coefficients for sand content and highly positive coefficients for clay content, exchangeable Mg, and Fed. These soil characteristics can be related to parent materials and clay formation, i.e. soils derived from mafic and/or clayey parent materials tend to exhibit fine-textured properties with high concentrations of exchangeable Mg and Fed through rapid mineral weathering and clay formation. Hence, the second component is denominated as the "texture" factor. The coefficients corresponding to the third component have highly positive or negative values for pH(H_2O), pH(KCl), and exchangeable Ca and Al, indicating that a close relationship exists between this component and soil acidity. This relationship can be denominated as the "acidity" factor. The fourth and fifth components are denominated "available P and K" and the "sodicity" factors, respectively, on the basis of the coefficients correlating each of the components and the soil variables (exchangeable K and available P, and exchangeable Na, respectively).

Table 2: Average values of measured soil variables in terms of parent materials or land uses

Variable	Averages for soils from different parent materials[1]				Averages for soils under different land uses[1]		
	Volcanic rocks	Granite and other plutonic rocks	Sedimentary and metamorphic rocks	Cenozoic rocks and deposits	Natural and matured secondary vegetation	Incipient fallow vegetation	Cropland
Number of samples	12(9)[2]	14	50(48)[2]	19	37(35)[2]	16	42(39)[2]
pH(H_2O)	5.91 ab	5.70 a	6.28 ab	6.43 b	6.34 a	6.14 a	6.04 a
pH(KCl)	5.18 a	4.86 a	5.49 a	5.56 a	5.53 a	5.32 a	5.26 a
pH(NaF)	9.04 b	7.95 a	8.04 a	8.05 a	8.12 a	7.99 a	8.22 a
EC (μS dm^{-1})	103.2 b	48.1 a	78.7 ab	63.9 a	87.2 b	35.9 a	77.6 b
CEC ($cmol_c$ kg^{-1})	29.5 b	6.93 a	12.5 a	13.3 a	14.3 a	8.6 a	15.7 a
Exch. Na ($cmol_c$ kg^{-1})	0.26 ab	0.07 a	0.11 a	0.39 b	0.11 a	0.12 ab	0.27 b
Exch. K ($cmol_c$ kg^{-1})	2.49 b	0.45 a	1.11 a	0.68 a	1.17 a	0.70 a	1.19 a
Exch. Mg (cmolc kg^{-1})	4.34 b	1.25 a	2.70 ab	3.12 b	2.95 a	2.53 a	2.72 a
Exch. Ca ($cmol_c$ kg^{-1})	11.60 b	1.68 a	5.87 ab	10.85 b	8.02 a	4.17 a	7.13 a
Exch. Al ($cmol_c$ kg^{-1})	0.18 a	0.28 a	0.25 a	0.07 a	0.22 a	0.28 a	0.18 a
Exch. bases ($cmol_c$ kg^{-1})	18.7 b	3.4 a	9.8 ab	15.0 b	12.2 a	7.5 a	11.3 a
Base satur. (%)	97.4 ab	88.5 a	95.4 ab	99.0 b	96.5 a	92.3 a	95.5 a
Sand (%)	36.2 a	73.9 b	64.7 b	70.2 b	66.2 a	69.0 a	59.1 a
Silt (%)	28.7 b	6.6 a	9.2 a	9.0 a	9.5 a	7.0 a	14.3 a
Clay (%)	35.1 a	19.5 a	26.1 a	20.8 a	24.3 a	23.9 a	26.5 a
Total C (g kg^{-1})	43.3 b	12.5 a	20.1 a	13.9 a	28.4 a	11.8 a	17.2 a
Total N (g kg^{-1})	3.40 b	0.98 a	1.42 a	0.87 a	1.98 a	0.80 a	1.33 a
Available P (gP_2O_5 kg^{-1})	0.431 b	0.044 a	0.128 a	0.112 a	0.154 a	0.067 a	0.180 a
Fe_o (g kg^{-1})	8.35 b	0.73 a	2.04 a	1.12 a	2.26 a	1.11 a	3.16 a
Al_o (g kg^{-1})	13.89 b	1.50 a	2.76 a	0.91 a	4.22 a	1.11 a	4.02 a
Si_o (g kg^{-1})	4.83 b	0.16 a	0.75 a	0.34 a	1.18 a	0.29 a	1.33 a
Fe_d (g kg^{-1})	40.2 b	13.2 a	26.5 ab	11.4 a	23.2 a	26.2 a	23.1 a
Al_d (g kg^{-1})	11.34 b	3.50 a	4.37 a	1.50 a	5.58 a	2.98 a	4.24 a
0.7 nm minerals (%)	75.8 a	80.4 a	73.3 a	63.0 a	73.9 a	79.6 a	68.3 a
1.0 nm minerals (%)	20.1 a	13.8 a	22.0 a	17.8 a	20.9 a	16.1 a	19.9 a
1.4 nm minerals (%)	4.2 a	5.8 a	4.7 a	19.2 b	5.2 a	4.3 a	11.8 a

[1] The values with the same letters are not significantly different by Tukey test ($p < 0.05$).

[2)] Parenthesis denotes the number of samples considered for XRD analysis (i.e. the percentage of 0.7, 1.0 and 1.4 nm minerals). Some samples were excluded from the analysis because of their X-ray amorphous natures.

Table 3: Factor pattern for the first four principal components (n = 95)

Variable	PC1	PC2	PC3	PC4	PC5
pH(H_2O)	-0.19	-0.07	-0.94	0.10	0.04
pH(KCl)	-0.05	-0.11	-0.95	0.10	-0.02
pH(NaF)	0.84	0.00	-0.05	0.29	0.05
CEC	0.57	0.51	-0.09	0.39	0.43
Exch. Na	-0.01	-0.01	0.04	0.08	0.93
Exch. K	0.01	0.32	-0.29	0.84	0.07
Exch. Mg	-0.03	0.62	-0.42	0.36	0.39
Exch. Ca	0.05	0.28	-0.64	0.38	0.47
Exch. Al	0.20	0.12	0.64	-0.08	0.07
Sand	-0.30	-0.83	-0.11	-0.35	-0.17
Silt	0.45	0.29	0.05	0.65	0.21
Clay	0.11	0.91	0.11	0.04	0.09
Total C	0.89	0.20	0.08	0.02	0.01
Total N	0.88	0.19	0.14	0.03	0.01
Avail. P	0.06	0.02	-0.26	0.87	0.04
Fe_o	0.62	0.32	0.18	0.57	-0.01
Al_o	0.97	0.00	0.13	0.05	0.01
Si_o	0.94	-0.07	0.07	0.10	0.04
Fe_d	0.09	0.86	0.15	0.13	-0.27
Al_d	0.87	0.24	0.26	-0.06	-0.08
Eigenvalue	5.98	3.43	3.14	2.92	1.60
Proportion (%)	29.9	17.2	15.7	14.6	8.0
	"SOM and amorphous compounds" factor	"Texture" factor	"Acidity" factor	"Available P and K" factor	"Sodicity" factor

Pedogenetic Conditions Determining the Distribution Patterns of Factor Scores for Each of the Principal Components

Figure 2 shows a scattergram of the factor scores of SOM and amorphous compounds and those of available P and K. Both factor scores were significantly higher in soils derived from volcanic rocks than in other soils, but no significant correlation was observed. The factor scores are plotted on the geological map, as shown in Figure 3. There are two representative volcanic areas in Tanzania, namely, Mount. Kilimanjaro and the surrounding region and the southern mountain ranging between the east of Mbeya and Lake Malawi. Generally, the scores of the factor for SOM and amorphous compounds were highest in the region of the southern volcanic mountain ranges, followed by some plots around Mt. Kilimanjaro (Fig. 3a), whereas the scores of the factor for available

P and K tended to be high in both volcanic regions (Fig. 3b). Msanya et al. (2007) indicated that the volcanic soils in the southern mountain ranges were rich in K, compared to several Japanese volcanic soils, most likely reflecting lithological differences among the parent materials. The predominantly high scores of the factor for SOM and amorphous compounds in the southern volcanic mountain ranges indicate a relatively incipient feature of soils after recent active volcanic events and potentially high soil fertility relating to SOM in these regions. In addition, soils located in those volcanic regions could be more fertile in terms of P and K nutrients supply from soils.

Figure 3c represents the distribution pattern of the factor scores of texture in terms of the geological conditions. There is a certain regional trend in these factor scores, though no statistical difference was observed in terms of the geological condition as a whole. Among the soils of volcanic origin, those in the northern volcanic regions exhibited higher scores in the texture factor, consistent with a previous report by Mizota et al. (1988), in which they postulated that these soils were in the advanced stages of weathering of volcanic materials. The scores were high for some soils originated from sedimentary and metamorphic rocks, which are mostly distributed in the western region around Kigoma and the hill slopes near Tanga. Otherwise, scores were in general low for soils originated from granite, except for those of the southern highland.

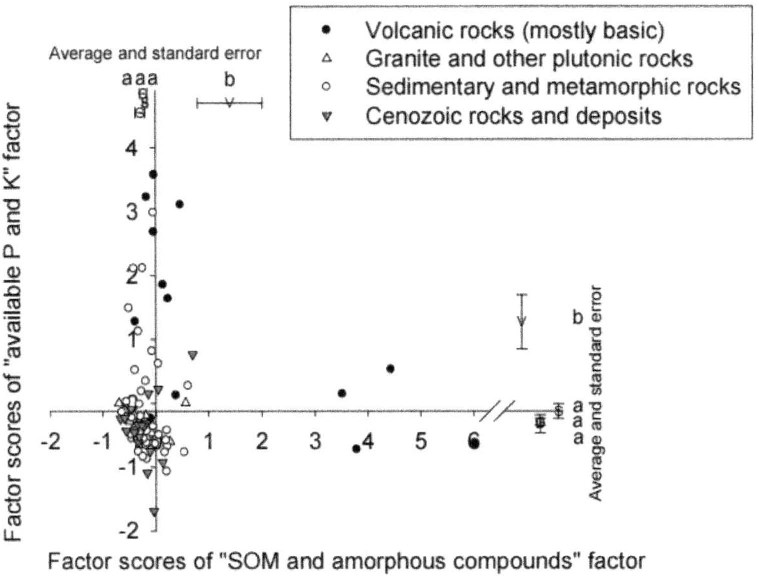

Figure 2: Relationship between the scores of the "SOM and amorphous" and "available P and K" factors

Figure 4 shows the influence of the amount of precipitation on the scores of selected factors. There was no significant relationship between the amount of precipitation and the factor scores of acidity or texture. Although the positive contribution of precipitation on mineral weathering might accompany soil acidification or the formation of clays and secondary Fe oxides, there was no correlation between those processes, which indirectly suggests that the influence of parent materials on soil properties is stronger than climatic factors among the soils studied.

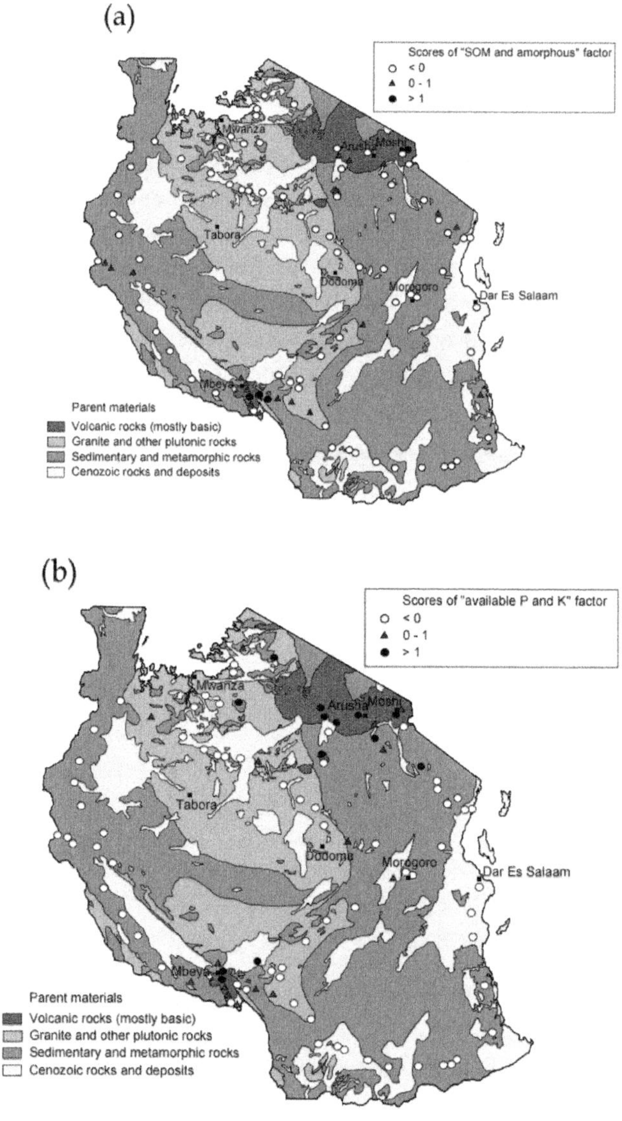

(c)

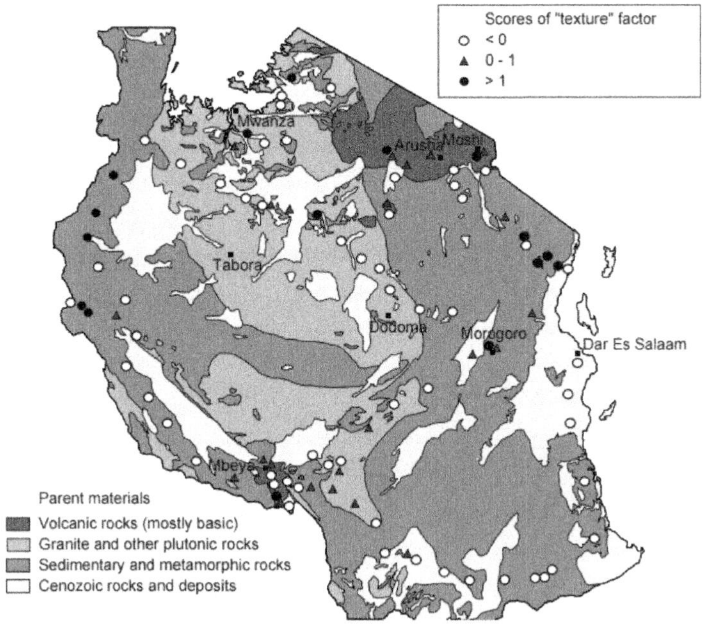

Figure 3: Distribution patterns of scores of (a) "SOM and amorphous," (b) "available P and K," and (c) "texture" factors in relation to geological conditions

(a)

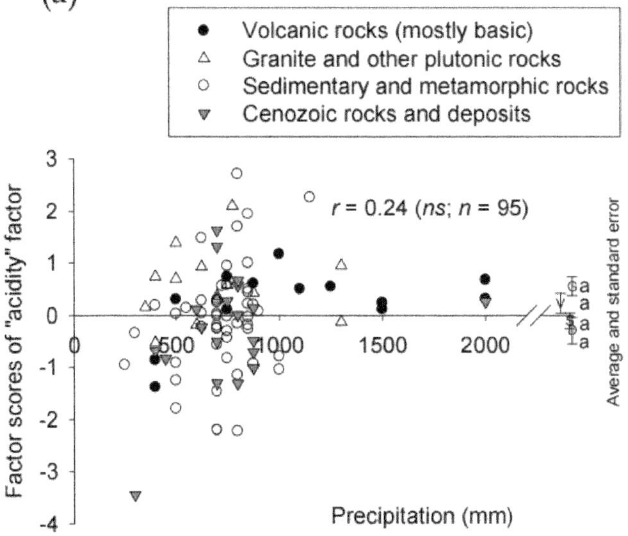

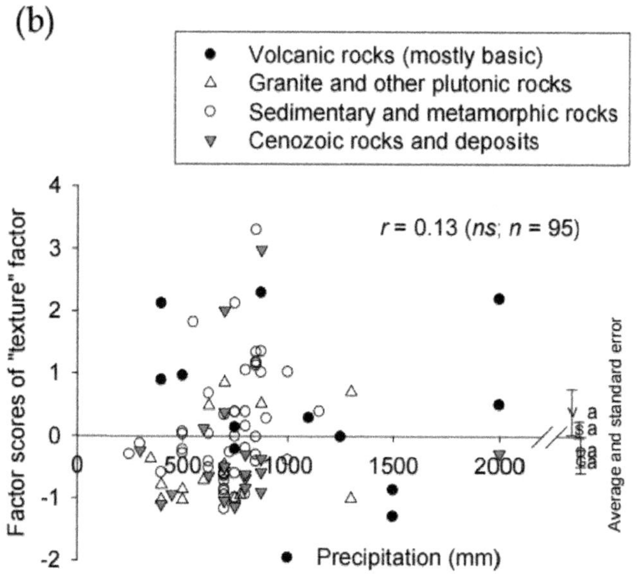

Figure 4: Relationships between precipitation and scores of (a) "acidity" and (b) "texture" factor

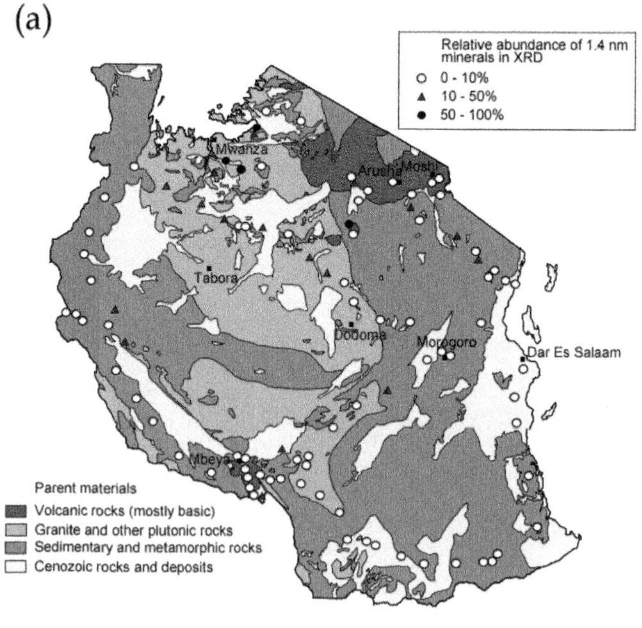

(b)

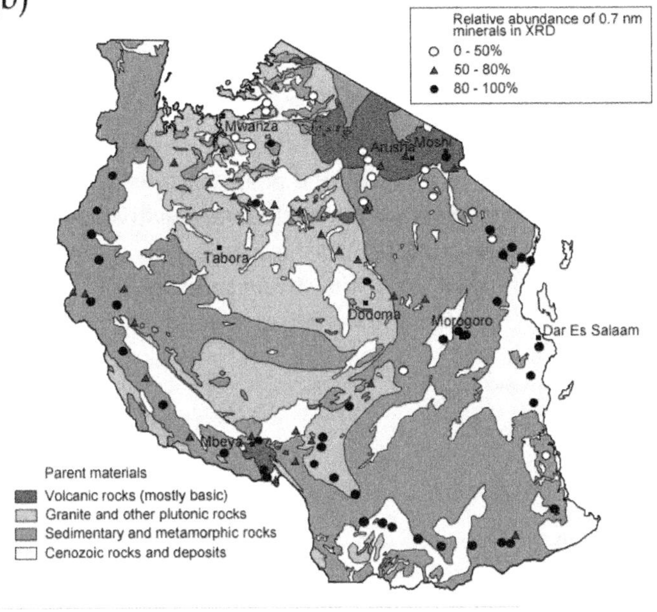

Figure 5: Distribution patterns of clay mineralogy in relation to geological or climatic conditions. Abundances of (a) 1.4 nm and (b) 0.7 nm minerals

Pedogenetic Conditions Determining the Clay Mineralogy of the Soils

Figure 5 shows the distribution patterns of the clay mineralogy in relation to the geological and climatic conditions. The relative abundance of 1.4-nm minerals was often higher in the northern region of the Great Rift Valley and around Lake Victoria. On the other hand, the abundance of 0.7-nm minerals tended to be lower in the central steppe, which has lower precipitation than other regions. These relationships are more clearly presented in Figure 6. Stepwise multiple regression indicated that the abundances of 1.4-nm minerals (mostly smectite) could be expressed by the following equation:

$$\begin{aligned}1.4-\text{nm minerals }(\%) &= 6.38 + 13.4\ (\text{sodicity factor}) - 9.78\ (\text{SOM / amorphous factor}) \\ &+ 3.17\ (\text{P / K factor});\quad r^2 = 0.58\ (p < 0.01, n = 90)\end{aligned} \tag{1}$$

The 1.4-nm minerals were probably formed under the strong influence of the high sodicity of the parent materials around the Great Rift Valley, and were often observed in the soils in the flat plains near Lake Victoria.

On the other hand, the abundances of the 0.7-nm minerals (kaolin minerals) can be expressed by the following equation:

$$0.7-\text{nm minerals } (\%) = -\ 56.2 + 19.5\ \ln(\text{precipitation in mm}) + 5.92(\text{acidity factor}) + 4.82\ (\text{texture factor}) - 11.2\ (\text{sodicity factor}) - 7.70\ (\text{P / K factor});\quad r^2 = 0.45\ (p < 0.01, n = 90) \tag{2}$$

From this equation, it can be stated that the kaolin formation is promoted under highly humid conditions with the positive influence of soil acidity and texture (or clayey parent materials) as well as the negative influence of sodicity. Hence, it can be inferred that the clay mineralogical properties of the soils studied herein were formed under the strong influence of the present climatic conditions as well as the parent materials on a countrywide scale in Tanzania.

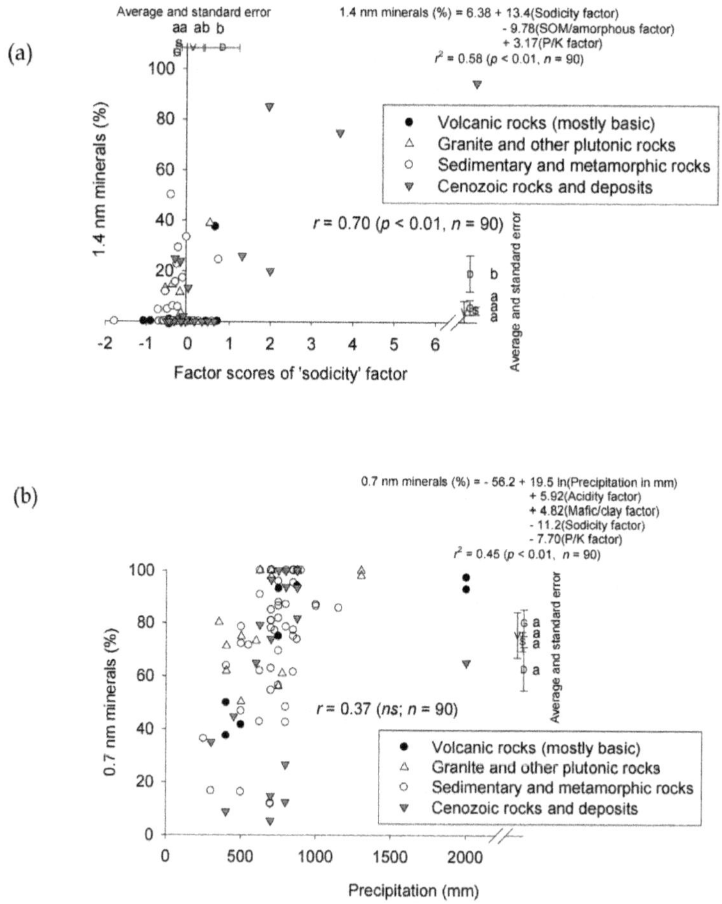

Figure 6: Relationships between clay mineralogy and soil and climatic factors. Abundances of (a) 1.4 nm and (b) 0.7 nm minerals

General Discussion on the Soil Conditions in Tanzania with Specific Reference to Potential Agricultural Development

As previously stated, soils can be considered as significantly fertile in the volcanic regions and areas around, due to the high SOM contents and the high P and K nutrient status. In addition, the soils around Lake Victoria are fertile due to the strong influence of the 1.4-nm minerals, which contributes to the retention of base cations. Both regions, namely, the volcanic regions and the regions around Lake Victoria, are included in the Great Rift Valley, which is the center of intensive agricultural activities of the country. However, in other areas of Tanzania, soils are generally low in SOM-related parameters and the 1.4-nm minerals are virtually absent, presumably due to consecutive mineral weathering under ustic soil moisture regime (Watanabe et al., 2006). The proportion of kaolin minerals increases with the precipitation; hence, soil fertility decreases in regions of high humidity. Soil fertility in terms of clay mineralogy is comparatively higher in dry regions than in humid regions because of the greater abundance of mica minerals. However, water availability decreases in such dry regions. Thus, the semiarid regions in Tanzania suffer from water scarcity, while the relatively humid areas have less fertile soil that predominantly contains kaolin minerals. In summary, high scores in SOM-related properties and the 1.4-nm minerals contribute to relatively high soil fertility in Great Lift Valley regions, whereas either water scarcity or low soil fertility are not favorable for agricultural production in the other regions of Tanzania. These conditions should be considered when studying the feasibility of agricultural development in different areas in the future.

CONCLUSION

From the principal component analysis of the collected soil samples, five individual factors—SOM and amorphous compounds, texture, acidity, available P and K, and sodicity—were determined which explained 85.4% of total variance. From the clay mineralogical composition and the relation between the geological conditions (or parent materials) and the annual precipitation and the scores of the five factors, the following conclusions can be summarized:

- The maximum scores of "SOM and amorphous compounds" were found at the volcanic center of the southern mountain ranges from the east of Mbeya to Lake Malawi.
- The scores of the "available P and K" were high in the volcanic regions around Mt. Kilimanjaro and in the southern volcanic mountain ranges.

- The abundance of 1.4-nm minerals (mostly smectite) can be expressed by the following equation (Equation 1):

$$1.4-\text{nm minerals }(\%) = 6.38 + 13.4\ (\text{sodicity factor}) - 9.78\ (\text{SOM / amorphous factor}) + 3.17\ (\text{P / K factor});\ r^2 = 0.58\ (p < 0.01, n = 90)$$

The 1.4-nm minerals were probably formed under conditions of high sodicity and were often observed in the soils near Lake Victoria.

- The abundance of 0.7-nm minerals (kaolin minerals) can be expressed by the following equation (Equation 2):

$$0.7-\text{nm minerals }(\%) = -56.2 + 19.5\ \ln(\text{precipitation in mm}) + 5.92\ (\text{acidity factor}) + 4.82\ (\text{texture factor}) - 11.2\ (\text{sodicity factor}) - 7.70\ (\text{P / K factor});\ r^2 = 0.45\ (p < 0.01, n = 90)$$

Equation 2 suggests that kaolin formation is promoted under highly humid conditions, which is also controlled by the acidity and texture of the soil (or parent materials). Hence, the results indicate that the formation of the soils studied in the present study was strongly influenced by climatic conditions and parent materials.

- In Tanzania, the volcanic regions and the Great Rift Valley region, where soil is generally more fertile than in other regions, are favorable to modernized agriculture. The semiarid regions in Tanzania suffer from water scarcity, while the relatively humid areas have less fertile soil that predominantly contains kaolin minerals. These conditions are not favorable for agricultural production and must be strongly considered when studying the feasibility of agricultural development in different areas in the future.

ACKNOWLEDGEMENTS

This study was supported by a Grant-in-Aid for Scientific Research (No. 17208028) from the Ministry of Education, Culture, Sports, Science and Technology, Japan.

REFERENCES

1. Araki, S., Msanya, B.M., Magoggo, J.P., Kimaro, D.N. & Kitagawa, Y. 1998. Characterization of soils on various planation surfaces in Tanzania. In: Summaries of 16th World Congress of Soil Science, Vol. I, pp. 310, Montpellier, France.
2. Bray, R.H. & Kurz, L.T. 1945. Determination of total organic and available forms of phosphorus in soils. Soil Science, 59, 39–45

3. Gee, G.W. & Bauder, J.W. 1986. Particle size analysis. In: Methods of Soil Analysis (ed. A. Klute), pp. 383–411. Soil Science Society of America, Madison, WI.
4. Kimaro, A.A., Timmer, V.R., Chamshama, S.A.O., Mugasha, A.G. & Kimaro, D.A. 2008. Differential response to tree fallows in rotational woodlot systems in semi-arid Tanzania: Post-fallow maize yield, nutrient uptake, and soil nutrients. Agriculture, Ecosystems & Environment, 125, 73–83.
5. McKeague, J.A. & Day, J.H. 1966. Dithionite- and oxalate-extractable Fe and Al as aids in differentiating various classes of soils. Canadian Journal of Soil Science, 46, 13–22.
6. Mehra, O.P. & Jackson, M.L. 1960. Iron oxide removal from soils and clays by a dithionitecitrate system buffered with sodium bicarbonate. Clays and Clay Minerals, 7, 317– 327.
7. Mizota, C., Kawasaki, I. & Wakatsuki, T. 1988. Clay mineralogy and chemistry of seven pedons formed in volcanic ash, Tanzania. Geoderma, 43, 131–141.
8. Msanya, B.M., Magoggo, J.P. & Otsuka, H. 2002. Development of soil surveys in Tanzania (Review). Pedologist, 46, 79–88.
9. Msanya, B.M., Otsuka, H. Araki, S. & Fujitake, N. 2007. Characterization of volcanic ash soils in southwestern Tanzania: Morphology, physicochemical properties, and classification. African study monographs (Supplementary issue), 34, 39–55.
10. Nandwa, S.M. 2001. Soil organic carbon (SOC) management for sustainable productivity of cropping and agro-forestry systems in Eastern and Southern Africa. Nutrient Cycling in Agroecosystems, 61, 143–158.
11. Olsen, S.R. & Sommers, L.E. 1982. 24. Phosphorus. In: Methods of Soil Analysis, Part 2, Chemical and Microbiological Properties, Second Edition (eds. A.L. Page, R.H. Miller & D.R. Keeny), pp. 403–430. American Society of Agronomy & Soil Science Society of America, Madison, WI.
12. SPSS 1998. SYSTAT 8.0. Statistics. SPSS, Chicago.
13. Surveys and Mapping Division, Tanzania 1967. Atlas of Tanzania. Dar es Salaam, Tanzania.
14. Szilas, C., Møberg, J.P., Borggaard, O.K. & Semoka, J.M.R. 2005. Mineralogy of characteristic well-drained soils of sub-humid to humid Tanzania. Acta Agriculturae Scandinavica, Section B - Plant Soil Science, 55, 241–251.

15. Thomas, G.W. 1982. Exchangeable cations. In: Methods of Soil Analysis, Part 2, Chemical and Mineralogical Properties (eds. A.L. Page, R.H. Miller & D.R. Keeney), pp. 159–165. American Society of Agronomy & Soil Science Society of America, Madison, WI.

16. Watanabe, T., Funakawa, S. & Kosaki, T. 2006. Clay mineralogy and its relationship to soil solution composition in soils from different weathering environment of humid Asia: Japan, Thailand and Indonesia. Geoderma, 136, 51–63.

Chapter 5

QUANTIFYING SOIL MOISTURE DISTRIBUTION AT A WATERSHED SCALE

Manoj K. Jha

North Carolina A&T State University USA

INTRODUCTION

Soil moisture content is a very vital component of the hydrological cycle. It is a key variable controlling water and energy fluxes in soils (Vereecken et al. 2007). It provides the plantavailable transpirable pool of water for vegetative life. In addition, the availability or retention of moisture in the soil controls the rainfall-runoff process. Despite its importance to vital lives and ecosystem, the distribution of soil moisture varies tremendously over the time and space. Spatial patterns of soil moisture are determined by a number of pysiographic factors that affect vertical and lateral redistribution of water in the unsaturated zone. These include topography and landscape position, slope aspect, vegetation, and texture. Temporal patterns depend on meteorological factors and their variation over the time. During the dry period (nonrainly periods), spatial variation in soil moisture is controlled by vegetation (Seyfried and Wilcox 1995). Different vegetation will have different impacts on soil moisture as their uptake will vary widely. Moisture content also exerts a strong control on soil biogeochemistry including microbial activity, nitrogen mineralization, and biogeochemical cycling of nitrogen and carbon (Turcu et al. 2005). Therefore, understanding the spatio-temporal distribution and quantity of available soil moisture that can be used without damaging the natural ecosystem are keys to sustainable development and prevention of ecosystem decline.

Soil moisture has been traditionally measured through point measurements, which is useful to understand field-scale soil water dynamics (Topp and Ferre 2002), and predominantly developed for applications in agriculture. Recent advancements in remote sensing technologies has developed capabilities that contribute to understanding of soil moisture distribution at very large scales such as large basins or continental or global scales; however, these prediction needs to be validated through a large number of ground based point measurements. It would be difficult to provide such information on a larger scale. Several techniques used in the past to represent spatial variation of soil moisture on a large scale using geostatistical anslyses tools such as kringing and semivariogram analysis, but these require a dense sampling character of the soil moisture field. The concept of temporal stability was able to capture spatial variation but limited to smaller scales (Brocca et al. 2010). Robinson et al. (2008) have extensively reviewed and summarized the challenges and opportunities for soil water content measurement in terms of laboratory, equipment, monitoring, remote sensing, and modelling challenges.

Recent advancement in watershed scale hydrology models have increasingly been adopted for soil and water management (Jha et al. 2007, 2010a, 2010b). These models provide a more holistic approach of modelling complex interconnected and nonlinear hydro-geological movement of water across all physical processes. This study used a watershed scale hydrologic model, called Soil and Water Assessment Tool (SWAT) (Arnold et al. 1998), to quantify long-term variation in spatial distribution of soil moisture on a medium-size watershed located in Midwestern USA. SWAT has been shown to perform well on both large river basins and small watersheds in terms of annual water and sediment yield (Arabi et al. 2006, Gitau et al. 2004, Spruill et al. 2000, and Jha et al. 2011, among may other studies). Gassman et al. (2007) has reviewed over a hundred of peer-reviewed SWAT related peerreviewed publications, which speaks of the magnitude and reliability of model use for hydrology and water quality analyses.

The combination of favourable climate and fertile soil makes the Midwest one of the most productive agricultural areas in the world. However, this brings an enormous application of fertilizers and manures on the cropland, unmanaged and overapplication, which led water quality problems in the local rivers and ultimately to larger ecosystems, e.g. hypoxia problem in the Gulf of Mexico (Rabalais et al. 1996). Many conservation practices have been proposed and implemented over decades. One such practice is the inclusion of winter cover crops in the traditional corn-soybean rotation. Winter cover crops can reduce nitrogen (N) leaching by extending the growing season and the uptake of N beyond that for corn and soybean (Shepherd and Webb 1999). These crops take up residual N, released by mineralization during fall and spring, and N released from fall-applied anhydrous ammonia. The cover crops then release this N as their residue decays the next spring or summer. While this practice was shown to have a tremendous potential for N reduction (Kaspar et al. 2005, Singer et al. 2011), it might have implication in soil moisture dynamics over a long period of time. This study analyzed the impacts of this conservation practice on spatial distribution of soil moisture. The main objective of this present study is to use SWAT model to quantify soil moisture distribution on a watershed scale and evaluate the impact of applying cover crop conservation practice on soil moisture content.

METHODS AND MATERIALS

Watershed Description

The Raccoon River Watershed (RRW) covers nearly 3,630 mi2 area in portions of 17 Iowa counties in west central Iowa (Figure 1). The North and Middle Raccoon Rivers flow through the recently glaciated (< 12,000 years old) Des Moines Lobe landform region, a region dominated by low relief and poor surface drainage. In contrast, the South Raccoon River drains an older (> 500,000 years old) Southern Iowa Drift Plain landscape region characterized by higher relief, steeply rolling hills, and well-developed drainage. The RRW is dominated by agricultural row crop production, with over 70% of the areas planted primarily in corn and soybeans. Other main land use includes grassland (16.3%), woodland (4.4%), and urban (4.0%). The grasses and trees generally are scattered throughout the South Raccoon basin on terrain difficult to cultivate. Figure 2 show the land use ypes in the watershed. As explained by the landorm region, north Raccoon is mostly tiled due to inadequate soil drainge property. Figure 3 depicts the tile drainage densitiy in the watershed that was very extenstively done in North Raccoon. The RRW stream system has been impacted by elevated levels of nitrogen, phosphorus, sediment, and bacteria pollutants during recent decades, primarily from nonpoint sources (Hatfield et al., 2009; Jha et al., 2010; Schilling et al., 2008).

Figure 1: Location of the study watershed

The modeling framework of the SWAT model for RRW was adapted from Jha et al. 2010. It has used SWAT vesion 2005 and relied on standard 12-digit watersheds (USGS 2009) as a basis for the subwateshed delineation. The process of watershed delineation and HRU creation was performed using the ArcView SWAT interface (AVSWATX). The resulting watershed configuration consisted of 112 subwatersheds. The hydrologica response unites (HRUs) were then created by overlaying Soil Survey Geographic (SSURGO) data (USDANRCS, 2008) and 2002 land cover data obtained from IDNR (2008). All together, a total of 3640 HRUs were created for modeling. Daily weather data was obtained from the National Weather Service COOP monitoring sites available through the Iowa Environmental Mesonet (www.mesonet.agron.iastate.edu). AVSWATX assigned the appropriate weather station information to each subwatershed based on the proximity of the station to the centroid of the subwatershed. Ten weather stations were used to provide the temperature and precipitation data for the entire simulation time frame. The SWAT model was run on a daily time step for the 1986 to 2004 period, with the first ten years (1986 to 1995) consisting of a model calibration period and a second nine year period (1996 to 2004) comprising a model validation period.

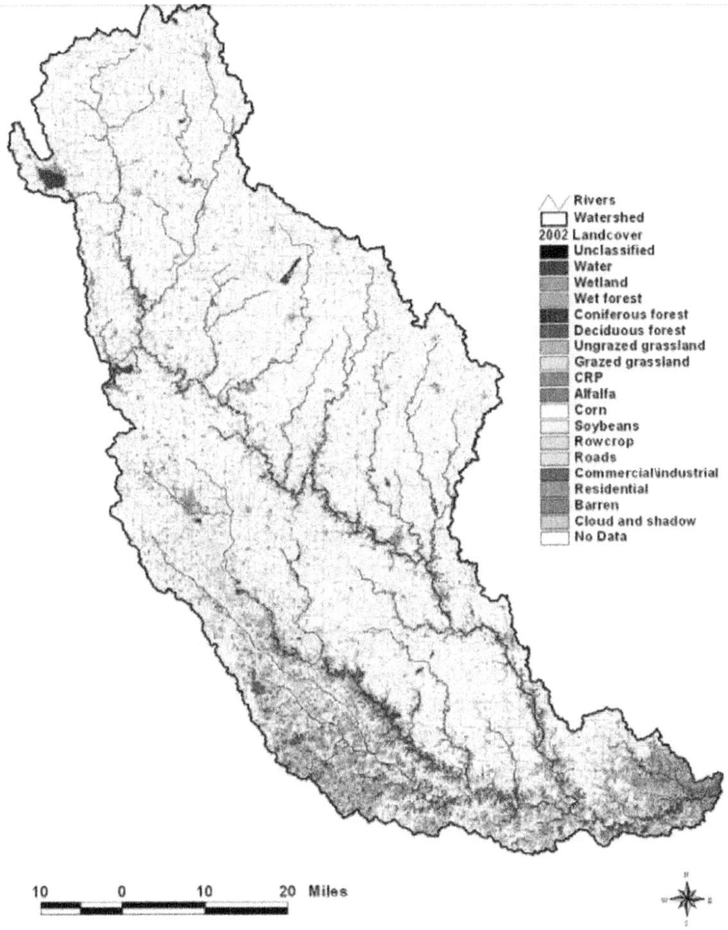

Figure 2: Land use pattern in the Raccoon River Watershed

The Penman-Monteith method was selected to estimate potential evapotranspiration and the Muskingum method was selected for channel flow routing simulation. Model calibration required varying model parameters within their ranges for match observed variables with the simulated variables. Figure 4 shows the monthly comparison of flow at the watershed outlet for both calibration and validation periods. Details on modeling setup can be found in Jha et al. 2010. Over the entire simulation period, the modeled average annual streamflow at the outlet (220 mm) was very close to the measured value (215 mm). Comparison of monthly values resulted in R^2 and E (NashSutcliffe's coefficient) values of 0.86 and 0.86 for calibraiton and 0.88 and 0.87 for validation.

The modeled average monthly streamflow (18.4 mm) closely matched the measured monthly average (17.9 mm) over the 228 months (19 years) simulation period. These statistical results can be viewed as quite strong for the resutls when viewed in the context of the suggested criteria by Moriasi et al. (2007).

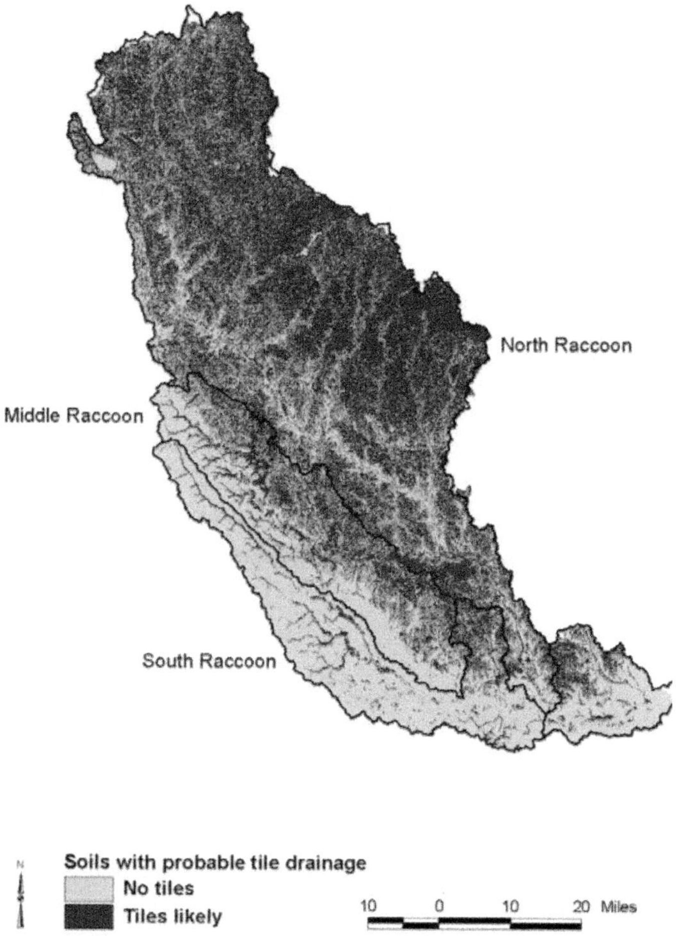

Figure 3: Soils with probable tile drainage in the watershed (adapted from Schilling et al. 2008)

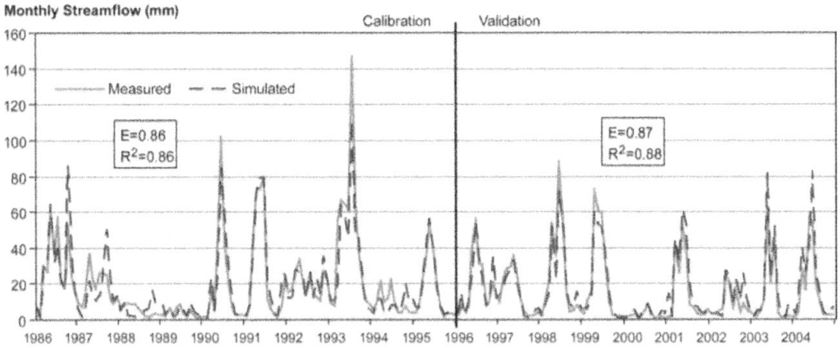

Figure 4: Long-term average (1986-2004) streamflow comparison at the watershed outlet (adapted from Jha et al. 2010)

Description of the Watershed Model, SWAT

The Soil and Water Assessment Tool (SWAT; Arnold et al. 1998) model is a watershed-based hydrologic and water quality model that operates on a daily times step and is capable of modeling the impact of different land use and management practices on hydrology and water quality of the watershed . It was developed by the U.S. Department of Agriculture (USDA) Agricultural Research Service (ARS) and has experienced continuous evolution since the first releases in the early 1990s. Major model components include hydrology, weather, soil temperature, crop growth, nutrient, bacteria, and land management. In SWAT, watersheds are divided into subwatersheds, which are further delineated by HRUs that consist of homogeneous soil, land use and management characteristics. The HRUs represent percentages of a subwatershed area and thus are not spatially defined in the model. The water balance of each HRU is represented by four storage volumes: snow, soil profile, shallow aquifer, and deep aquifer. Flow generation, sediment yield and pollutant loadings are summed across all HRUs within a subwatershed, and the resulting alues are then routed through channels, ponds, and/or reservoirs to the watershed outlet. The model has several options to estimate potential evapotranspiration including Hargreaves method, PenmanMonteith method, and others. Two options are available to simulate channel routing: variable storage method and Muskingum method. SWAT simulates a complete plant growth process and model nutrient dynamics throughout several interconnected nutrient pools.

Water that enters the soil profile may move along one of several different pathways. The water may be removed from the soil by plant uptake or evaporation. It can percolate past the bottom of the soil profile and ultimately become aquifer recharge. A final option is that water may move laterally in the profile and contribute to streamflow. Of these different pathways, plant uptake of water removes the majority of water that enters the soil profile. Two stages of water content are recognized: field capacity (water held at a tension of 0.033 MPa) and permanent wilting point (water held at a tension of 1.5 MPa). The amount of water held in the soil between field capacity and permanent wilting point is considered to be the water available for plant extraction. SWAT directly simulates saturated flow only. The model records the water contents of the different soil layers but assumes that the water is uniformly distributed within a given layer. This assumption eliminates the need to model unsaturated flow in the horizontal direction. Unsaturated flow between layers is indirectly modelled with the depth distribution of plant water uptake (Equation 1) and depth distribution of soil water evaporation (Equation 2). Depth distribution of plant water uptake:

$$w_{up,z} = \frac{E_t}{[1-\exp(-\beta_w)]} \cdot [1 - \exp(-\beta_w \cdot \frac{z}{z_{root}})_0] \tag{1}$$

Where $w_{up,z}$ is the potential water uptake from the soil profile to a specified depth, z, on a given day (mm), E_t is the maximum plant transpiration on a given day (m), β_w is the wateruse distribution parameter, z is the depth from the soil surface (mm), and z_{root} is the depth of root development in the soil (mm). The potential water uptake from any soil layer can be calculated by solving above equation for the depth at the top and bottom of the soil layer and taking the difference.

Depth distribution of soil water evaporation:

$$E_{soil,ly} = E_{soil,zl} - E_{soil,zu} \tag{2}$$

Where $_{Esoil,ly}$ is the evaporative demand for layer l_y (mm), $E_{soil,zl}$ is the evaporative demand at a lower boundary of the soil layer (mm), and $E_{soil,zl}$ is the evaporative demand at the upper boundary of the soil layer (mm).

Design Experiment for Soil Moisture Analyses

The calibrated SWAT model was examined for predicting the hydrological response at a subwatershed level. The level of spatial detail framed in this study is the size of the subwatershed (total number of which is 112 in the Raccoon River watershed with an average area of about 83.5 km^2). Various hydrological processes including precipitation, water yield, evapotranspiration, and soil water content were looked at from the perspective of spatial distribution across the watershed on a long-term average annual basis. While the spatial distribution of precipitation was derived from historical climatic observation from 10 weather stations located in and around the watershed, other parameters are simulated outcomes from the calibrated SWAT model.

It is hypothesized that the total water yield (surface runoff and baseflow) is very close (if not equal) to the difference between precipitation and evapotranspiration, while soil moisture content remains unaffected over a long-period of time. This hypothesis was tested at a subwatershed level to evaluate the model's ability to predict hydrological processes at smaller spatial scales. There is no set specific criterion to evaluate the hypothesis, but it was assumed that the model performance would be considered acceptable if the bias was found to be less than or equal to 10%. Model prediction of soil moisture was not directly validated by comparing with actual measurement due to the lack of available data on such a large scale (a motivation of this study). However, the reasonable prediction of other hydrological parameters by the model satisfied the validity of the model's ability to replicate hydrological response of the watershed through prediction of hydrological processes.

After the model validation, it was used to evaluate the effect of incorporating winter cover crops into standard corn soybean rotation in the watershed. In this scenario, rye was planted after the corn and soybean harvest. Harvest of the rye crop was not simulated but was simply plowed in prior to corn or soybean planting. This scenario provided an opportunity to assess the impact of adoption of this practice on soil moisture content on a long-term basis. Winter cover crops provide ground cover on cultivated cropland after the growing season. Rye, oats, and alfalfa have been used as cover crops in cropland areas in the Midwest for number of years, and continuously increasing. It has shown a promise of significant reduction in N losses from agricultural lands (Kaspar et al. 2004) thereby protecting local streams from nonpoint source pollution, and contributing positively to regional ecosystems. Implementation of this practice into vast majority of traditional corn and soybean rotation in the Midwest has potential to reduce N loss significantly, and ultimately reducing the concern of delivering significant nutrient loadings from Iowa and Illinois watersheds into the Mississippi and ultimately to the Gulf of Mexico.

RESULTS AND DISCUSSION

Meteorological input to the modelling system was from 10 weather stations located in and around the watershed. Spatial distribution of the most important hydrological driver precipitation is shown in Figure 5. It can be seen that the distribution does not vary significantly over the watershed spatially, and values range from 805 to 885 mm on a longterm average annual basis over the period of 19 years (1986-2004). Based on the input on temperature, other meteorological data, and information on land cover, SWAT estimated evapotranspiration (ET) using Penman-Monteith method (Figure 6). Spatial distribution of ET ranged from 470 to 660 mm with higher values in north and central portion of the watershed. Average ET among subwatersheds was found to be 564 mm with standard deviation of 36.

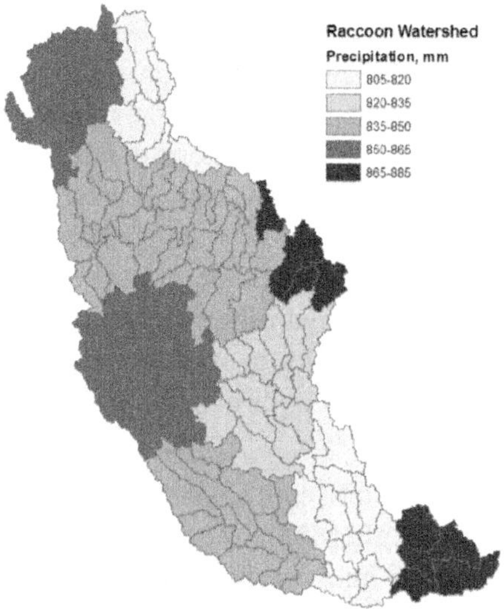

Figure 5: Spatial representation of precipitation on a long-term annual basis

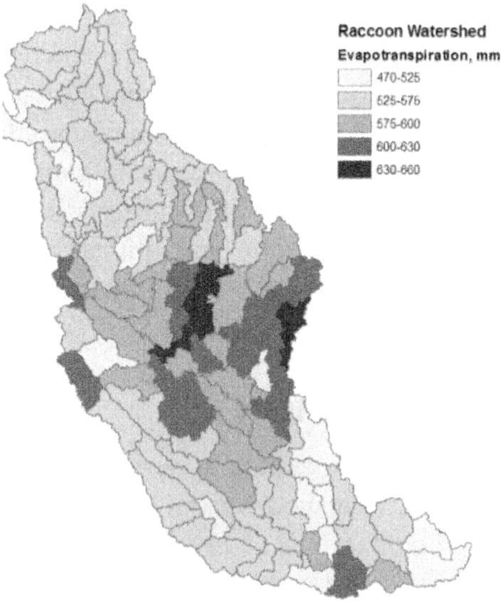

Figure 6: Estimated evapotranspiration (ET) over a subwatershed scale on a long-term annual basis

Hydrological model performed daily water balance on scale much finer than subwatershed (at HRU or response unit level). The total water yield (sum of surface runoff and baseflow) calculated at each response unit were aggregated at subwatershed level. The distribution of water yield at the subwatershed level is show in Figure 7. This was achieved after the model was calibrated for overall watershed hydrology and then for time-series data of streamflow at the watershed outlet. Our hypothesis about water yield be equal to precipitation minus evapotranspiration on a long term basis, was tested for each subwatershed individually for the calibrated model. It was found that the absolute deviation of water yield values as compared with the difference in precipitation and evapotranspiration values were very small (mean = 3 mm, standard deviation = 3 mm, and values range from +6 to -10 mm) over the entire watershed. This is the error of less than 1% in predicting water yield on a longterm basis on such a large scale. This validates the accuracy of model prediction on a longterm average annual basis. The resulting soil water content and its spatial distribution are shown in Figure 8. Its value ranges from 164 to 300 mm with an average value of 250 mm and standard deviation of 25mm. Higher moisture content was seem to exist mostly in the eastern portion of the watershed.

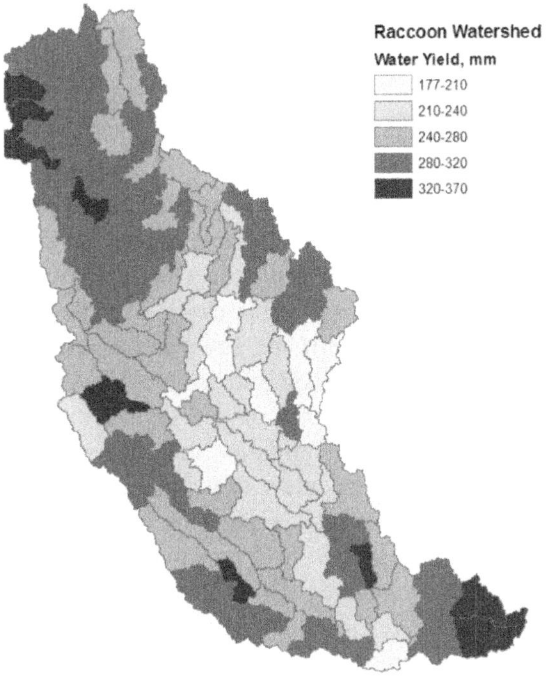

Figure 7: Total water yield distribution as predicted by SWAT on a long-term basis

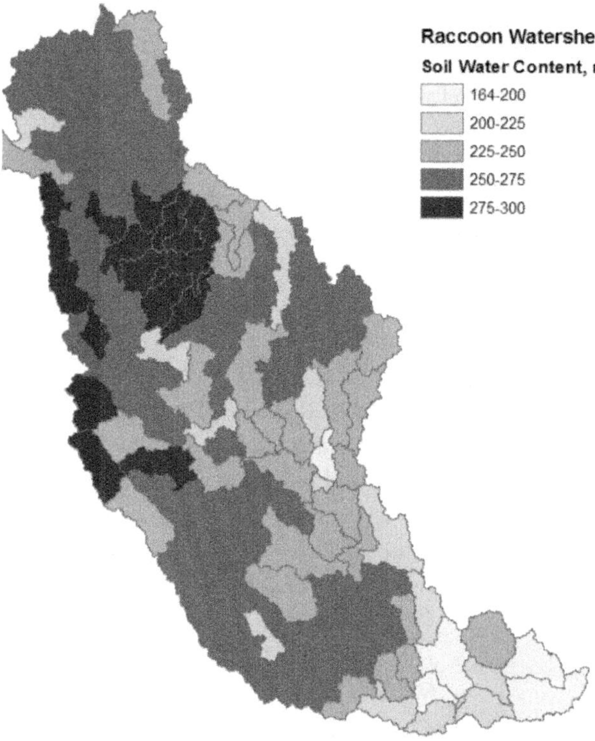

Figure 8: Soil moisture content predicted by SWAT at subwatershed scale on a long-term basis

Once the model was successfully tested to predict soil moisture content, a scenario was conducted to examine the impact on soil moisture content for a promising land management practice: inclusion of winter cover crops into cropland (corn and soybean in this case). A winter cover crop, rye, was simulated to be planted after corn and soybean harvest each year. While this practice is well known for both soil and water quality and conservation, this study attempts to quantify its impact on soil moisture content. The modelling setup was run with cover crop simulation included into the original baseline condition, and soil moisture content was predicted at each subwatershed. The long-term impact of this management practice on soil moisture content is reflected as shown in Figure 9. Soil moisture content was found to reduce significantly across the watershed with a new mean of 167 mm and standard deviation of 21. The range of values across subwatershed was found to be 116 to 207 mm, while compared to the baseline condition which was 164 to 300 mm. Spatial distribution of soil moisture was consistent with the original baseline condition where Eastern part of the watershed had higher moisture content. Moreover, the reduction in moisture content was found to be consistent on a spatial scale. The magnitude of reduction was found significant as evident by reduction in mean by 67%. Even though it is an outcome of a simulation model, the signal of impact is very high. Figure 10 show the spatial distribution of reduction in soil water content due to inclusion of winter cover crops in standard corn-soybean rotation on a long-term basis.

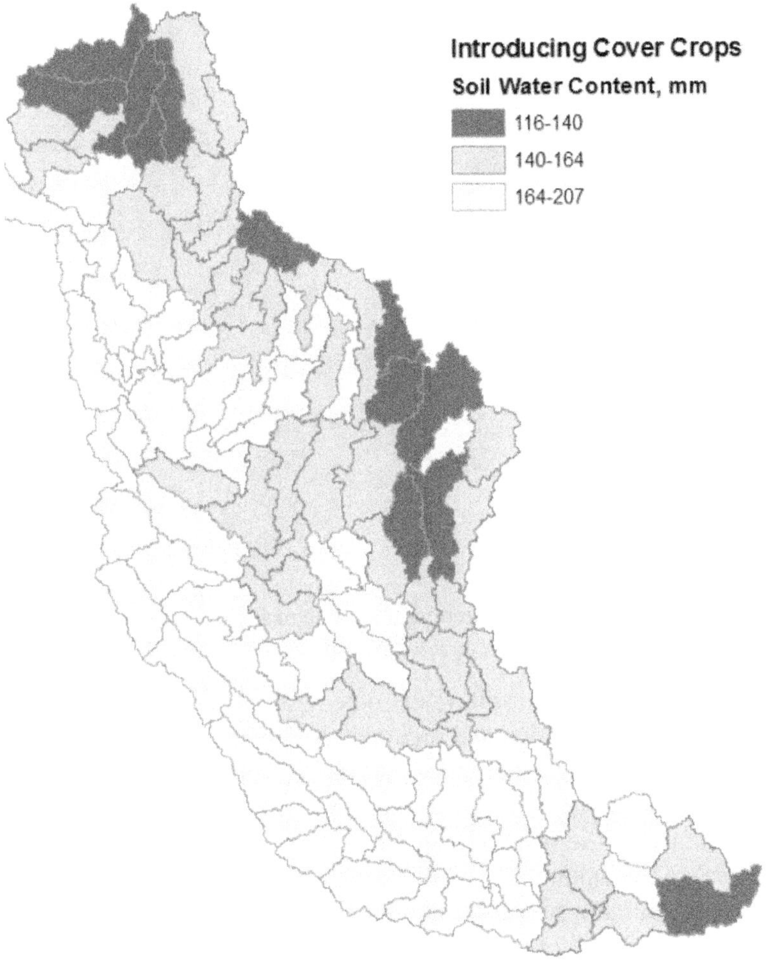

Figure 9: Soil moisture content (after introducing winter cover crop) as predicted by SWAT at subwatershed scale on a long-term basis

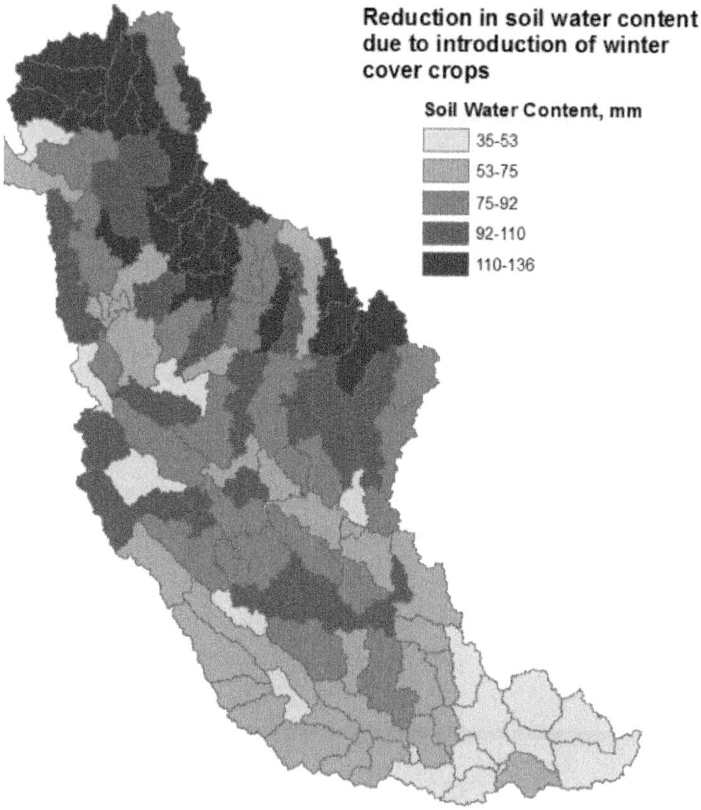

Figure 10: Reduction in soil moisture content due to inclusion of winter cover crops in standard corn-soybean rotation on a long-term basis

Significant reduction in soil water content raises the sustainability concern of the future crop production and regional ecosystem. As soil water content is very vital for crop growth and other ecosystem variables, it is imperative that it needs to be conserved. Added to that, the uncertainties in climate change with a certain increase of temperature and uncertain changes (may increase or decrease) in the amount of precpitation pose more threat to the sustainable agriculture system. It is warranted that the large scale implementation of winter cover crops should be examined with caution for changes in soil moisture content and its impact on future use of the land for agricultural production.

CONCLUSION

Understanding the spatio-temporal distribution and quantity of available soil moisture that can be used without damaging the natural ecosystem are keys to sustainable development and prevention of ecosystem decline. This study attempted to quantify the distribution of soil moisture content on a 3,630 mi2 Raccoon River Watershed located in the Midwest United States through the use of a watershed scale hydrologic model SWAT. After a successful test of SWAT's ability to predict soil moisture content, it was used to quantify the impact of introducing winter cover crops in standard corn-soybean rotation in the Midwest. The unit of analyses was at a subwatershed scale; a finer unit with total number of 112 comprise the entire watershed. Successful calibration of the SWAT modelling setup for the watershed input parameters and databases was found to produce total water yield very accurately (less than 1% error) which lead to the accurate estimation of soil moisture content at a subwatershed scale. While introducing winter cover crops has shown to be effective positively for both soil quality as well as water quality, this modelling study on the impact of this change in soil moisture found to have an adverse impact on a long-term basis. Soil moisture content was found to reduce significantly across the watershed with a mean of 167 mm and standard deviation of 21. The range of values across subwatershed was found to be 116 to 207 mm, while compared to the baseline condition which was 164 to 300 mm. The magnitude of reduction was found significant as evident by reduction in mean by 67%. Even though it is predicted by simulating a well calibrated model, signal of the impact is very high. It is warranted that the large scale implementation of winter cover crops should be examined with caution for changes in soil moisture content and its impact on future use of the land for agricultural production.

REFERENCES

1. Arabi, M.; Govindraju, R.S. & Hantush, M.M. (2006). Role of watershed subdivision on evaluation of long-term impact of best management practices on water quality, Journal of American Water Resources Association, Vol. 42, No. 2, pp. 513-528.
2. Arnold, J.G.; Srinivasan, R.; Muttiah, R.S. & Williams, J.R. (1998). Large area hydrologic modeling assessment: Part 1. Model development, Journal of American Water Resources Association, Vol. 34, No. 1, pp. 73-89.
3. Brocca, L.; Morbidelli, R.; Melone, F. & Moramarco, T. (2007). Soil moisture spatial variability in experimental areas of central Italy, Journal of Hydrology, Vol. 333, No. 2–4, pp. 356–373.
4. Gassman, P.W.; Reyes, M.; Green, C.H. & Arnold, J.G. (2007). The Soil and Water Assessment Tool: Historical development, applications, and future directiosn, Transactions of the ASABE, Vol. 50, pp. 1211-1250.

5. Gitau, M.W.; Veith, T.L. & Gburek, W.J. (2004). Farm-level optimization of BMP placement for cost-effective pollution reduction, Transactions of the ASABE, Vol. 47, No. 6, pp. 1923-1931.

6. Hatfield, J.L.; McMullen, L.D. & Jones, C.S. (2009). Nitrate-nitrogen patterns in the Racoon River Basin related to agricultural practices, Journal of Soil and Water Conservation, Vol. 64, No. 3, pp. 190-199.

7. Doi: 10.2489/jswc.64.3190. IDNR (2008). Natural resources geographic information systems library. Iowa City, Iowa: Iowa Department of Natural Resources, Geological Survey. http://www.igsb.uiowa.edu/nrgislibx/, accessed July 2008.

8. Kaspar, T.C.; Jaynes, D.B.; Parkin, T.B. & Moorman, T.B. (2007). Rye Cover Crop and Gamagrass Strip Effects on NO3 Concentration and Load in Tile Drainage, Journal of Environmental Quality, Vol. 36, No. 5, pp. 1503-1511.

9. Jha, M.K.; Arnold, J.G. & Gassman, P.W. (2007). Water quality modeling for the Raccoon River Watershed using SWAT, Transactions of the ASABE, Vol. 50, No. 2, pp. 479- 493.

10. Jha, M.K.; Wolter, C.F.; Schilling, K.E. & Gassman, P.W. (2010a). Assessment of total maximum daily load implementation strategies for nitrate impairment of the Raccoon River, Iowa, Journal of Environmental Quality, Vol. 39, pp. 1317-1327. Doi: 10.2134/jeq2009.0392.

11. Jha, M.K.; Schilling, K.E.; Gassman, P.W. & Wolter, C.F. (2010b). Targeting land-use change for nitrate-nitrogen load reductions in an agricultural watershed, Journal of Soil and Water Conservation, Vol. 65, No. 6, pp. 342-352. Doi: 10.2489/jswc.65.6.342.

12. Jha, M.K. (2011). Evaluating hydrologic response of an agricultural watershed for watershed analysis. Water, Vol. 3, No. 2, pp. 604-617.

13. Moriasi, D.N.; Arnold, J.G.; Van Liew, M.W.; Binger, R.L.; Harmel, R.D. & Veith, T. (2007). Model evaluating guidelines for systematic quantification of accuracy in water simulations. Transactions of the ASABE, Vol. 50, No. 3, pp. 885-900.

14. Rabalais, N. N.; Turner, R.E.; Justic, D.; Dortch, Q.; Wiseman, Jr., J. W. & Sen Gupta, B.K. (1996). Nutrient changes in the Mississippi River, Estuaries, Vol. 19, No. 2B, pp. 385- 407.

15. Robinson, D.A.; Campbell, C.S.; Hopmans, J.W.; Hornbuckle, B.K.; Jones, S.B.; Knight, R.; Ogden, F.; Selker, J. & Wendroth, O. (2008). Soil moisture measurement for ecological and hydrological watershed-scale observations: A review, Vadose Zone Journal, Vol. 7, pp. 358-389.

16. Doi: 10.2136/vzj2007.0143. Schilling, K.E.; Wolter, C.F.; Christiansen, D.E.; Schnoebelen, D.J. & Jha,M.K. (2008). Water quality improvement plan for Raccoon River, Iowa. TMDL Report. Watershed Improvement Section, Iowa Department of Natural Resources, pp. 202. Available at: http://www.iowadnr.gov/ water/ watershed/ tmdl/ files/ final/raccoon 08tmdl .pdf Seyfried, M.S. & Wilcox, B.P. (1995). Scale and the nature of spatial variability: Field examples having implications for hydrologic modeling, Water Resources Research, Vol. 31, pp. 173–184.
17. Shepherd, M. A. & Webb, J. (1999). Effects of overwinter cover on nitrate loss and drainage from a sandy soil: Consequences for water management? Soil Use Management, Vol. 15, No. 2, pp. 109-116.
18. Singer, J.W.; Malone, R.W.; Jaynes, D.B. & Ma, L. (2011). Cover crop effects on nitrogen load in tile drainage from Walnut Creek, Iowa, using root zone water quality model (RZWQM), Agricultural Water Management, Vol. 98, No. 10, pp. 1622-1628.
19. Spruill, C.A.; Workman, S.R. & Taraba, J.L. (2000). Simulation of daily and monthly stream discharge from small watersheds using the SWAT model, Transactions of the ASBE, Vol. 43, No. 6, pp. 1431-1349.
20. Topp, G.C. & Ferre, P.A. (2002). Th ermogravimetric method using convective oven-drying. p. 422–424. In J.H. Dane and G.C. Topp (ed.) Methods of Soil Analysis: Part 4. Physical methods. SSSA, Madison, WI. Turcu, V.E.; Jones, S.B. & Or, D. (2005). Continuous soil carbon dioxide and oxygen measurements and estimation of gradient-based gaseous flux, Vadose Zone Journal, Vol. 4, pp. 1161–1169.
21. USDA-NRCS. (2008). Overview and history of hydrologic units and the watershed boundary dataset (WBD). Washington, D.C.: USDA Natural Resources Conservation Service. http://www.ncgc.nrcs.usda.gov/products/datasets/watershed/history.html, accessed July 2008.
22. USGS. (2009). Federal guidelines, requirements, and procedures for the National Watershed Boundary Dataset. U.S. Geological Survey, U.S. Department of the Interior, Reston, VA and Natural Resources Conservation Service, U.S. Department of Agriculture, Washington, D.C. Available at: http://pubs.usgs.gov/tm/tm11a3/.
23. Vereecken, H.; Kamai, T.; Harter, T.; Kasteel, R.; Hopmans, J. & Vanderborght, J. (2007). Explaining soil moisture variability as a function of mean soil moisture: a stochastic unsaturated flow perspective, Geophysical Research Letters, Vol. 34, L22402. Doi: 10.1029/2007GL031813.

Chapter 6

FIRE IMPACT ON SEVERAL CHEMICAL AND PHYSICOCHEMICAL PARAMETERS IN A FOREST SOIL

Andrea Rubenacker, Paola Campitelli, Manuel Velasco and Silvia Ceppi

Departamento de Recursos Naturales, Facultad de Ciencias Agropecuarias, Universidad Nacional de Córdoba, Córdoba Argentina

INTRODUCTION

Cordoba is a Mediterranean State, with semiarid climate, dry autumn and winter, in which the wild fire can take place, especially at the end of dry season. Forest fires happen frequently in the mountain zones of the province of Córdoba, Argentina, which are located at west and south-west region. The vegetation, in the south-west zone, are principally Pinus halepensis Mill.; Pinus elliottii implanted and the native vegetation cover is Stipa caudata, Piptochaetium hackelii, P. napostaense y Briza subaristata, between others. Taxonomically the soil corresponds to an Ustorthent.

In the west area the native vegetation is principally Acacia caven, Festuca hieronymi, Stipa, Poa stukerti, between others. The soil is an Argiustoll. Forest wild fires constitute a serious environmental problem, not only due to the destruction of vegetation but also because the degradation that may be induced in a soil as a consequence of the change produced in its properties. Wild fire can strongly modify the abiotic and biotic characteristics of soil, altering its structure, chemical and physicochemical properties, carbon content and macronutrient levels. The degree of the alteration produced depends on the frequency and intensity of fire, all these modifications being particularly important in the surface horizons.

Organic matter is a key factor for forest soil. It has a direct and /or an indirect influence on all physical and chemical characteristics of the soil. While low severity fires, such as those prescribed for forest management, have been reported to have transient but positive effect on soil fertility, severe wildfire result in significant losses of soil organic matter, and nutrient, and deterioration of the overall physical-chemical properties of soil that determine its fertility, such as porosity, structure among others (Certini, 2005).

Fire may directly consume part or all of the standing plant material and litter as well as the organic matter in the upper layer of the soil. One of the most important soil change, during the burning is the alteration in the organic matter content therefore, the nutrient contained in the organic matter are either more available or can be volatilized and lost from the site. The soluble nutrient would be loss for erosion or leaching if they are not immediately absorbed by plants or retained by soil.

Humic substances are one of the most important fractions of the organic matter and are considered the most abundant organic component in nature and largely contribute to soil structuring and stability, to its permeability for water and gases, to its water holding capacity, to the nutrient availability, to the pH buffering and to the interaction with metal ions (Schnitzer 2000; Hayes & Malcom, 2001, Campitelli & Ceppi 2008)

Depending on the fire severity the organic matter may change not only their level, but also, their different fractions, i.e. the humic substances (humic and fulvic acids) content and their principal characteristics (Vergnoux et al., 2011a, 2011b; Duguy & Rovira 2010). The fire could induce transformation in the solubility of humic substances in different media, alkali or acid, and of their fraction (humic and fulvic acids). Thus, it could means that the humic and fulvic acids suffer structural modification, provably of the peripheral chains and the oxygenated moieties.

The study of the burned soil is necessary to analyze the soil degradation, not only to estimate the modification of the nutrient content but either the physicochemical characteristics of the organic matter, specially analyzing the humic substances and their fractions.

Humic acids have an important role in soil structure and nutrient capacity due to their surface charge development which may change due to the fire event. Some researchers have mainly shown changes in the aromaticity and in the oxygen-containing functional groups (Almedros et al., 2003; Gonzalez-Vila & Almendros 2009; Kniker et al., 2005; Vergnoux et al., 2007).

Most of the researches are focusing in the nutrient content, carbon content and their fraction, the fate of nutrient after fire, the effect of erosion, in general, the effect of fire disturbance of the soil properties. Moreover, a single parameter is insufficient to give an accurate evaluation of soil alteration. That is why several parameters need to be taken into consideration (Vergnoux et al., 2009).

Because the abundance and importance in soil of the organic matter and their fraction, mainly the humic acids, is necessary to focus the study not only onto the principal nutrient content, but also, on the characteristics of the humic acids and compare it with those extracted from the same unburned soil.

The objective of this research was to study, through different analytical techniques, i) several soil parameters related to chemical soil fertility and ii) chemical and physicochemical properties of organic carbon and their different fractions, focusing mainly, on the humic acids extracted from the forest soil after the fire event in order to compare it with the humic acids from the unburned soil.

MATERIALS

Study site and Sampling

The area selected was the south-west area of the province of Cordoba, named as San Agustín (Departamento de Calamuchita).

The annual average rainfall is about 600-800 mm. The mean temperature in the area fluctuates from 9 0C in winter to 20 0C in summer.

The tree stratum are principally Pinus halepensis Mill.; Pinus elliottii implanted and native vegetation cover is Stipa caudata, Piptochaetium hackelii, P. napostaense y Briza subaristata, between others. Three composite samples (10 subsamples) were taken from the upper layer (0-10 cm) of soil, some days after the wildfire occurrence and before any rainfall event. The samples were taken from the burned (BS) and adjacent unburned soil (UBS), at the same sampling moment. Litter and the ash were not removed from the soil surface before sampling in the unburned and burned soil, respectively. The soil was taxonomically characterized as Ustorthent. The samples were air-dried, crushed and passed through a 2 mm sieves before all the analytical analysis The humic acids (HA) analyzed were extracted from the burned (HA-BS) and unburned soil (HA-UBS)

Methods

The samples of burned and unburned soil were analyzed for pH at a rate 1:2.5 (w:v), electric conductivity (EC), total nitrogen content (TN) by the Kjeldahl method, phosphorus available (P) by Bray & Kurtz method (1945), total organic carbon (TOC) by combustion at 540 0C for 4 h (Abad, et al., 2002) and oxidable carbon (Cox) by the methodology proposed by de Richter & Von Wistinghausen (1981). Organic light fraction (OLF) were determined according the method proposed by Janzen et al., 1992, the C and N content of the OLF by dry combustion using a Perkin Elmer CN Elemental Analizer.

The carbon content of humic substances (CHS), humic acids (CHA) and fulvic acids (CFA) were determined according to the technique proposed by Syms & Haby 1971. The carbon content of each fraction (CHS, CHA and CFA) were calculated as percentage of the TOC, therefore, the % CHA correspond to the Humification Index (HI) (Roletto et al., 1985; Ciavatta et al., 1988).

The apolar or free lipidic fraction (FLF) were extracted with petroleum ether (40-60 0C) in 250 ml Soxhtel loaded with 50 g of soil; the extraction phase was renewed every 12h. The total extract was dehydrated with anhydrous Na_2SO_4 evaporated under reduced pressure to approximately 50 ml, dried under N_2 stream at room temperature (20-25 0C), and finally weighted, following the methodology proposed by Zancada et al., 2004.

The spectroscopy characteristics of the alkaline extract of both soil samples, the absorbance at different wavelength (280, 470 y 664 nm), were determined by the methodology proposed by Sapeck & Sapeck (1999). The ratio E2/E6, E4/E6 and E4/E6 were calculated from the corresponding absorbance value of the alkaline extract. The measures were determined using Spectronic 20 Genesys Spectrophotometer.

Humic Acids isolation

HA from burned and unburned soil were extracted with NaOH 0.1 mol L^{-1}, purified with HCl:HF (1:3) and dried at low temperature until constant weight, according to the procedure recommended by Chen et al. (1978). All solutions were prepared with tridistilled water and all the reagents were ACS reagent grade.

Humic Acids analyses

HA ash content was measured by heating it at 550 ^{0}C for 24 h. The elemental composition for C, H, N, S was determined by an analyzer instrument Carlo Erba 1108, using isothiourea as standard. Oxygen was calculated by difference: O%= 100 - (%C+%H+%N+%S) (ash and moisture-free basis).

Spectroscopic characteristics

The absorbance of the extracted HA were measured on a solution containing 3.0 mg of each HA in 10 mL of 0.05 mol L^{-1} $NaHCO_3$ at different wavelength (280, 470 y 664 nm) according the methodology proposed by Kononova, 1982; Zbytniewski & Buszewski, 2005; Sellami et al., 2008. From the absorbance value then were calculated the E2/E4, E2/E6 and E4/E6 ratio

Potentiometric Titration

Potentiometric titrations were carried out according to the technique proposed by Campitelli et al. (2003), which is briefly: HA solution of each samples were prepared by dissolving HA (≈ 50 mg) with minimum volume of NaOH solution (0.1 mol L^{-1}) and adding water up to the final volume (50 ml). An aliquot containing the desired amount of HA (≈7–8 mg) was transferred to the titration flask containing 10 mL of tridistilled water. The titrant (HCl =0.05 mol L^{-1}) was added from an automatic burette (Schott Geräte T80/20) at a titrant rate of 0.1 ml/40 s. This rate was chosen taking into consideration that the variation of pH values should range between 0.02 and 0.04 pH units. The pH values were measured with an Orion Research 901 pH meter equipped with a glass-combined electrode (Orion 9103 BN). All titrations were performed in KCl 0.01 mol L^{-1} as background electrolyte. The same titration was followed in absence of HA (reference or blank titration) for each titration curve, in order to subtracts it from the raw data titration, and thus obtain the charge developed by the HA sample. Each HA solution, with the corresponding blank solution, was titrated by triplicate and the reported data representing the average values. All the reagents were ACS reagent grade.

Capillary zone electrophoresis

Capillary zone electrophoresis (CZE) experiments were performed on an Agilent Technology Capillary electrophoresis system equipped with a diode array. Operation of the instrument, data collection and analysis were controlled by Agilent ChemStation software. The polarity was negative, voltage of -30kV, temperature 25 °C, total run time 30 min (for time migration higher than 30 min no significant peak were observed). The samples were injected hydrodynamically using pressure of 5000 Pa for 20 s. The absorbance was monitored at four different wavelengths (210 nm, 230 nm, 260 nm and 450nm) and 260 nm was selected to report.

Each HA electropherogram was carried out by triplicate and the reported data representing the average values. The dimensions of the fused-silica capillary were 75 μm internal diameter; 73.0 cm total length and 64.5 cm effective length. All the solutions and background electrolyte (BGE) were prepared from analytical (p.a. or HPLC) chemicals and ultra-pure water. BGE was buffer borate 20 mmol L^{-1} at pH=9.3, the concentration of the HA solutions were 1000 ppm. At the beginning of daily work, the capillary was washed for 5 min with 0.1 mol L^{-1} NaOH solutions, followed by 5 min washing with ultra pure water and 20 min with BGE at 25°C and 104 Pa. At the end of the daily work, the capillary was rinsed with BGE for 5 min and water for 10 min, at the same temperature and pressure condition. The capillary was treated before each sampling injection, as following, pre-condition: 2 min with NaOH 0.1 mol L^{-1} at 10^4 Pa, followed by washing with BGE for 3 min at 10^4 Pa, and finally waiting for 1 min. Post-run conditions were: 1 min with NaOH 0.1 mol L^{-1} at 10^4 Pa, followed by 5 min with water at the same pressure.

RESULTS AND DISCUSSION

Soil Characterization

Main Properties

The results of the principal chemical parameters are shown in table 1. The concentration of the cations, such as Na+ and K+ were not altered by fire, Ca^{2+} content slightly increase after fire, probably due to their release from the litter layer and Mg^{2+} decrease. The increase observed in the availability of Ca^{2+}, may be remarkably in a fire event, but ephemerally (Certini, 2005).

Table 1: Principal chemical characteristic of burned and unburned soil

Sample	pH	EC	TN	P	TOC	Cox	TOC/TN	Na	K	Ca	Mg	CIC
Unburned soil (UBS)	6.20a	0.60a	6.6a	23.0a	105a	24.2a	15.9a	0.22a	1.15a	10.25a	2.25b	23.5a
Burned soil (BS)	6.53a	1.19b	7.1a	52.4b	128b	24.9a	18.0b	0.22a	1.03a	11.1a	1.5a	23.6a

EC: dSm^{-1}; P: mg kg^{-1}; TOC, Cox: g kg^{-1}; CIC, Na, K, Ca Mg: cmol kg^{-1} Different letters (a–b) in the same column indicate significant differences ($p<0.05$) according to Tukey test.

The effect of burning onto soil Total Nitrogen (TN) content present a paradox, which have been debated for years (Neary et al., 1999; Knicker & Skjemstad,, 2000). Fisher & Binkley (2000) found that the immediate response of soil N to heating is a decrement because some loss through volatilization; Certini (2005), suggested that organic N could be volatilizes and in part mineralized to ammonium. Santin et al. (2008) found that the TN after fire increase and González-Vila et al. (2009) suggest that wildfire promote the accumulation of recalcitrant organic-N forms. The N, would be as NH_4^+ or NO_3^- , the NH_4^+ could be adsorbed onto negative charge of mineral and/or organic surface, but with time transformed to NO_3^- . Nitrate, without any plant uptake, will be lost from the ecosystem either by denitrification or leaching (Certini, 2005; Knicker, 2007).

The TN content increased slightly in the burned soil, but this change is not statistically significant; this behavior may be due to the nitrogen supplied by the burned litter and/or the ash contained in the sample.

The forest fires have not necessarily the same impact on soil P as on N, because losses of P through volatilization or leaching are small. The combustion of vegetation and litter causes modification on biogeochemical cycle of P. Burning convert the organic pool of soil P to orthophosphate, which is the form of P available to biota. Furthermore, the peak of P bioavailability is around pH 6.5. These could be the reason for which an enrichment of P is observed in the studied burned soil, but this enrichment will decline soon, because it precipitates as slightly available mineral forms (Certini, 2005; Cade-Menun et al., 2000). In agreement with this suggestion, the increase in the available P content in this burned soil could be due to the soil pH value (table 1).

Cation Exchange Capacity (CIC), on average, decrease after a fire event due to the loss of organic matter (Certini 2005; Badía & Martí, 2003), in this soil CIC was not changed, probably because the Cox content is the same before and after fire.

In general the soil pH increase by soil heating as a result of organic acids denaturalization, this increase take place when the temperatures are higher than 450 or 5000C, in coincidence with the complete combustion of fuel and the bases release (Arocena & Opio 2003; Knicker, 2007; Certini 2005). For the soil analyzed, the increasing observed in soil pH after the fire event was slight (around 5%), this is in agreement with the cation (Na, K, Ca, Mg) content which were not largely modified by the heating soil, suggesting that the temperature did not raise up to 450^{0}C or greater.

The electric conductivity (EC) increase in the burned soil, it could be assigned to the release of inorganic ions from the combusted organic matter present as litter or ash; this increase could be temporary (Kutiel & Imbar 1993; Hernandez, et al.,1997; Certini, 2005).

Organic Matter

The most intuitive expected change in the soils during a fire event is the loss of organic matter. This change depends on the fire severity, vegetation type, soil texture and even slope. The impact on the organic matter consist of slight distillation (volatilization of minor constituents), charring or complete oxidation. Substantial consumption of organic matter begins in the 200-250 ^{0}C range to complete at around 450-500 ^{0}C (Fernandez et al., 1997; Giovannini et al., 1988; Certini, 2005; Knicker, 2007).

The influence of fire on the organic matter content have been reported a wide range of effects, showing even contrasting results (Gonzalez-Perez et al., 2004; Czimczik et al., 2005, Dai et al., 2005; Knicker et al., 2005; Alexis et al., 2007)

The oxidable organic carbon (Cox) content was not altered by fire, but, total organic carbon (TOC) increase around 21%, this behavior could be attributed to the accumulation of recalcitrant hydrophobic fraction of organic matter (Gonzalez-Perez et al., 2004; Santin et al., 2008).

The organic fraction extracted with petroleum ether, the soil free lipids, represents a diverse group of hydrophobic substances ranging from simple compounds such as fatty acids, to more complex substances as sterols, terpenes, polynuclear hydrocarbons, chlorophylls, fats, waxes and resins. The hydrophobic fraction extracted (FLF) from the sample after fire was greater ($\approx$ 38%) than that quantified for the sample of the control soil (table 2), in agreement with those found by Almendros et al. (1988), for a soil under Pinus pinea. Although, such compounds occur in fire unaffected soil, their abundance is increased by fire due to greater stability of lipids and lignin derivatives but also due to the neoformation of aromatic polymers (Almendros et al., 2003; Fernandez et al., 2004; Knicker et al., 2005a).

The high TOC content before and after fire event could be due to the sampling methodology, taking the soil sample with all the litter and grass in soil before fire and litter from decaying fire affected vegetation. The increase in the TOC content suggests that this fire event contributes to an enhancement of the organic matter, through the incomplete combust vegetation and thus contributes to a soil TOC increase. With the time residence in the soil this unstable organic matter could be incorporated to the stable pool of organic matter, this behavior is related to the process of accumulation of organic compounds in soil controlled by their chemical affinity with the native organic matter. The randomness of the process and the heterogeneity of the organic molecules, probable produced by fire, lead to the accumulation of organic matter in which hydrophilic association may be contiguous with hydrophobic domains or contained in one other, and thus the native organic matter pool could behave as sink of the decaying fire affected vegetation. (Spaccini et al., 2000; Santin et al., 2008; Gonzales-Perez et al., 2004; Knicker et al., 2005).

The increase in the TOC/TN ratio (Table 1) after the fire event is due, principally, to the TOC increase more than to the TN change after fire. This could confirm the accumulation of incompletely burnt plants necromass, or a post-fire enhancement of the litter from decaying fire-affected vegetation production (Knicker et al., 2005a; 2007; Gonzalez-Perez et al., 2004; Santin et al., 2008).

The light fraction (LF) content (Table 2), which represents all residues, with a density value lower than 1.7 g ml^{-1}, on the top soil before and after the wildfire event, could be the reason for the high value of the TOC observed. This fraction (LF) increases after fire in the same way as the TOC, around 28%;

which represent one possible source of organic material (incomplete burnt plants) that would be incorporated to the native pool of soil organic matter and thus a way to a progressive stabilization of the different organic compounds produced by the fire effect, such as, aliphatic compounds, polysaccharides, peptides of plant and microbial origin and other organic compound generated by fire. The carbon content slightly increases and nitrogen content decreases significantly (≈ 27%) after fire in the LF. The C/N ratio indicate that this fraction is formed by an unstable organic fraction, composed by debris with incomplete combustion, thus, it could produce a nitrogen immobilization during the stabilization and the incorporation to the native soil organic matter.

Organic Matter Fractions Analysis

The carbon content of each fraction (CHS, CHA and CFA) were calculated as a percentage of the TOC, therefore, the % CHA correspond to the Humification Index (HI) (Roletto et al., 1985; Ciavatta et al., 1988).

Vergnoux et al.(2011a, 2011b), found that the different fraction of the humic substances decrease after fire, in agreement with Almendros et al.(1990); Fernandez et al.(1997); Gonzalez-Perez et al.(2004); Kincker et al.(2005). Other studies suggest that during the wildfire a humic-like fraction can be produced from burned plant biomass and thus it would be extractable in alkaline solution. In general, medium heating, i.e. temperatures not higher than 250°C, leads to increase complexity of the organic matter: newly formed compounds, oxidation and thermal fixation of alkyl moieties, etc. (Almendros et al., 1992; Gonzalez-Perez et al., 2004).

The organic carbon content of each fraction (CHS, CHA, and CFA) of the burned and unburned soil is shown in table 2.

Table 2: Carbon content in each humic substances fraction (CHS, CFA, and CHA), carbon light fraction content (LF), nitrogen and carbon content in the carbon light fraction, and the free lipidic fraction (FLF) content in burned and unburned soil

Sample	CHS	CFA	CHA(HI)	CHA/CFA	LF	N%	C%	C/N	FLF
Unburned soil (UBS)	1.92a	0.68a	1.23b	1.80b	56.7a	1.54b	17.35a	11.2a	0.24a
Burned soil (BS)	1.84a	0.79b	1.06a	1.34a	66.9b	1.12a	18.37a	16.4b	0.39b

CHS, CFA, CHA, LF and FLF: expressed as % in function of 100 g of TOC Different letters (a–b) in the same column indicate significant differences ($p < 0.05$) according to Tukey test.

The variation in the CHS content after the wildfire is not statistically significant; this could be due to the original humic materials transformations into an alkali-insoluble macromolecule material (Gonzalez-Perez et al., 2004;

Fernandez et al., 2004), which is in agreement with the amount of hydrophobic fraction (FLF) found in both soil samples (table 2).

The CFA increases around 15% after fire and CHA decrease around 12% in the soil exposed to high temperatures. The increase of the CFA content indicate the newly formed compounds, with more aliphatic chains, in general, with less molecular size, produced by the breakup of the more aggregated structures of the humic acids and thus, the carbon humic fraction decrease. The Humification Index (HI) (Table 2) is reduced about 12% indicating, also, the alteration in the humic substances by wildfire. The ratio CHA/CFA (Table 2), also known as "degree of polymerization or polymerization index", decrease around 25% in the burned soil, reflecting the breakdown of the complex and more aggregated structures of unheated soil humic fraction, indicating that the wildfire lead to an important change in the structure and the properties of the humic substances fraction (Debano et al., 2000; Shakeesby &Doerr 2006).

Spectroscopic Properties of Soil Alkaline Extracts

The scattering of monochromatic light in a diluted solution of macromolecules or colloidal particles is closely related to weight, size, aggregation and interaction of particles in solution. The UV-Visible absorption of humic substances was used to evaluate the condensation degree of the aromatic compounds (Chen el al., 1977; Stevenson, 1982; Polak et al., 2009).

Sutton & Sposito (2005), suggest that the apparent size of humic materials do not change due to tight coiling (or uncoiling), but instead change due to disaggregation (or aggregation) of clusters of small molecules.

The absorption at 280 nm was also introduced to represent total aromaticity, because the π- π^* electron transition occurs in this UV region, for phenolic arenes, benzoic acids, aniline derivatives polyenes and polycyclic aromatic hydrocarbon with two or more rings (Uyguner & Bekbolet, 2005).

The absorption at 470 nm is related with the fragment produced for the depolimerization or disaggregation of the supramolecular structure or material with a low humification degree (Sellami et al., 2008; Zbytniewski & Buszewski, 2004).

The absorbance at 664 nm is characteristics of high oxygen content, aromatic compound, strongly humified material with a high degree of condensed groups (Sellami et al., 2008). Lipski et al.(1999) defined E2/E4 ratio (the ratio of absorbance at 280 and 400 nm) to characterize the degradation of phenolic/quinoid core of humic acids to simpler carboxylic aromatic compounds. This ratio may represent an alternative parameter for the elucidation of the photocatalytic degradation efficiency.

The value of the quotient E4/E6 (the ratio of absorbance at 400 and 665nm) and E2/E6 (the ratio of absorbance 280 and 665 nm) coefficient are related with aromatic condensation; suggest the aggregation level, phenolic and benzene-carboxylic group content, among other characteristics. A low ratio reflects a high degree of aromaticity, aggregation and high humification level; large values are associated with the presence of smaller size organic molecules, more aliphatic structures, high content of functional groups, high disaggregation level (Chen et al., 1977; Pertusati & Prado, 2007, Zbytniewski & Buszewski, 2004). The value of the coefficient E2/E4 for CHS and CFA (Table 3) obtained in the alkaline extracts for the burned and unburned soil, don't have a great variation, suggesting that the degradation of core structure of humic substances, depolymerization or the disaggregation of the supramolecular structure was not significant, probably several aggregate disruption was produced by heating the soil (Uyguner & Bekbolet, 2005; Sutton & Sposito, 2005). The values of the quotient E2/E6 and E4/E6 are around 20-30% greater for both fraction (CHS and CFA) in the burned soil than in the unburned (Table 3). This variation suggest that the temperatures reached during the fire event, probably around 250-300°C, produced some degree of disaggregation effect and also, the increasing in the quotient value could be due to the newly organic compounds produced by the litter and vegetal residues burned during the wildfire.

Table 3: Alkaline extracts Absorbance ratio of burned and unburned soil samples

Sample	E2/E4	E2/E6	E4/E6		E2/E4	E2/E6	E4/E6
	CHS	CHS	CHS		CFA	CFA	CFA
Unburned soil (UBS)	7.7a	0.67a	5.2a		28.65a	0.63a	18.1a
Burned soil (BS)	7.5a	0.84b	6.33b		29.4a	0.81b	23.9b

Different letters (a-b) in the same column indicate significant differences ($p<0.05$) according to Tukey test.

The greater content obtained for the CFA (table 2) is in agreement with the disaggregation observed through the E2/E6 and E4/E6 values after the fire event.

Humic Acids Characterization

Elemental Composition

Elemental composition (ash and moisture-free basis) O/C, H/C (atomic ratios) and E2/E4, E2/E6 and E4/E6 ratio of the HA extracted from unburned and burned soil are shown in Table 4.

The increase in the carbon content after fire could be produced by the incorporation of the incompletely burned necromass to the original supramolecular structure. The decrease in the oxygen content after fire suggests that the environment could have reducing properties.

The atomic ratio of O/C and H/C are often used to monitor structural changes of humic substances (Gonzalez-Perez et al., 2004; Adani et al., 2006).

Table 4: Elemental composition (ash and moisture-free basis) O/C, H/C (atomic ratios) and E2/E4, E2/E6 and E4/E6 ratio of the HA studied

Sample	C	H	N	O	S	O/C	H/C	E280	E460	E660	E4/E6
HA-UBS	49.67a	5.46a	4.97a	39.46b	< 0.4a	0.59b	1.32b	1.78a	0.34a	0.09a	3.77a
HA-BS	53.89b	5.28a	4.93a	35.45a	< 0.4a	0.49a	1.18a	2.15b	0.47b	0.12b	3.92a

Different letters (a-b) in the same column indicate significant differences ($p<0.05$) according to Tukey test.

The decrease in the atomic H/C ratio observed for HA-BS, suggest a diminution in the peripheral aliphatic chains with low thermal stability and thus, an increase in the aromaticity because this domains was found resistant to the effects of fire. The decrease in the O/C ratio indicates a substantial loss of oxygen-containing functional groups. The mains change observed in HA heated in laboratory or in natural fire are the dehydration and decarboxyilation which explain the progressive alteration in the colloidal properties of soil affected by fire (Gonzalez-Perez et al., 2004).

Spectroscopic Properties

UV-Visible spectra were recorded for both HA analyzed, the specific absorbance decreases steadily with increasing wavelength. The spectra are close to those presented in other studies related to the chemical nature of humic acids (Senesi et al., 1989; Fuentes et al., 2006). The absorption properties are conventional and versatile for the characterization and were used to evaluate the condensation degree of the humic aromatic nuclei. Various absorption wavelengths at 270, 280, 300, 400, 465 nm, among other, and their ratios have been cited for the spectral differentiation of humic substances (Sellami et al., 2008; Uyguner et al., 2005). By analyzing the absorption spectrum of UV-Visible, three important regions were observed at 280, 460 and 660 nm. The absorbance at 280 nm (E280) is related to lignin, aniline derivatives, polyenes and polycyclic aromatic hydrocarbon with two or more rings (Uyguner &

Bekbolet, 2005). The absorbance at 460 nm (E460) is the result of organic macromolecules with a low polymerization degree, and the absorbance at 660 nm (E660) is characteristic of high oxygen content, aromatic compound, high size and molecular weight (Sellami et al., 2008; Uyguner et al., 2005).

The absorbance of the HA extracted from the burned soil is greater than the absorbance of the HA isolated from the unburned soil, similar to those obtained for Vergnoux et al.(2011a). This behavior indicate that the HA isolated from the soil exposed to high temperatures have greater content of different fraction of organic compounds. The increase of the absorption at 280 nm (Table 4) indicate the presence of fraction like lignin derivatives and compounds with aliphatic chains; the absorption at 460 (Table 4) suggest the increment of compounds with a low polymerization degree or less condensed structural domains and the increment of the absorption at 660 nm (Table 4) suggest the increase of aromatic compounds with great microbial and /or chemical resistance, structures that have refractory character (Vergnoux et al. 2011a; Sellami et al., 2008; Santin et al 2008; Gonzalez-Perez et al., 2004). The growth observed in the content of all these fractions could be due through the incorporation of the compounds produced by an incomplete combustion of the vegetation, and therefore, a considerable amount of newly formed C forms were adding together to the thermal modified C forms previously existing in the ecosystem (Cofer et al., 1997; GonzalezPerez et al., 2004). Through the E4/E6 value for both HA, burned and unburned HA, (3.92 and 3.77 respectively), in general, is possible to suppose that the nuclei of the macromolecule of HA, the aromaticity, the size, the weight were not disrupt by the temperature reached in this event fire, instead, the wildfire could have enough energy to produce a disruption onto the linkage which retain together the small fraction of the supramolecular structure and thus a disaggregation could take place; this behavior is shown through the increment of the absorbance values.

Potentiometric titration: Acid base properties and charge evolution

The charges-pH curves (-Q versus pH) of the HA isolated, between pH 3 and 11, obtained from potentiometric titration, corrected for blank solution and fitted with sixth degree polynomial according to Machesky (1993) and Campitelli & Ceppi (2008), are shown in the Figure 1a. This smoothing function was selected for their simplicity.

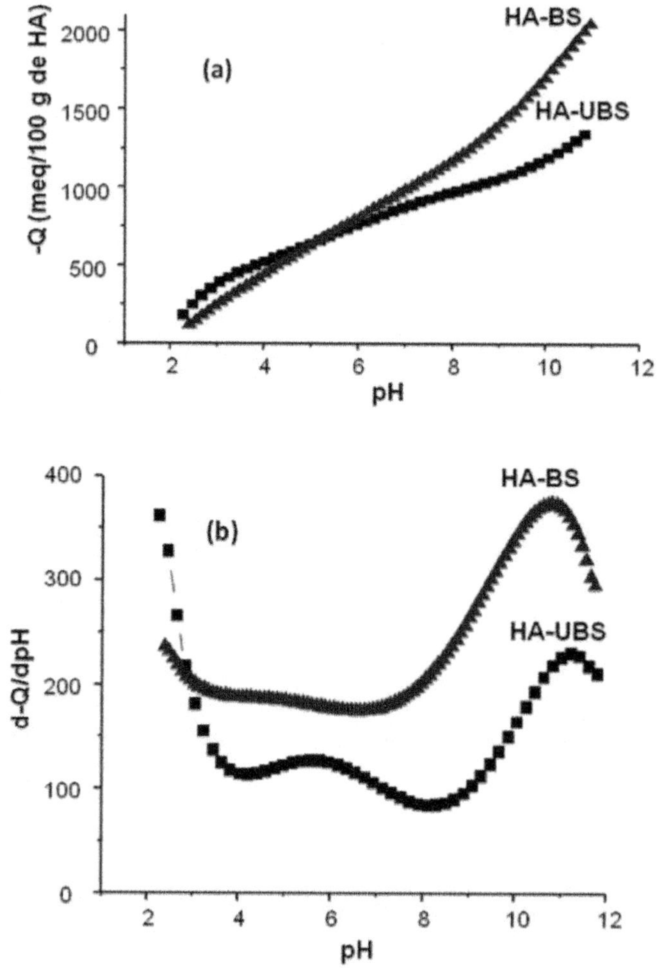

Figure 1: (a): Charge-pH curves of humic acids extracted from burned (HA-BS) and unburned soil (HA-UBS). The charge developments were calculated on the basis of the sixth polynomial equation (with R^2 values exceeding 0.999 in all cases). (Charge development were calculated taking into account the ash content); (b): Apparent proton-affinity distribution of humic acids extracted from from burned (HA-BS) and unburned soil (HAUBS) obtained from the first derivatives through charge-pH curves [d(−Q)/d (pH)] smoothing with sixth degree polynomial equation through the experimental data in the range of 3–10

The charge development of HA isolated from the burned soil (HA-BS) is greater than for the humic acids extracted from unburned soil (HA-UBS) in the region of pH 6 to 11 and lower at the more acidic region (3 to 6). Total acidity is about 60% greater in the HA extracted from burned soil than those of the unburned soil. In the acidic pH region (3 to 6) the lower charge development for HA isolated from burned soil could be due to the loss of strong acidic sites produced by the disruption of the supramolecular structure.

The disaggregation produced by temperature could be the reason for the increment of the negative charge development up to pH 6, because the negative charged groups increase as the size of the fractions decrease (Tombacz, 1999). This behavior is in agreement with that observed through the spectroscopic analysis.

Through the first derivative of the –Q versus pH curves (-dQ/dpH) obtained from the titration curves smoothed with the polynomial equation (Figure 1b), is possible: i) to obtain the average of apparent proton-dissociation constant (pKaap) of each set of acidic groups, ii) to analyze the chemical heterogeneity of each class of acidic group present in the HA macromolecule, iii) to estimate the concentration of each set of acidic groups by the calculus of the area under each peak and iv) to estimate the buffer capacity developed by each class of acidic site (Nederlof et al., 1994; Koopal et al., 2005; Campitelli et al., 2006; Campitelli & Ceppi, 2008). In this way, is possible to follow how the acid-base characteristics, i.e, the evolution in quantity and quality for the principal acidic groups (carboxylic and phenolic), were changed for the fire event.

The number of site classes (set of acidic groups) is then equal to the number of peaks and the peak position could be used as an average of the apparent dissociation constant (pKaap)(de Wit et al.,1993) The samples of HA extracted from unburned soil (HA-UBS) show two main peaks, the first would be assigned to the carboxylic groups (strong acidic sites) and the second to the phenolic groups (weak acidic sites). For the HA isolated from burned soil (HA-BS) the first peak is only a shoulder and the second peak is well defined. In both HA samples (HA-BS and HA-UBS), also is observed, a small or developing peaks at more acidic pH values (≤ 4), indicating, probably, a presence of stronger acidic sites; this behavior is more clear in HABS. This is in agreement with previous results obtained studying HA extracted from soil (Campitelli et al., 2006; Campitelli & Ceppi, 2008).

HA isolated from burned soil (HA-BS) presents the first peaks or shoulder, not well defined, with the maxima at around pH 3.5 and the second with a maximum at pH 10.8; the first could be assigned to strong acid sites (carboxylic groups) and the second to weak acidic sites (phenolic groups). The peak at pH=3.5 was wider than the peak at pH=10.8. The minimum was not well defined, and the partial overlapping of peaks indicate that there is no significant differences among the acidic sites in the surface, in terms of proton dissociation strength. This results suggest a large chemical heterogeneity on the HA present or the production of small organic compounds during the fire event.

These small organic compounds could be produced by the incomplete combustion of the vegetation present; Knicker et al. (2007) suggested that around 250 °C new molecular structures are produced; the principal structures could be aliphatic C; phenol and/or furan C; Sharma et al. (2004) suggested that some decarboxylation could occur at higher temperatures (>250°C) but the aromatic rings still remain essentially intact. This behavior could justify the decrease in the negative charge development at pH values lower than 6 and their increase at higher pH values (pH > 6).

HA isolated from unburned soil (HA-UBS) have two well defined peaks, the first with the maximum at pH 5.6 and the second at pH 11.2, these values are similar to other obtained for soil derived humic acids (Campitelli et al., 2006; Campitelli & Ceppi 2008).

The pKaap for the carboxylic and phenolic groups in the HA derived from the burned soil (HA-BS) are lower than the corresponding for HA extracted from unburned soil (HA-UBS), this could be due to the disruption of the supramolecular structure of the humic acids, and in this way the carboxylic groups that remains in the surface are those with very strong acidic characteristics, probably those in the aromatic structures, like o-COOOH or in greater fractions, and the phenolic groups are those in the small fraction produced by the disaggregation (Table 5) (Knicker et al., 2007; Sharma et al., 2004). For both type of acidic groups (o-COOH and OH-Phenolic), the contribution could be from the partial combustion of vegetation and then extracted with the alkaline media, without discrimination (Adani et al., 2004).

Table 5: Acidic functional groups content (o-carboxylic, carboxilic and phenolic) content calculated by integration of the area under each maximum of the curves (d-Q/dpH) obtained through the first derivative of smoothed experimental data. The pKaap values correspond to the maximum of each peak

Humic acids	o-COOH	pKaap	-COOH	pKaap	phenolic-OH	pKaap
HA-UBS	320a	2.3	473	5.6	567a	11.2
HA-BS	588b	3.5	---	---	1318b	10.8

Acidic groups: cmol kg^{-1} Different letters (a-b) in the same column indicate significant differences ($p<0.05$) according to Tukey test.

The fire event altered the concentration of acidic sites (Table 5) and therefore the buffer capacity. For the burned soil, the buffer capacity of HA was neglectable at soil pH value around 6 (Table 1) and for pH value ranging between ≈ 3 – 7. This can be attributed to the great heterogeneity of HA in this pH range and to the lost of carboxylic groups with pKaap values around 5.

In the zone up to pH 8 (weak acidic sites) the buffer capacity is greater than that observed for HA from unburned soil, but this groups, in both cases, are not dissociated at soil pH values, thus they have not a significant contribution to the soil buffer capacity. The fire event produced important changes in the acid-base properties, principally in the buffer capacity of the HA.

The loss of carboxylic groups onto this HA structure produced by fire event (Table 5), i.e. the decrease of negative charge development below pH 6, cause a deficiency of charged site to make linkage between the inorganic and organic fraction through cation-bound; and thus, the formation of soil aggregates. In this way, this characteristic could be the key factor promoting soil erosion (Mill & Fey 2004). The fire event could generate important modification in the physicochemical properties of the HA

At the lowest pH measured (sites domains below 4), the HA-UBS shows a developing peak (Fig 1b) indicating that very acidic sites could be present in the macromolecule, in HA-BS it seems that this sites are the only present (Table 5). The minimum around pH 4, which could be considered as a separation of both type of acidic sites (like COOH) from the very acidic sites (like o-COOH), is clearer in the HA extracted from unburned soil (HA-UBS) than in the HA from the burned soil (HA-BS), this indicate, also, the heterogeneity of the acidic groups present in the HA extracted from soil exposed to high temperatures, due to the disaggregation produced by the temperature developed during the wildfire.

Capillary zone electrophoresis

The main characteristics of HA are the occurrence of acidic site with different strength, the principal groups are the strong (carboxilic groups) and weak (phenolic groups) acidic site.

For these HA analyzed the average Pkaap value are around 3.5 – 5.5 for the carboxylic groups and 10 – 11 for the phenolic groups (Table 5).

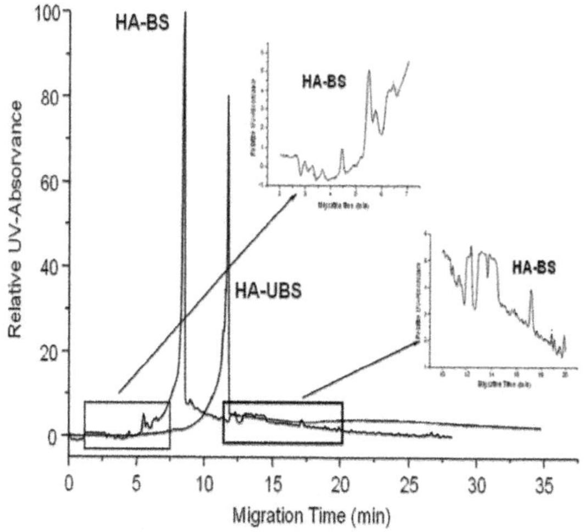

Figure 2: Electropherograms of acids extracted from burned (HA-BS) and unburned soil (HAUBS) in buffer borate 20 mmol L^{-1} (pH=9.3), temperature 25°C, the concentration of the AH solutions were 1000 ppm. CZE conditions: voltage of -30kV, injection hydrodynamic 5000 Pa for 20 s, detection at 260 nm, fused-silica capillary, 73 cm total length, 75 μm i. d. (effective length 64.5 cm). Total run time 30 min (for time migration higher than 30 min no significant peak were observed)

At the experimental condition (pH ≈ 9) all of the strong acidic groups and approximately, the half of the weak acidic groups of HA are deprotonated (negatively charged). The presence of negative charges permit to separate HA by electrophoresis in an electrical field (+) to (-) in which the EOF (electro osmotic flow) is responsible for the movement of the analyte (Peuravouri et al., 2004).

The electropherogram of HA extracted from unburned soil (HA-UBS) (fig 2) shows a principal and well defined peak at time migration 11.73 min and the characteristic hump at time migration around 7 – 8 min, just before the main peak, is shown as a tail; at migration time higher than 12 min no peaks were distinguished.

The electropherogram of HA isolated from burned soil (HA-BS) presents the main peak at lower time migration (8.40 min) than in HA of unburned soil (HA-UBS), and several peaks are detected before and after the main peak (fig 2); at migration time higher than 25 min no peaks were distinguished.

The peak at 11.73 min observed in the electropherogram corresponding to the HA-UBS, could indicate that in these experimental conditions the macromolecule migrate as a unbroken entity, the tailing observed at lower time migration, could be assigned to some structure with low mass/charge ratio difficult to be separated; i.e. the macromolecule is not easy to be separated in subfraction with different electrokinetic properties, similar behavior was observed for Fetch & Havel (1998); Pokorna et al.(2000); Peuravouri et al.(2004). The different time migration for the principal peak of the HA from burned soil (HA-BS) and the peaks detected at both side of the peak at 8.40 min could indicate changes in the macromolecule structure and the presence of subfraction.

The BGE, borate, could react with phenols, phenols carboxylic, polycarboxilic acids, dihidroxy or perihydroxy groups present in the solution and thus the separation of each fraction would be improved (Fetsch & Havel 1998). The phenolic groups present in HA isolated from burned soil (HA-BS) is greater than that quantified in HA extracted from unburned soil (HA-UBS), this characteristic could produce the interaction between the BGE and this acidic groups and enhance the separation.

The electropherogram profile of the HA extracted from burned soil (AH-BS) indicates the presence of distinct subfraction, which could be produced by the disaggregation of the macromolecule of HA and/or the formation of newly small carbon compounds after heating, suggesting that the temperature reached during the fire event, breaks, disaggregates or creates new structure, with lower and higher mass/charge ratio and diverse electrokinetic mobility. This behavior confirms the large heterogeneity, the disaggregation and the new carbon compound produced for the wildfire and are in agreement with those observed through the other different analytical techniques used to study these HA.

CONCLUSIONS

The temperature reached in the fire event was enough to produce several changes in the organic matter characteristics, i.e. changes in the quantity and/or quality of their fraction: light fraction, humic acids, fulvic acids, free lipidic fraction. The fire event produced important changes in the structure of the macromolecule of humic acids, like break and/or disaggregation which generate compound with lower size, weight, mass/charge ratio and/or newly formed carbon compounds originated by the incomplete combustion of the vegetal materials. The fire event could generate important modification in the physicochemical and acid-base properties of the HA.

The amount of acidic functional group was changed: the COOH sites were decreased and the OH phenolic sites were increased by the fire event. The pKaap values were modified, in general, the acidic site are stronger after fire than in the unburned soil. The COOH groups with pKaap value about 5 were lost after fire. The buffer capacity is lower or practically missing at soil pH (≈

6) after fire. The negative charge development decrease significantly at field pH (≈ 6) after the fire event, producing a deficiency on sites to make linkage between the organic and inorganic soil fraction, and in this way a reduction of aggregates formation. This characteristic could be the key factor promoting soil erosion.

ACKNOWLEDGEMENTS

SeCyT-UNC are gratefully acknowledged for financial support.

REFERENCES

1. Abad, M., Noguera, P., Puchades, R., Maqueira, A. & Noguera, V., 2002. Physico-chemical properties of some coconut coir dusts for use as a peat substitute for containerised ornamental plants. Bioresour Technol., 82: 241-245
2. Adani, F & G. Ricca 2004 The contributionof alkali soluble (humic acid-like) and unhydrolyzed-alkali soluble(core-humic acid-like) fraction extracted from maize plant to the formation of soil humic acid Chemosphere 56: 13-22.
3. Adani, F., Ricca, G., Tambone, F. & Genevini, P., 2006. Isolation of the stable fraction (the core) of humic acids. Chemosphere 65:1300-1307.
4. Alexis, M. A., Rasse, D. P., Rumpel, C., Bardoux, G., Pechot, N., Schmalzer, P., Drake, B., & Mariotti, A. 2007 Fire impact on C and N losses and charcoal production in a scrub oak ecosystem. Biogeochemistry 82: 201-219
5. Almendros, G., Gonzalez-Vila F. J., Martin, F., Frund, R. & Ludemann H. D. 1992 Solid state NMR studies of fire induced changes in the structure of humic substances. Sci Total Environment 117-118: 63-74 Almendros, G., Gonzalez-Vila F.J. & Martin, F. 1990 Fire-induced transformation of soil organic matter from an oak forest: an experimental approach to the effects of fire on humic substances. Soil Science 149: 158-168.
6. Almendros, G., knicker, H. & Gonzalez-Vila F. J. 2003 Rearrangement of carbon and nitrogen forms in peat after progressive isothermal heating as determined by solidstate 13C and 15N-NMR spectroscopy. Organic geochem 34: 1559-1568
7. Almendros, G., Martin, F. & Gonzalez-Vila, F. J. 1988 Effects of fire on humic and lipid fraction in a Dystric Xerochrept in Spain Geoderma 42:115-127

8. Arocena, J. M. & Opio, C. 2003 Prescribed fire-induced changes in properties of sub-boreal forest soil Geoderma 113:1-16
9. Badía, D. & Martí, C. 2003 Plant ash and heat intensity effects on chemical and phisical properties of two contrasting soils. Arid Land Res Management 17:23-41
10. Cade- Menun, B. J., Berch, S. M., Preston, C. M. & Lavkulic, L. M. 2000 Phosphorus forms and related soil chemistry of Podzolic soils on northern Vancouver Island. II. The effects of clear-cutting and burning. Can J Forest Research 30:1726-1741
11. Campitelli, P, A., Velasco, M, I. & Ceppi, S, B., 2006. Chemical and physicochemical characteristics of humic acids extracted from compost, soil and amended soil. Talanta, 69:1234–1239.
12. Campitelli, P. & Ceppi, S., 2008. Effects of composting technologies on the chemical and physicochemical properties of humic acids. Geoderma, 144:325–333.
13. Campitelli, P., Velasco, M. & Ceppi, S., 2003. Charge development and acid-base characteristics of soil and compost humic acids. J Chil Chem Soc., 48:91-96
14. Certini, G. 2005 Effects of fire on properties of forest soil: a review. Oecologia 143: 1-10
15. Ciavatta, C., Antisari, V. & Sequi, P. 1988 A first approach to the characterization of the presence of humified materials in organic fertilizers. Agrichimica 32:510-517
16. Cofer III W. R., Koutzenogii, K. P., Kokorin & A. Ezcurra, A. 1997 Biomass burning emission and the atmosphere. In Clark J. S., Cachier, H., Goldammer J. G., Stocks, B. editors. Sedimient records of biomass burning and global change. NATO ASI Serie, Vol I Berlin Germany:
17. Springer Czimczik, C. I., Schmidt, M. W. I. & Schulze, E. D. 2005 Effects of increasing fire frecuency on black carbon and organic matter in Podzols of Siberian Scots pine forest. European J of soil Sci. 56: 417-428
18. Chen, Y., Senesi, N. & Schnitzer, M., 1977. Information provide on humic substances by E4/E6 ratio. Soil Sci Soc Am J. 41: 352-358.
19. Chen, Y., Senesi, N. & Schnitzer, M., 1978. Chemical and physical characteristics of humic and fulvic acids extracted from soils of the Mediterranean region. Geoderma, 20:87- 104.
20. Dai, X., Boutton, T. W., Glaser, B., Ansley, R. J. & Zech, W. 2005 Black carbon in a temperate mixed-grass savanna. Soil Biology and biochemistry 37: 1879-1881

21. Debano, l. F. 2000 The role of fire and soil heating on water repellence in wildland environment: a review. J Hydrol 231: 195-206
22. Duguy, B. & Rovira, P. 2010 Differential thermogravimetry and differential scanning calorimetry of soil organic matter in mineral horizons: Effect of wildfire and land use. Organic Geochemestry 41: 742-752
23. De Wit, J. C. M., Van Riemsdijk, W. H. & Koopal L. K. 1993 Chemical heterogeneity and adsorption models. Environm. Sci. Technol. 27: 2015-2022
24. Fernandez, I., Cabaneiro, A. & Carballas, T. 1997 Organic matter changes immediately after a wildfire in an Atlantic forest soil and comparison with laboratory soil heating. Soil Biology and Biochemestry 29: 1-11
25. Fernandez, I., Cabaneiro, A. & Gonzalez-Prieto, S. J. 2004 Use of 13C to monitor soil organic matter transformations caused by a simulated forest fire. Rapid Commun Mass Spectrom 18:435-442
26. Fuentes, M., Gonzalez-Gaitano, G. & García-Mina, J. M. 2006 The usefulness of UV-vis and fluorescencespectroscopies to study the chemical nature of humic substances from soil and compots. Org. Geochem 37:1949-1959
27. Giovannini, G. Lucchesi, S. & Giachetti, M. 1988 Effect of heating on some physical and chemical parameters related to soil aggregation and erodibility. Soil Sci 146: 255-261
28. Golzalez-Vila, F. J. & Almendros, G. 2009 Thermal transformation of soil organic matter by natural fires and laboratory controlled heating. In: Natural and Laboratory Simulated Thermal Geochemical Processes. R. Ikan (ed.) Kluver Academic Publisher. Netherlands.
29. Gonzales-Perez, J. A., Gonzalez-Vila, F. J., Almendros, G. & Knicker, H. 2004 The effect of fire on soil organic matter: a review Environmental International 30: 855-870
30. Hayes, M. H. B. & Malcom, R. L. 2001 Considerations of the compositions and of aspects of the structure of humic substances. In: Clapp, C.E., Hayes, M. H. B., Senesi, N., Bloom, P. R., Jardine, P. M. (Editors) Humic substances and chemical contaminations. Soil Science Society of America Inc. Madison (pp 3-39)
31. Hernandez, T., Garcia, C. & Reinhardt, I. 1997 Short-term effect of wildfire on the chemical, biochemical and microbiological properties of Mediterranean pine forest soil. Biol. Fertil. Soil 25:109-116
32. Janzen, H. H., Campbell, S. A., Brand, S. A., Laford, G. P. & Townley-Smith, A. 1992 Light fraction organic matter in soil from long-term crop rotation. Soil Sc. Soc of Am J. 56:1799-1806

33. Kincker, H., Gonzalez-Vila, F. J., Polvillo, O., Gomzalez, J. A. & Almendros, G. 2005 Fire induced transformation of C and N forms in different organic soli fractions from a Dystric Cambisol under Mediterranean pine forest (Pinus pinaster). Soil Biol. Biochem. 37: 701-718

34. Knicker, H. 2007 How does fire affect the nature and stability of soil organic nitrogen and carbon? A review. Biogeochemistry 85-118

35. Knicker, H., Gonzalez-Vila, F. J., Plovillo, O. & Gonzalez, J. A., Almendros, G. 2005a Foreinduced transformation of C and N forms in different organic soil fraction from a Dystric Cambisol under a mediterranean pine forest (pinus pinaster). Soil Biol Biochemestry 37:701-718

36. Knicker, H. & Skjemstad, J. O. 2000 carbon and nitrogen functionality in protected organic matter of some Australian soils as revealed by solid-state 13C and 15N NMR . Australian J of Soil Sc, 38: 113-127

37. Kononova, M. M. 1982. Materia Orgánica del Suelo. Vilassar de Mar. Barcelona. España Koopal, L. K., Saito, T., Pinheiro., van Riemsdijk, W. H. 2005 Ion binding to natural organin matter: General considerations and the NICA-Donnan model. Colloids and Surface A: Physicochem. Eng aspects 265: 40-54.

38. Kutiel, P. & Imbar, M. 1993 Fire impact on soil nutrients and soil erosion in a mediterranean pine forest plantation. Catena 20:129-139

39. Lipski, M., Slawinski, D. & Zych, D. 1999 Chnages in the luminescent properties of humic acid induced by UV radiation. J Fluorescence 9: 133-138

40. Machesky, M., 1993. Calorimetric acid-base titrations of aquatic and peat derived fulvic and humic acids. Environm Sci Technol., 27: 1182-1198.

41. Mill, A. J. & Fey, M. V. 2004 Frequent fire intensity soil crusting: physicochemical feedback in the pedoderm of long-term burn experiments in South Africa. Geoderma 121: 45- 64

42. Neary, D. G., Klopatec, C. C., deBano, L. F. & Fgolliott P. F. 1999 Fire effects on belowground sustainability: a review and synthesis. Forest Ecol manag. 122:51-71

43. Nederlof, M. M., van Riemsdijk, W. H. & Koopal, L. K., 1994. Heterogeneity analysis for binding data using adapted smoothing spline techniques. Environ Sci Technol., 28: 1037-1047.

44. Pertusatti, J. & Prado, A. G. S., 2007. Buffer capacity of humic acid: Thermodynamic approach. J Colloid and Interface Sci., 314:484-489.

45. Polak, J., Bartoszek, M. & Sulkowski, W. W. 2009 Comparison of some spectroscopic and physico-chemical properties of humic acids extracted from sewage sludge and bottom sediments. J. of Molecular Structure. 924-926: 309-312

46. Richter, M. & Von Wistinghausen, E. 1981 Unterscheidbarkut von humusfraktione in boden be unterschiedlicher Bewirtschaftung, Z. Pflanzenernaehr Bodenk 144:395-406

47. Roletto, E., Barberis, R., Consignlid, M. & Jodice, R. 1985 Chemical parameters for evaluation of compost maturity. Biocycle March: 46-48

48. Santín, C., Knicker, H., Fernandez, S., Mendez-Duarte, R. & Alvarez, M. A. 2008 Wildfire influence on soil organic matter in an Atlantic mountainous regiôn (NW of Spain). Catena 74: 286-295

49. Sapeck, B. & Sapeck, A. 1999. Determination of optical propierties in weakly humified samples. In: Dziadowiec, H., Gonet, S.S. (Eds.), The Study of Soil Organic Matter-the Methodical Guide. Warszawa, Poland.

50. Schnitzer, M. 2000 A lifetime perspective on the chemical of soil organic matter. Adv. Agron. 68: 3-58

51. Sellami, F., Hachicha, S., Chtourou, M., Medhioub, K. & Ammar, E., 2008. Maturity assessment of compostded olive mill waste using UV spectra and humification parameters. Bioresour Technol., 99:6900-6907.

52. Senesi, N., 1989. Composted materials as organic fertilizer. Sci Total Environm., 81:521-542.

53. Shakeesby, R. A. & Doerr, S. H. 2006 Wildfire as a Hydrological and geomorphological agent. Earth Science Rev 74:269-307

54. Sharma, R. K., Wooten, J. B., Baliga, V. L., Lin, X., Chan, W. G. & Hajaligol, M. R. 2004 Characterization of chars from pyrolysis of lignin. Fuel 83: 1469-1482

55. Sims, J. R. & Haby, V. A., 1971. Simplified colorimetric determination of soil organic matter. Soil Sci., 112:137-141.

56. Spaccini, R., Piccolo, G., Haberhauer, G. & Gerzabek, M. H. 2000 Transformation of organic matter from maize residues into labile and humic fraction of three European soil as reveled by 13C distribution and CPMAS-NMR spectra. European J soil Sci. 51: 583- 594

57. Stevenson, F. J., 1982. Humus chemistry. Genesis, composition, reactants. John Wiley and Sons N. York.

58. Sutton, R. & Sposito, G., 2005 Molecular structure in soil humic substances: the new view. Environm Sci Technol., 39: 9009-9015.

59. Tombacz, E. 1999 Colloidal properties if humic acids and spontaneos

changes of their coloidal state under variable solution conditions Soil Sci 164:814-824

60. Uyguner, C. S., Bekbolet, M. 2005 Evaluation of humic acid photocatalytic degradation by UV-vis and fluorescence spectroscopy. Catalysis Today 101: 267-274

61. Vergnoux, A., Di Rocco, R., Domeizel, M., Guiliano, M., Doumenq, P., Theraulaz, F. a) 2011 Effects of fire on water extractable organic matter and humic substances from Mediterranean soil: UV-vis and flourecsence spectroscopy approaches Geoderma 160:434-443

62. Vergnoux, A., Dupuy, N., Guiliano, M., Vennetier, M. Theraulaz, F. & Doumenq, P. 2009 Fire impact on forest soil evaluated using near-infrared spectroscopy and multivariate calibration. Talanta 80: 39-47

63. Vergnoux, A., Guiliano, M. Di Rocco, R., Domizel, M., Theraulaz, F. & Doumenq, P. b) 2011 Quantitative and mid-infra-red changes of humic substances from burned soils. Environmental Research 111: 193-198

64. Vergnoux, A., Malleret, L., Domeizel, M. Theraulaz,F. & Doumenq, P. 2007 Effect of forest fire on water extractabke organic matter and polycyclic aromatic hydrocarbon in soil. Progerss in Environmental Science abd Technology. Beiijing, China Zancada, M. C., Almendros, G., Sanz, J., & Romám, R. 2004 Speciation of lipids and humuslike coloidal compounds in a forest soil reclaimed with municipal solid waste compost. Waste management and research 22: 24-34

65. Zbytniewski, R. & Buszewski, B. 2004 Characterization of natural organic matter (NOM) derived from sewage sludge compost. Part 1: chemical and spectroscopic properties. Bioresource technology 96: 471-478

66. Zbytniewski, R. & Buszewski, B. 2005 Characterization of natural organic matter (NOM) derived from sewage sludge compost. Part 2: multivariate techniques in the study of compost maturation. Bioresource technology 96: 479-484

Chapter 7

FOREST PRESERVATION, FLOODING AND SOIL FERTILITY: EVIDENCE FROM MADAGASCAR

Bart Minten[1] and Claude Randrianarisoa[2]

[1]International Food Policy Research Institute, Addis Ababa, Ethiopia

[2]United States Agency for International Development (USAID), Madagascar

INTRODUCTION

In several developing countries, forest preservation programs have been put in place with an economic justification based on the local ecological services that they provide (Pagiola et al., 2002). It is argued that the presence of forests preserve the hydrological balance; reduce soil erosion due to increased soil stability; reduce flooding and regulate flows (Perrot-Maitre and Davis, 2001; Johnson et al., 2002; Pattanayak and Kramer, 2001a,b). However, other authors dispute the domestic benefits of forests and state that natural scientists often overvalue forests (Chomitz and Kumari, 1998; Aylward and Echeverria, 2001; Calder, 1999). Assuming that an externality costs of deforestation exists, policy makers have started to look at how to correct for this and how a workable system can be put in place to pay for ecological services locally. Increasing attention is going towards the direct payment for environmental services (Ferraro and Simpson, 2002; Durbin, 2002; Pagiola et al., 2002).

We look at this issue in a case study in Madagascar. Multiple studies have shown the high and accelerating deforestation rate in Madagascar (McConnell, 2002). Causes of deforestation are multiple and have been linked to poverty (Zeller et al., 2000), conversion of forest land to pastures (McConnel, 2002), use of wood for charcoal (Casse et al., 2004), wood exports or household fuel consumption (Minten and Moser, 2003), slash-and burn agriculture (Barrett, 1999; Keck et al., 1994; FOFIFA, 2001; Casse et al., 2004; Terretany, 1997), rural insecurity (Minten and Moser, 2003), and land tenure problems (Freudenberger, 1999). While deforestation threatens the unique eco-system of

Madagascar, it has also been linked to higher incidences of flooding and greater soil erosion and damages therefore the agricultural resource base domestically (Freudenberger, 1999, Kramer et al., 1997). Overall, it is estimated that the damage of soil erosion in Madagascar is high (Kramer et al., 1997; World Bank, 2005) although the numbers that have been suggested might have been exaggerated (see f.ex., Larson, 1994).

In this analysis, we study the potential domestic benefits of forests on lowland agriculture. While we do not try to establish explicit linkages between deforestation and sedimentation off-site, we do look at the effects of flooding and sedimentation downstream as perceived by rice farmers. The analysis is based on a small-scale survey in Northern Madagascar where we try to monetize the cost to farmers of flooding and sedimentation on their rice fields downstream.1 If the link between forest cover and flooding would exist in this area and if the link is strong, a positive willingness to pay might then justify investments in conservation measures upstream.

We contribute to the literature in two ways. First, we show that an important percentage of rice farmers benefit from flooding and sedimentation (as shown in higher land values after sedimentation and refusal for contribution towards conservation) and that current economic returns to investment in forest preservation, largely beneficial because of averted rice productivity declines, might thus be overestimated.2 Second, in the rural scarcely monetized settings of developing countries where land transactions are rare, we develop an alternative to the hedonic price analysis of land values using willingness-to-accept scenario's explicitly allowing for uncertainty.

The structure of the paper is as follows. First, we discuss the conceptual framework. Second, the methodology, data sources and the structure of the survey are presented. Third, we look at descriptive statistics describing households as well as sedimentation and flooding incidence. Fourth, the determinants of land values, incorporating the impact of sedimentation, and the results of a willingness to pay question to avoid flooding and sedimentation are discussed. We finish with the conclusions.

CONCEPTUAL FRAMEWORK

Assume an expenditure minimization problem where expenditures are minimized subject to the constraint that utility equal or exceed some stated level, U^0. The solution to this minimization problem is the restricted expenditure function

$$e = e(p^0, T^0, U^0, \varepsilon^0)$$

where p^0 can be thought of as a vector of prices, T^0 is land availability to the household and ε^0 represents uncertain factors not reflected in p^0, T^0 and U^0.

In a first offer, the household is asked to sell land for a total payment of P^1. In a second offer, the household is asked to pay for conservation for a total payment of P^2. The change from T^0 to T^i in either of the two scenarios will result in a new expenditure function with a new set of prices and environmental and resource flows, i.e. $e = e(p^1, T^1, U^0, \varepsilon^1)$ in the first scenario and $e = e(p^2, T^2, U^0, \varepsilon^2)$ in the second scenario. It seems reasonable if you take away land or income, and given imperfect markets, that the shadow prices and wages are likely to change, i.e. we do not assume the price vector to be independent in the two scenarios.

In such a set-up, the welfare change - the Hicksian compensating surplus - is defined as the difference between the two expenditure functions,

$$e(p^i, T^i, U^0, \varepsilon^i) - e(p^0, T^0, U^0, \varepsilon^0)$$

where i is 1 (scenario 1) or 2 (scenario 2). The value of the welfare change is established by using contingent valuation measures and the Willingness to Accept/Pay (W) at the farm household level might be represented by W_j for household j

$$W_j = e(p^i, T_j^i, U_j^0, \varepsilon_j^i, X_j) - e(p^0, T_j^0, U_j^0, \varepsilon_j^0, X_j) + \eta_j$$

where X_j is a vector of socio-economic characteristics for household j and η_j is an error term. Such a model can be further refined to allow for dynamic behavior (Holden and Shiferaw, 2002). If we let W_2 represent the subjective present value of future land productivity gains by switching from no interventions to conservation efforts in the uplands, the following equation holds in the case of the maximization of an expected intertemporal utility function:

$$U_j^0 (C_j^0) - U_j^0 (C_j^0 - W_j) = \sum_{t=1}^{\infty}(1+\delta_j)^{-t} EU_j^t(C_{1j}^t - C_{0j}^t)$$

Where δ_j is individual j time preference, EU_j^t is the expected utility for individual j in time t, and $U_j^t(C_{1j}^t - C_{0j}^t)$ is the utility gain in time t when switching from no interventions to conservation efforts in the uplands. Nonseparability in a dynamic context implies that intertemporal markets do not work well and that W would then vary over time with household discount rates that can be very high for poor liquidity constrained households.3 W can then be specified as a random variable which is a continuous function of observational variables that appear in the expenditure function such as farm, technology and socio-economic characteristics. W can thus be written as

$$W_j = Z_j\beta + \mu_j$$

where $\mu_j \sim (0, \sigma^2)$

where Z is a vector of explanatory variables, μ_i is the error term and σ is the standard deviation.

METHODOLOGY AND DATA

An agricultural household survey was organized in November 2001 in an area northwest of Maroantsetra, in the northeast of Madagascar. The area was selected on the basis of the high diversity in watershed forms and areas and the perceived clear link between upstream activities and lowland impacts. First, a census of all the watersheds was done. In total, 65 watersheds were identified. Due to logistical reasons, only 52 watersheds were sampled. In each watershed, a stratified sample of rice plots was done. Rice plots were stratified based on the distance to the main river. In each watershed, around six fields were sampled, depending on the size of the watershed. In total, data on 268 rice farmers were obtained. The questionnaire that was implemented consisted of four parts. The first part dealt with plot characteristics (including a land valuation question), the second with questions on the rice harvest of last year on that plot, and the third on the overall structure of the agricultural firm. The final part described a willingness to pay scenario where households were asked to value their desire to avoid flooding and sedimentation in their rice fields.

Instead of the widely used and recommended dichotomous choice valuation question (Arrow et al., 1993), a stochastic payment card method (Wang and Whittington, 2005) was implemented for different reasons: (1) Given logistical constraints, a relatively small sample had to be relied upon. The payment card format gives the benefit of having extra information beyond the yes/no question (For papers that discuss the benefits of information beyond dichomotous choices, see Blamey et al. (1999) and Ready et al. (2001)). (2) Whittington (1998) and Wang and Whittington (2005) show that a main problem in contingent valuation studies is that the range that is offered is often not large enough to allow for a robust estimation of the valuation function. Moreover, as we had little a priori knowledge about the valuation function, we had to make sure that extreme levels were included in the bids on the payment card. Given the small sample, this could not have been achieved in the dichotomous choice variable format. (3) Uncertainty (for example on the future price evolution of agricultural products) and imperfect information (household chief had to answer immediately during the interview and could not consult with family members and/or village leaders) is allowed for in this format. Wang (1997),

Wang and Whittington (2005) and Alberini et al. (2003) show the benefits of the explicit modeling of uncertain responses in contingent valuation data.

Two valuation questions were asked. The valuation questions were set up in such a way to reduce as much as possible the problem with starting point bias and with yea-saying: therefore, it started with an open-ended question (no starting point bais) followed by a payment card (additional information). In the case of the land valuation, a willingness to accept scenario was described where a certain monetary payment was given in exchange for the plot studied. As previous surveys in Madagascar had shown the reluctance of farmers to give a sales price for land - they would often report they would be unwilling to sell the plot whatever happened - it was made clear from the beginning that this was a hypothetical situation where we like to know their approximate financial value of the plot in their farming enterprise. The respondent was presented with a payment card in local currency but with references to values of local rice units, bikes, and value of livestock. On this payment card, the enumerator proceeded to fill in for every amount that was mentioned a code corresponding to 1. Accept to pay for sure; 2. A little bit in doubt but would say yes; 3. Not yes or no, do not know; 4. A little bit in doubt but would say no; 5. Will not pay for sure.

In the case of the question on willingness to pay for reduction in flooding and sedimentation, the valuation scenario was constructed as follows. Respondents were first asked if they thought if flooding and sediments had a negative, neutral or positive influence on rice productivity, in general and on the specific plot that was studied. A scenario was then described in the following way:

"Suppose *that we leave the situation as it is and we leave damage as it is without any intervention to limit deposits on this rice field or to reduce the frequency of flooding on this field. In a second situation, actions will be undertaken in the watershed upstream of your fields. In this case, you will not suffer anymore from problems of flooding and sediments. However, you know that these actions will cost money. We would like to know how much you would be willing to pay for these actions, taking into account your possibilities. If you do not pay as much than what you would be really able to pay, actions will not be sufficient to reduce flooding and sedimentation. On the other hand, if you give a level that is higher than you can afford, functional interventions cannot be agreed upon. How much would you be willing to pay? x sobika of rice?*"

The question was formulated in local units of rice as this measure was easily recognizable by farmers. To finish the valuation section, a question was asked to the farmers on where they would get the rice from for the amount that they were willing to contribute. It was hoped that this would remind them of their budget constraint. Corrections on the payment card were allowed for afterwards.

While non-responses were not a problem in the plot valuation question, about one third of the respondents did not answer the willingness to pay question to avoid flooding or sedimentation. The characteristics of the respondents that refused to answer are not randomly distributed and might therefore cause inconsistency and inefficiency in the estimation of the coefficients in the regression of the willingness to pay question. A common method to control for non-responses to the willingness to pay question is to estimate a sample selection model (Messonier et al., 2000; Mekkonen, 2001), usually referred to as the Heckman two stage approach (Heckman, 1997). In this case, we estimate:

$$Y^* = \beta' X + \varepsilon$$

$$Y = 0 \text{ if } Y^* \leq 0$$

$$\text{and } Y = Y^* \text{ otherwise}$$

$$Z = \alpha' V + \mu$$

$$Z = 1 \text{ if } Z^* > 0$$

$$\text{and } Z = 0 \text{ if } Z^* \leq 0$$

where Y is willingness to pay (censored at 0); X is a vector of explanatory exogenous variables that explain Y; Z is 1 when there is a valid response and 0 otherwise; V is vector of explanatory exogenous that influence the probability of giving a valid response; α and β are parameters to be estimated; ε and μ are disturbances; Y^* and Z^* are latent variables.

DESCRIPTIVE STATISTICS

The Maroantsetra area in the Northeast of Madagascar is a humid area characterized by two types of agriculture: slash-and-burn cultivation (“tavy”) on the hillsides and lowland rice cultivation. The area is isolated from the rest of Madagascar and is highly dependent on agriculture for income. The region is also still highly forested and is one of the largely untouched areas in Madagascar. Table 1 shows the basic descriptive statistics of the households in the survey. The head of households have a low average level of education, i.e. only three years. 10% of the households are female headed and these are mostly poorer households (Razafindravonona et al., 2000). The average size of the household is six members. Almost all the households are natives from the region and all the households report to depend on agriculture for their livelihood.

An average household in the sample possesses 62 ares4 of lowland and 73 ares of upland. As in most of Madagascar, the main staple is rice. The average production is just below 1 ton which is estimated to be sufficient for subsistence by almost 70% of the population. However, most households - even some that declare to be self-sufficient in rice - reduce overall consumption during the lean period. The average length of this lean period is estimated to be three months. A household possesses on average 2 zebus. Total annual monetary household income is estimated at 2.7 M Fmg[5], i.e. around 415$US, i.e. low but consistent with the high poverty levels and the low GNP of Madagascar (Razafindravonona et al., 2001).

Tables 2 presents the descriptive statistics of the rice plots that will be analyzed in more detail later on. The average plot size is small, 2.1 ares, with a range between 1 and 25 ares. Most of the plots are reported to be irrigated through a dam (96%). When asked about production problems in the last agricultural year, 28% of the farmers complained of droughts, 21% of sedimentation problems, and 14% of floods. Average yields during the previous agricultural year were estimated at 3.3 ton per hectare, high compared to the rest of the country but consistent with the excellent country-wide production conditions in 2001.6

Table 1: Descriptive statistics of household variables

variable	Unit	N	mean	median	min	max
size of household	number of people	268	5,65	5	1	14
education level head of hh	years	268	3,13	3	0	12
age	years	268	45,55	44	15	81
gender	man=1	268	0,90	1	0	1
native of region	yes=1	268	0,99	1	0	1
lowland	ares	268	61,87	50	0	340
upland	ares	268	73,40	50	0	1000
forest savoka	ares	268	33,09	0	0	500
primary forest	ares	268	30,06	0	0	600
zebus	number	268	1,75	0	0	18
total production of rice	kg	268	913,46	720	60	4500
total income	1000 Fmg	268	2695,57	1635	0	30100
rice production is enough	yes=1	268	0,27	0	0	1
length of lean period	number of months	268	2,81	3	0	10
potential access to credit	1000 Fmg	268	706,03	100	0	25000

Two major cyclones hit the area in the last five years: Huddah in 2000 and Gloria in 1997. The majority of the farmers state that production of plots was not affected by these events. Even when plots were affected, the perceived impact was reported to be small. Only 12% and 3% of the farmers declare that these cyclones had an impact on their rice yields in 2000 and 1997 respectively. Of these farmers, only 3% and 1% state that the impact on rice yield had been very high. Hence, it seems that the direct overall impact of these cyclones has been very small. This might be because the cyclones normally hit outside the regular growing period in Maroantsetra.7

Table 2: Descriptive statistics parcel, flooding, and sedimentation

variable	Unit	N	mean	median	min	max
Parcel characteristics						
area	ares	268	2,16	1,2	0,1	25
distance from house	minutes	268	15,40	10	1	90
isolated parcel	yes=1	268	0,04	0	0	1
parcel along river	yes=1	268	0,14	0	0	1
traditional perimetre	yes=1	268	0,82	1	0	1
parcel far from river	yes=1	268	0,57	1	0	1
parcel in terras	yes=1	268	0,17	0	0	1
parcel close to river (<100m)	yes=1	268	0,15	0	0	1
parcel between 100 and 200m of river	yes=1	268	0,10	0	0	1
interior of bend of river	yes=1	268	0,04	0	0	1
exterior of bend of river	yes=1	268	0,19	0	0	1
parallel to river	yes=1	268	0,56	1	0	1
irrigated by rainfall	yes=1	268	0,04	0	0	1
irrigated by dam	yes=1	268	0,96	1	0	1
distance river parcel	meters	268	103,45	40	0,2	1200
height difference parcel river	meters	266	2,51	2	0,2	20
order in irrigation (rank)	number	268	9,51	5	1	99
soil depth	cm	267	26,34	20	3	120

Sedimentation and flooding						
no deposits	yes=1	268	0,44	0	0	1
deposits of clay	yes=1	268	0,26	0	0	1
deposits of sand	yes=1	268	0,30	0	0	1
Cyclone Huddah 2000						
length flooding	days	218	1,66	1	0	30
maximal depth of water	cm	202	116,46	100	0	600
no impact on yields	yes=1	268	0,56	1	0	1
little impact on yields	yes=1	268	0,05	0	0	1
medium impact on yields	yes=1	268	0,04	0	0	1
strong impact on yields	yes=1	268	0,03	0	0	1
Cyclone Gloria 1997						
length flooding	days	144	1,31	1	0	17
maximal depth of water	cm	122	117,37	100	0	500
no impact on yields	yes=1	268	0,41	0	0	1
little impact on yields	yes=1	268	0,01	0	0	1
medium impact on yields	yes=1	268	0,01	0	0	1
strong impact on yields	yes=1	268	0,01	0	0	1
This harvest						
problems with flooding	yes=1	268	0,14	0	0	1
problems with drought	yes=1	268	0,28	0	0	1
problems with deposit sand	yes=1	268	0,21	0	0	1

Runoff and erosion happen often during rare events such as cyclones and heavy, intense rainfall (Kaimowitz, 2000; Brand et al., 2002). While direct impact on productivity might be small, long-term impacts through increased sedimentation might be large. In the next section, we will evaluate the values these rice farmers attach to sedimentation and flooding. We will estimate these through well-established methods in environmental economics: (1) an indirect valuation method using the hedonic pricing methodology and (2) a direct valuation method using the contingent valuation technique.

REGRESSION RESULTS

Land Valuation

To evaluate to what extent farmers incorporate physical and environmental amenities in land valuation, a modified hedonic pricing analysis was done. Given that land sales are rare in the region and good land valuations are therefore more difficult to get at, a stochastic payment card method was implemented to arrive at approximate land valuations of the rice plot in the sample. The stated price at which households are willing to sell their plot for sure is used as dependent variable in the regression analysis. The results of this regression are shown in Table 3.

The results illustrate that farmers are well aware of the effect of the physical characteristics on the value of their plots. As expected, area is shown to be a significant determinant of value (see Figure 1). A doubling in area increases the value of the plot by only 0.54, i.e. significantly different from one. This result indicates that larger plots are relatively less valuable than smaller plots, controlling for physical characteristics. On first sight, this implies that there are potential profits to be made by repacking plots in smaller units.8 While returns to scale would result in relatively higher values for larger plots, a potential explanation might be that farmers prefer different smaller plots compared to one big plot as in this way, farmers are able to diversify their risk.9 The likelihood that small plots, that are spatially segregated, are all hit by calamities at the same time - such as flooding, drought, sedimentation problems or plant diseases - is less than for one big plot. This risk averseness, typical for poor small farmers, might be an important explanation of the concave land price relationship.

Table 3: Hedonic price regression (dep. var. = log (value of land); robust standard errors)

variables	Unit	Coefficient	t-value	P>|t|
plot characteristics				
area	log(ares)	0,503	**7,590**	**0,000**
parcel in terras	yes=1	-0,343	**-1,840**	**0,067**
parcel along river	yes=1	-0,617	-1,470	0,142
tradional perimeter	yes=1	-0,215	-0,530	0,596
interior bend of river	yes=1	-0,371	-1,600	0,111
exterior bend of river	yes=1	-0,045	-0,300	0,765
distance river parcel	log(meters)	0,027	0,700	0,488
height difference parcel river	log(meters)	0,040	0,300	0,762
soil depth	log(cm)	0,260	**2,140**	**0,034**
irrgation directly from river	yes=1	0,216	**1,700**	**0,091**
irrigated by dam	yes=1	-0,305	-1,000	0,320
clay deposit after cyclones	yes=1	0,299	**1,990**	**0,048**
sandy deposits after cyclones	yes=1	0,429	**2,980**	**0,003**

household characteristics				
education head of household	years	0,016	0,740	0,463
age of head of household	years	0,001	0,130	0,893
gender head of household	man=1	-0,237	-1,110	0,267
annual monetary income	log(Fmg)	0,041	1,330	0,185
length of lean period	months	0,017	0,690	0,492
potential access to credit	log(Fmg)	-0,010	-0,870	0,384
owned number of zebus	log(number)	0,300	**3,460**	**0,001**
owned agricultural land	log(ares)	-0,070	-0,950	0,341
intercept		12,248	**14,700**	**0,000**
Number of observations	256			
F(21, 234)	9,62			
Prob > F	0			
R-squared	0,3929			
Root MSE	0,9256			

Most of the physical variables turn out not significant at the conventional statistical levels, indicating that these are not major determinants of sales prices. However, there are a few exceptions. Plots in terraces, at the top of the river, are less valuable. This might be because these plots are more likely to be affected by drought. The impact is shown to reduce the value of the plot by around 34%. The perceived cultivable soil depth is a highly important determinant of land prices. A doubling of soil depth increases the value of rice land by 26%. Agronomic evidence suggests that soil depth is crucial for root development which has been shown to be an important constraint on rice production in Madagascar.

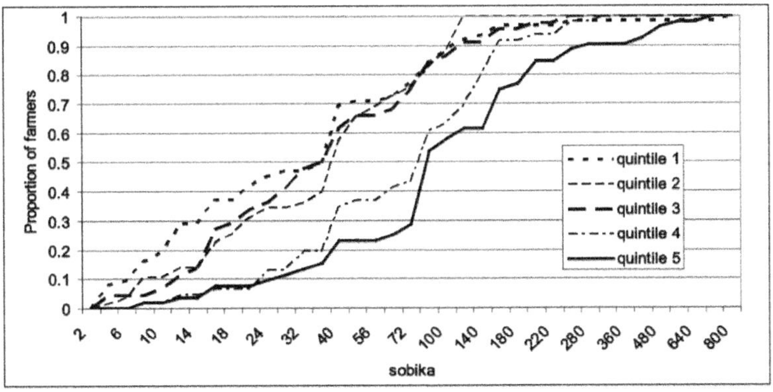

Figure 1: Willingness-to-accept the sales price 'for sure' (by plot size quintile)

In line with de Janvry et al. (1991), we assume imperfect or missing markets where farm households are the decision makers and production and consumption decision are not separable. This implies that land prices would also depend on household characteristics and they were thus included in the regression. Few of these variables come out significant. Only the ownership of cattle leads to significant higher land values. This seems linked to the importance of ownership of cattle to access to manure, an important lasting fertility and land quality enhancing input in these environments (Minten et al., 2007; Barrett et al., 2002).

To measure the effect of sedimentation, we created dummies for clay and sand deposits during recent floods. Compared to soils without deposits during floods, these plots are estimated to be significantly more valuable. The plots affected by soil and sand deposits are estimated to be respectively 30% and 43% more valuable. The latter results might seem surprising at first sight. However, sand deposits come usually together with organic material that might significantly improve the fertility of soils. Farmers also often remove the more damaging sand from the plot. These results indicate overall that sedimentation does not reduce the value of the plot per se, ceteris paribus. We discuss this in more detail below.

Willingness-to-pay to Avoid Sedimentation and Flooding

All sedimentation is not perceived to be bad for rice productivity. In fact, erosion and heavy rainfall might induce runoff of the good topsoil of the uplands that ends up in the lowland ricefields (Chomitz and Kumari, 1998). This seems also to be the case in the lowlands of the Maroantsetra region. When asked about the perceived effect of flooding and sedimentation on rice yields overall, 53% of the farmers reported that they thought this relation was negative (Table 4). However, 38% of the farmers thought that it was actually good for rice yields (while 9% thought its effect was neutral). In a follow-up question, it was asked what the rice farmers expected of the effect of sedimentation and flooding on the rice plot in the sample. Farmers were evenly divided on the question: 37% thought that the effect would be negative, 38% expected a positive effect and 26% reported to expect a neutral effect.

Table 4: Perceived effect of sedimentation/flooding

variable	Unit	N	mean
Overall effect sediment/flooding on rice yield...			
positive	yes=1	268	0,38
neutral	yes=1	268	0,09
negative	yes=1	268	0,53
Effect on studied parcel of sediments/flooding on rice yield...			
positive	yes=1	268	0,37
neutral	yes=1	268	0,26
negative	yes=1	268	0,38

Finally, farmers were asked what they were willing to pay to avoid flooding and sedimentation. Figure 2 illustrates, for the respondents that were willing to pay, how the willingness to pay varies for the different levels that were offered to the respondent. We see that the median willingness to pay (at 95% for sure) to avoid flooding is just over 2 sobika, the local unit for a rice basket containing 12 kgs on average per household per year. This amounts to around 4$. This implies that if a vote would be held in the region, more than 4$ would not be accepted by a majority of the population. 50% of the farmers would refuse to pay more than 4.5 sobika for sure. On average, this corresponds to 7% of their total rice production of last year.

The number of farmers that were undecided about accepting or refusing the offer is largest in the middle of the graph, as could be expected (see Wang (1997; p. 223)). For some bids, the indecision domain contains up to 15% of the farmers. This high number indicates the importance of allowing farmers to convey information beyond the simple yes/no format in contingent valuation studies as has been shown by other authors (Blamey et al., 1999; Ready et al., 2001; Alberini et al., 2003).

Regressions were run to look at the determinants of the willingness to pay to avoid flooding and sedimentation on the plot in the sample. These results serve to validate the WTP answers. A two-step approach was used. In a first step, a selection equation was run to explain the characteristics of the households that are willing to contribute to avoid flooding and sedimentation. In this step, variables are included that are potential determinants of the likelihood of the plot to be subject to flooding and sedimentation. In a second step - controlling for the characteristics of the plot and the household which explain if it is willing to contribute - economic variables are included in the regression to measure to what extent they are able to contribute, taking into account their socio-economic background. A selectivity coefficient was then included in the second-stage willingness to pay regression.

This set-up would allow us to obtain efficient and unbiased estimates in the second stage regression.

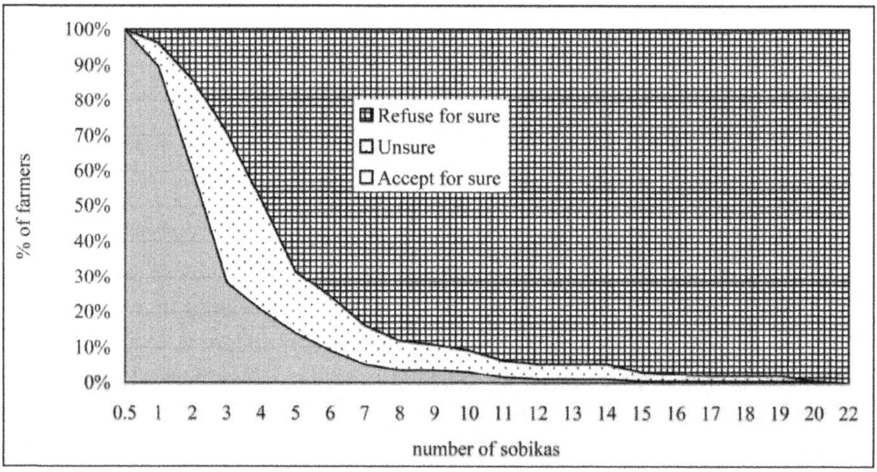

Figure 2: Willingness to pay to avoid flooding/sedimentation

The results are largely conforming to expectations. The coefficient that measures the expected effect of sedimentation and flooding on the plot and the household perceived effect of sedimentation are significant determinants of the probability that the household is willing to contribute.10 Households with plots on the exterior bend of the river, bigger soil depth and irrigated by dams are significantly more willing to contribute. These plots might be more exposed to risk or are more valuable. It is interesting to note that negative experiences with the last two cyclones make the household less likely to contribute. These households might believe that there is not much that can be done or, alternatively, that this type of adversity might easily be overcome.

The results on the amount that households are willing to contribute – the second stage regression - suggest that wealthier households are willing to pay more. Different measures of wealth were included. A doubling of the area of lowland in possession would increase the willingness to pay significantly by 11%. A lean period that lasts one month longer as measured by the period that they do not have sufficient rice, an indicator of poverty of the household (Barrett and Dorosh, 1998; Minten and Zeller, 2000), reduces the willingness to pay of the household by 6%. Potential access to credit increases the willingness to pay significantly. However, its coefficient is small. Overall income and the number of zebus owned by the household show the expected positive sign but are not significant at the 10% level. Household characteristics, such as level of education, gender and age of the head of household, do not influence the willingness to pay significantly.

Table 5: Willingness to pay to avoid flooding on rice plot (Heckman selection model)

variables	Unit	Coef.	z	P>z
dep. var.: willingness to pay in log(Fmg)				
area of the plot	log(ares)	-0,014	-0,230	0,816
education head of household	years	-0,016	-0,820	0,410
age of head of household	years	-0,006	-1,320	0,187
gender head of household	man=1	-0,108	-0,610	0,542
owned lowland	log(ares)	0,112	**1,950**	**0,051**
annual monetary income	log(Fmg)	0,036	1,090	0,276
length of lean period	months	-0,058	**-2,460**	**0,014**
potential access to credit	log(Fmg)	0,021	**1,910**	**0,057**
owned number of zebus	log(number)	0,030	0,400	0,691
intercept		0,116	0,220	0,822
selection equation				
expected effect of sedimentation on plot	1=pos; 2=neutral; 3=neg.	0,683	**4,990**	**0,000**
area of the plot	log(ares)	-0,161	-1,440	0,150
parcel along river	yes=1	-1,074	-1,500	0,134
traditional perimetre	yes=1	-0,704	-1,060	0,287
parcel in terras	yes=1	-0,442	-1,460	0,144
parcel close to river (<100m)	yes=1	-0,448	-1,500	0,134
parcel between 100 and 200m of river	yes=1	-0,030	-0,070	0,943
interior of bend of river	yes=1	-0,216	-0,400	0,689
exterior of bend of river	yes=1	0,836	**3,200**	**0,001**
irrigated by dam	yes=1	0,970	**1,960**	**0,050**
distance river parcel	meters	0,005	0,080	0,939
height difference parcel river	meters	0,113	0,810	0,420
order in irrigation (rank)	number	-0,087	-0,910	0,362
slope (distance one sees w/o obstacle)	meters	0,056	0,640	0,525
soil depth	cm	0,464	**2,740**	**0,006**
estimated age of plot	years	-0,003	-0,410	0,680
little impact on yields of cyclone 1	yes=1	-0,087	-0,180	0,859
medium impact on yields of cyclone 1	yes=1	-1,182	**-1,780**	**0,075**
strong impact on yields of cyclone 1	yes=1	-1,642	**-2,510**	**0,012**

little impact on yields of cyclone 2	yes=1	-0,333	-0,390	0,695
medium impact on yields of cyclone 2	yes=1	-0,087	-0,060	0,949
strong impact on yields of cyclone 2	yes=1	0,434	0,290	0,770
overall perceived impact of sedimentation	1=pos; 2=neutral; 3=neg.	0,385	**2,580**	**0,010**
intercept		-3,186	1,018	-3,130
rho		-0,744	0,119	
sigma		0,778	0,050	
lambda		-0,578	0,118	
LR test of indep. eqns. (rho = 0): chi2(1) = 13.88 Prob > chi2 = 0.0002				
Number of obs		265		
Censored obs		82		
Uncensored obs		183		
Wald chi2(9)		21,33		
Prob > chi2		0,011		
Log likelihood		-296,3996		

The coefficients on the explanatory variables show that the stated amount is consistent with economic logic. Households with access to liquidity and who perceive to suffer from flooding are willing to pay more. To further test for robustness, regressions were run without the selectivity coefficient and with the refusal to pay for sure as dependent variable. The coefficients obtained - but not reported - confirm the results discussed earlier. We end this section with a final note on the significance of these results at the national level. There are two main differences of the surveyed farmers with the rest of the country. First, the watersheds in this area are small and sedimentation downstream can easy be linked to upstream activities. This is however not the case in the rest of Madagascar and the link between sedimentation downstream and corrective measures upstream are more difficult to make as watersheds are larger.11 Second, the rice harvest in the Maroantsetra region is at the end of the year, i.e. before major cyclones hit the country. This might reduce the willingness to pay for a reduction of floods. In the rest of the country, the main harvest is in the beginning of the year and it might thus more directly be affected by rice losses due to floods and submersion. The run-off of good soils might then only affect the subsequent harvests. Based on the data of the national household survey of 2001, it is found that 15% of agricultural households cultivate lowland, 21% upland and 64% both. This compares to respectively 25% and 75% of the households that we interviewed in the Maroantsetra area (because of the study subject, only rice farmers were selected). The majority of the households in Madagascar and in our dataset cultivate thus both uplands and lowlands and it seems that farmers that might cause erosion and those that suffer from it are often the same households.12 This makes a compensation mechanism cumbersome.

Secondary sources of information further seem to indicate that siltation and erosion might be a relatively minor problem in Malagasy agriculture overall and this despite the high recent deforestation rates in Madagascar.13 The 2001 national household survey asked farmers about the biggest constraints they faced to improved agricultural productivity. The same question was asked in the 2004 national household survey, based on a different sampling frame and with a bigger sample. Respondents had to rank options from 'not important' to 'very important'. The results are presented in Table 6 ordered in decreasing percentage of households that identified the constraint as 'quite' or 'very' important. Answers were strikingly consistent between the two surveys, three years apart and with a different sample. The most and least frequently cited constraints were common to both surveys. Access to agricultural equipment, access to cattle for traction and transport and access to labor are ranked among the top four constraints in both surveys. The clear pattern in these answers is that inputs that complement labor and boost its productivity are most limiting in farmers' opinion (Minten et al., 2007). By contrast, less than 40 percent of households identify the siltation of land as an important constraint and it is more commonly identified as not a constraint on agricultural productivity. Farmers were further asked for each plot in the national household survey of 2001 about the production problems in the year preceding the survey. Siltation was mentioned as a problem on less than 1% of the rice plots.

CONCLUSIONS

Flooding and sedimentation downstream are often linked to deforestation upstream. While the debate is on-going and results seem to be variable and site specific (Chomitz and Kumari, 1998; Calder, 1999), policy makers are looking for ways to solve this externality problem to ensure sustainable financing for ecological services of conservation efforts such as reforestation and soil conservation measures. Based on interviews with almost 300 rice farmers - users of land downstream - in the Northeast of Madagascar, this paper tries to shed light on the willingness to pay for ecological services for forests, in this case to avoid flooding and sedimentation. The results of our analysis show that the rice farmers are clearly aware of the effect of sedimentation on production. Sedimentation is not perceived to be unambiguously bad for lowland productivity. Policy interventions that focus on only correcting the perceived negative relationship are therefore misguided. A hedonic pricing analysis on riceland values shows that farmers take sedimentation into consideration in the valuation of their rice plots but that rice plots with sedimentation are valued significantly higher, ceteris paribus.

Table 6: Farm households' reported constraints on improved agricultural productivity

Variables	Percentage of households that state this constraint is ... important: not	a bit	quite	very
Constraints to overall agricultural productivity				
EPM 2001, 2470 agricultural households				
Access to agricultural equipment	19	18	27	35
Access to land	27	19	29	25
Access to cattle for traction and transport	24	23	29	24
Access to labor	22	28	30	20
Access to credit	36	19	23	22
Degradation of irrigation infrastructure due to environmental problems	29	31	22	18
Access to agricultural inputs (e.g. fertilizer)	34	26	19	21
Access to cattle for fertilizer	42	23	19	16
Land tenure insecurity	44	26	22	8
Silting of land	46	29	18	7
EPM 2004, 3543 agricultural households				
Access to agricultural equipment	11	14	32	43
Access to irrigation	13	21	29	37
Access to cattle for traction and transport	16	20	35	29
Access to labor	17	22	37	24
Avoid droughts	20	19	27	34
Access to agricultural inputs (e.g. fertilizer)	24	20	26	30
Phyto-sanitary diseases	19	25	30	26
Avoid flooding	25	20	26	29
Access to cattle for fertilizer	28	22	25	25
Access to credit	31	23	22	24
Silting of land	33	29	23	15
Land tenure insecurity	38	24	23	15

The results of the survey further show that, while 10% of the farmers believe that flooding and sedimentation has no effect, a significant part of the farmers (almost 40% of the rice farmers in the sample) feels that their plots actually benefit from flooding and sedimentation. This seems related to the fact that flooding occurs outside the main harvest period and thus therefore not seem to cause any large immediate production damage. Damage depends then on the type of deposits as flooding can actually cause valuable soils and organic material to be transported to the ricefield and to be beneficial for rice productivity. The negative or positive effect of flooding seems to depend on spatial determinants, i.e. location with respect to the main river that irrigates the rice fields matters.

However, a significant part of the farmers also realize the bad effects that sedimentation can have on their rice production. Therefore, they are willing to contribute to avoid flooding and sedimentation on their fields. These farmers are willing to contribute 4$ per household per year. The magnitude of the amount that they are willing to pay corresponds to spatial as well as economic rationales. Households that are richer, not credit constrained, and that suffer less from seasonality problems are willing to pay significantly more to avoid this flooding and sedimentation damage. Given beneficial effects of sedimentation for some farmers and small willingness to pay by other farmers, our results overall thus suggest that current economic rates of return on forest preservation projects in Madagascar, largely beneficial because of across-the-board domestic agricultural benefits on lowlands, might be overestimated.14

ACKNOWLEDGEMENT

We would like to thank Jurg Brand, Tim Healy, and Andy Keck for help with the set-up of survey, helpful discussions and comments on preliminary results. The field work for this research was financed by ONE (Office National de l'Environnement), PAGE (Projet d'Appuie à la Gestion de l'Environnement) and by the Ilo program. The two last projects were financed by USAID-Madagascar. Any remaining errors are solely the authors' responsibility

REFERENCES

1. Alberini, A., Boyle, K., Welsh, M., Analysis of Contingent Valuation Data with Multiple Bids and Response Options allowing Respondents to express Uncertainty, Journal of Environmental Economics and Management, 2003, 45: 40-62.
2. Arrow, K., Solow, P., Portney, P. Leamer, E.E., Radner, R., Schuman, K., Report of the NOAA Panel on contingent valuation, NOAA, 1993
3. Aylward, B., Echeverria, J., Synergies between livestock production and hydrological function in Arenal, Costa Rica, Environment and development economics, 6, 2001, pp. 359-381
4. Barrett, C.B., Stochastic Food Prices and Slash-and-Burn Agriculture, Environment and development economics, vol. 4, no. 2, 1999, pp. 161-176
5. Barrett, C.B., Place, F., Aboud, A., Natural Resource Management in African Agriculture: Understanding and Improving Current Practices, CAB International, 2002

6. Barrett, C.B., Dorosh, P., Farmers' welfare and changing food prices: Nonparametric evidence from rice in Madagascar, American Journal of Agricultural Economics, 78, 1996, pp. 656-669
7. Blamey, R.K., Bennett, J.W., Morrison, M.D., Yea-Saying in Contingent Valuation Studies, Land Economics, February 1999, 75(1), pp. 126-141
8. Blarel, B., Hazell, P., Place, F., Quiggin, J., The economics of farm fragmentation: Evidence from Ghana and Rwanda, World Bank Economic Review, vol. 6, no. 2, may 1992, pp. 233-254
9. Brand, J., Minten, B., Randrianarisoa, C., Etude d'impact de la deforestation sur la riziculture irriguée, Cahier d'études et de recherches en économie et sciences sociales, No. 6, December 2002, FOFIFA, Antananarivo Calder, I.R., The Blue Revolution: Land Use and Integrated Water Resources Management, 1999, Earthscan publications Ltd, London
10. Casse, T., Milhoj, A., Ranaivoson, S., Randriamanarivo, J.R., Causes of deforestation in southwestern Madagascar: What do we know?, Forest Economics and Policy, 2004, 6(1): 33-48
11. Chomitz, K.M., Kumari, K., The domestic benefits of tropical forests: A critical review, World Bank Research Observer, Vol. 13, No. 1, February 1998, pp. 13-35
12. Dasgupta, P, An Inquiry into Well-Being and Destitution, Clarendon Press, 1993, Oxford, England.de Janvry, A., Fafchamps, M., Sadoulet, E. (1991), Peasant household behavior with missing markets : Some paradoxes explained, The Economic Journal, 101: 1400-1417
13. Durbin, J., 'The potential of conservation contracts to contribute to biodiversity conservation in Madagascar', 2001, mimeo. Ferraro, P.J., Simpson R.D., The cost-effectiveness of conservation payments, Land Economics, 2002, 78(3): 339-353 FOFIFA, Culture sur brûlis: vers l'application des résultats de recherche, actes de l'atelier de EPB-BEMA, 2001
14. Freudenberger, K., 1999, Flight to the forest: a study of community and household resource management in the commune of Ikongo, Madagascar Heckman, J., Sample selection bias as a specification error, Econometrica, 47, 1979, pp. 153- 161
15. Holden, S.T., Shiferaw, B., Peasants' willingness to pay for sustaining land productivity, in Barrett, C., Place, F., Aboud, A., Natural Resources Management in African Agriculture: Understanding and improving current practices, CAB international, 2002

16. Holden, S.T., Shiferaw, B., Wik, M., Poverty, market imperfections and time preferences: of relevance for environmental policy?, Environment and Development Economics, 1998, 3:105-130

17. Johnson, N., White, A., Perrot-Maître, Developing Markets for Water Services from Forests: Issues and Lessons for Innovators, Forest Trends, 2002

18. Kaimowitz, D., Useful Myths and Intractable Truths: The politics of the link between forests and water in Central America, CIFOR, 2000, mimeo

19. Keck, A., Sharma, N.P., Feder, G., Population growth, shifting cultivation, and unsustainable agricultural development: a case study from Madagascar, World Bank Discussion Paper, No. 234, Africa Technical Department Series, The World Bank, Washington DC, 1994

20. Kramer, R.A., Richter, D., Pattanayak, S., Sharma, N., Ecological and economic analysis of watershed protection in Madagascar, Journal of Environmental Management, 49, 1997, pp. 277-295

21. Larson, B.A., Changing the economics of environmental degradation in Madagascar: Lessons from the national environmental action plan process, World Development, Vol. 22, No. 5, May 1994, pp. 671-689

22. Lin, T., Evans, A.W., The relationship between the price of land and size of plot when plots are small, Land economics, August 2000, pp. 386-394

23. McConnell, W.J., Madagascar: Emerald Isle or Paradise Lost?, Environment, October 2002, vol. 44, No. 8, pp. 10-14 Mekkonen, A., Valuation of community forest in Ethiopia: A contingent valuation study of rural households, Environment and Development Economics, 2001, 5:289-308

24. Messonier, M.L., Bergstrom, J.C., Cornwell, C.M., Teasley, R.J., Cordell, H.K., Survey response-related biases in contingent valuation: concepts, remedies, and empirical application to valuing aquatic plant management, American Journal of Agricultural Economics, 83, May 2000, pp. 438-450

25. Minten, B., Moser, C., Forêts: Usages et menaces sur une ressource, in (eds) Minten, B., Randrianarisoa, J., Randrianarison, L., Agriculture, pauvreté rurale et politiques économiques à Madagascar, Programme Ilo/ Cornell University, 2003, pp. 86-89

26. Minten, B., Zeller, M., Beyond market liberalization: welfare, income generation and environmental sustainability in rural Madagascar, Ashgate, 2000 Minten, B., Randrianarisoa, J., Barrett, C., Productivity in Malagasy rice systems: Wealthdifferentiated constraints and priorities, Proceedings of the IAAE conference, 2007.

27. Pagiola, S., J. Bishop and N. Landell-Mills, Setting Forest Environmental Services: Marketbased Mechanisms for Conservation and Development, London: Earthscan, 2002.
28. Pattanayak, S.H., Kramer, R.A., Worth of watersheds: a producer surplus approach for valuing drought mitigation in Eastern Indonesia, Environment and Development Economics, 6, 2001, pp. 123-146
29. Pattanayak, S.H., Kramer, R.A., Pricing ecological services: Willingness to pay for drought mitigation from watershed protection in eastern Indonesia, Water resources research, Vol. 37, No. 3, 2001, pp. 771-778
30. Pender, J.L., Discount rates and credit markets: Theory and Evidence from rural India, Journal of Development Economics, 1996, 50:257-296
31. Perrot-Maître, D., Davis, P., Case studies of markets and innovative financial mechanisms for water services from forests, 2001, Forest Trends, mimeo Razafindravonona, J., Stifel, D., Paternostro, S., Changes in poverty in Madagascar: 1993- 1999, Instat, Antananarivo, 2000.
32. Ready, R.C., Navrud, S., Dubourg, W.R., How do respondents with uncertain willingness to pay answer contingent valuation questions?, Land economics, August 2001, vol. 77, no. 3, pg 315-326
33. Terretany, Cahier Terretany no. 6, 1997, Un système agro-écologique dominé par le tavy: la region de Beforona, Falaise Est de Madagascar Wang, H., 'Treatment of don't know responses in contingent valuation surveys: a random valuation model', Journal of Environmental and Economic Management, 1997, 32: 219-232.
34. Wang, H., Whittington, D., 2005, 'Individuals' valuation distributions using a stochastic payment care approach, Ecological Economics, 55, pp. 143-155
35. Whittington, D., Chapter 16: Environmental Issues, in Eds. Grosh, M., Glewwe, P., Designing Household Survey Questionnaires for Developing Countries: Lessons from Ten Years of LSMS Experience, World Bank, 1998
36. World Bank, Review of agricultural and environmental sector, 2003, Washington.
37. World Bank, Environmental Program 3, 2005
38. World Bank, Madagascar: The impact of public spending on irrigated perimeters productivity (1985-2004), Economic and Sector Work, 2005, Washington DC.

39. Zeller, M., Lapenu, C., Minten, B., Randrianaivo, D., Ralison, E., Randrianarisoa, C., Rural development in Madagascar : Quo vadis ? Towards a better understanding of the critical triangle between economic growth, poverty alleviation and environmental sustainaibility, Quarterly Journal of International Agriculture, 1999, Vol.2

Chapter 8

EFFECT OF SALINITY ON SOIL MICROORGANISMS

Celia Maria Maganhotto de Souza Silva and Elisabeth Francisconi Fay

Embrapa Environment Brazil

INTRODUCTION

Whilst the majority of countries have criteria to evaluate the quality of the air and water, the same does not occur for the quality of the soil. Traditionally soil quality is associated with productivity, but recently it has been defined in terms of sustainability, that is, the capacity of the soil to absorb, store and recycle water, minerals and energy such that the production of the crops can be maximized and environmental degradation minimized. Thus preservation of soil quality is a critical factor for environmental sustainability.

A significant decline in soil quality has occurred throughout the entire world as a result of adverse changes in its physical, chemical and biological properties. According to Steer (1998), in the last decades of the last century, about 2 billion of the 8.7 billion agricultural lands, permanent pastures, forests and wild native lands were degraded. The global grain production growth rate fell from 3% in the seventies to 1.3% in the period from 1983-1993, and one of the main reasons for this decline was inadequate soil and water management. Inventories carried out on the soil productive capacity in the last decade indicated that 40% of the degradations of arable land were induced by man as a result of soil erosion, atmospheric pollution, intensive cultivation, over-grazing, deforestation, salinization and desertification (Oldeman, 1994).

Soil degradation processes constitute a serious problem on a worldwide basis, with significant environmental, social and economic consequences. As the world population increases, so does the need to protect the soil as a vital resource, particularly for food production, .

The soil is a dynamic medium, constituting the habitat of abundant biodiversity, with unique genetic patterns where one can find the greatest amount and variety of living organisms, which serve as a nutrient reservoir. One gram of soil in good conditions can contain 600 million bacteria belonging to 15,000 or 20,000 different species. These values decrease to 1 million bacteria encompassing from 5000 to 8000 species in desert soils (Informativo Capebe, 2010). Depending on the amount of organic matter present in the soil, the biological activity eliminates pathogenic agents, decomposes organic matter and other pollutants into simpler components (frequently less noxious), and contributes to maintaining the physical and biochemical properties required for soil fertility and structure. However, soil is not an inexhaustible resource and consciousness of this, allied to knowledge of the need to maintain or increase the capacity of this agro-ecosystem, directing its multiple functions in an adequate way, is increasing, as also changes in the overall perception of its importance as an environmental component.

As an open system, the soil is dynamic and in constant interaction with the atmosphere, the hydrosphere, the biosphere and the lithosphere. Depending on the intensity with which these factors act, soil can present differentiated characteristics which define its potentialities for exploitation by man. Its structure is defined as the aggregation of primary particles in compound particles, separate from adjacent aggregates. However its structure implies in an arrangement of the primary particles (sand, silt, clay) which, by way of cementing agents, can group together forming aggregates with certain structural patterns, which necessarily include porous space.

Alterations in the chemical conditions of cultivated soils, such as the concentrations and types of ions in solution in the soil, variations in pH and in the critical flocculation concentration of the particles, can cause modifications in the dispersion of the clay fraction, degrading the original soil fraction. The sodium ion, being monovalent, increases the width of the diffuse double layer on the surface of the clays, reducing the attractive forces between them with a consequent increase in particle dispersion. The consequence of this dispersion of the clay is also shown by a reduction in stability of the soil aggregates, which are thus easily transported by rain or irrigation (Almeida Neto, 2007).

Although soil structure is not considered as a plant growth factor, it exerts an influence on the air and water supplies to the roots of the crops, on nutrient availability, on the penetration and development of the roots, as also on the movement of the soil macro-fauna. It also influences the loss of agrochemicals by way of erosion and leaching and can have considerable importance on the negative environmental impact of some agricultural practices. According to

Machado (2002), inadequate soil use has been causing a gradual loss in their productive capacity.

The main threats to soil are erosion, mineralization of the organic matter, reduction in biodiversity, contamination, water proofing, compacting, salinization, and the degrading effects of floods and landslides. Soil degradation produces deterioration of the plant covering and the hydric resources. In addition, by mean of a series of physical, chemical, biological and hydrological processes, it causes destruction of both the biological potential of the land and of their use to sustain the population connected to it.

SOIL SALINIZATION

Soil salinity is part of the natural ecosystem in arid and semi-arid regions and an increasing problem in agricultural soils the world over. In temperate, moist climates salinity occurs on a smaller scale, principally in salt water marshes, at the side of highways and in salty effluent discharges (Pathak & Rao, 1998; Keren, 2000; Qadir et al, 2000; Wichern, 2006). Salinization consists of an accumulation of water soluble salts in the soil. These salts include the ions potassium (K^+), magnesium (Mg^{2+}), calcium (Ca^{2+}), chloride (Cl^-), sulfate (SO^{42-}), carbonate (CO^{32-}), bicarbonate (HCO^{3-}) and sodium (Na^+). Sodium accumulation is also called sodification. High sodium contents result in destruction of the soil structure which, due to a lack of oxygen, becomes incapable of assuring plant growth and animal life.

Salt affects crop germination and density, as also vegetative development, reducing productivity and, in the most serious cases, leading to generalized plant death, limiting nutrient absorption and reducing the quality of the available water. For example, elevated salinity weakens plants due to the increase in osmotic pressure and the toxic effect of the salts. In addition, salinization affects the metabolism of the organisms present in the soil, drastically reducing soil fertility and increasing water proofing of the deeper layers, impeding cultivation of the land. In an indirect way, soil salinization can adversely affect plant growth, due to destruction of the soil structure and its consequent compacting. This occurs due to a dispersion of the clay particles caused by substitution of the calcium (Ca^{+2}) and magnesium (Mg^{+2}) ions present in the complex by sodium (Na^+), resulting in an increase in soil sodicity, that is, in the percentage of exchangeable sodium (PES), which, in the last instance, is the main factor responsible for the deterioration of the physical properties of salt-affected soils (sodic, or alkaline, and saline-sodic). Saline soils show the following physical-hydric characteristics: low permeability, low hydraulic conductivity and aggregate instability (Freire, 2009).

On a world wide scale, the production by approximately 400 million hectares of arable land is being severely restricted by salinity (Bot et al., 2000). In the European Union salinization affects about 1 million hectares, mainly in the Mediterranean countries, constituting one of the main causes of desertification (Iannetta & Colonna, 2011). The most affected soils are situated in Hungary, Romania, Greece, Italy and the Iberian Peninsula (Agricultura…., 2011). About 8.1 million hectares are salinized in India, of which 3.1 million are in coastal regions (Triphati et al., 2007). In Nordic countries, the use of salt to remove ice from highways produces localized salinization phenomena (Agricultura…., 2011). Considering the increasing temperatures and decrease in pluviosity which have characterized climates in recent years, the salinization problem has increased.

Salinization results from natural or anthropogenic factors, constituting a process of soil degradation which, in some cases, is responsible for irreparable losses in their productive capacity, with great extensions of arable land becoming sterile.

Natural soil salinization and sodification factors

The natural factors influencing soil salinity are:

- geological phenomena which increase the salts concentration in groundwater and consequently in the soil;
- natural factors capable of bringing groundwater containing elevated salt contents to the surface;
- infiltration of groundwater in below sea-level zones (micro-depressions with reduced or absent drainage); • drainage of waters from zones with geological substrates capable of liberating large amounts of salts;
- action of winds, which, in coastal zones, can transport moderate amounts of salts to the interior.

The weathering of primary minerals (which make up the rocks or the original soil material) is the indirect source of nearly all the salts present in soils, although there are only a few cases in which this results in sufficient accumulation of salt (primary or pedogenetic salinization) to form saline soils. The areas of lands salinized by such natural processes, from which salt-affected soils could arise, such as Planossolo Solódico, Solonetz Solorizado, Solonchack Solonétzico, do not increase so drastically when compared to the increasing growth intensity of the extension of land salinized by anthropic activity.

In general, the salinization process occurs in soils situated in regions of low rainfall and which have a water-bearing stratum near the surface. In coastal zones, salinization could be associated with the over-exploitation of groundwaters due to the demand induced by increased urbanization, or by industry and agriculture. The over-extraction of groundwaters can result in a lowering of the normal water-bearing stratum levels, leading to the intrusion of sea water.

Secondary factors leading to soil salinization and sodification

The most influential anthropogenic factors are:

- irrigation with water containing elevated salt contents;
- rise in phreatic water level due to human activities (infiltration of water from unlined channels and reservoirs, irregular distribution of irrigation water, deficient irrigation practices, inadequate drainage);
- use of fertilizers and other production factors, namely for intensive agriculture in land with low permeability and reduced possibilities for leaching;
- irrigation with residual waters with high salt contents;
- elimination of residual waters with high salt contents by way of the soil;
- contamination of the soil with industrial water and sub-products with high salt contents.

Soils affected by salts commonly appear in irrigated areas due to inadequate management of the irrigation and other practices, such that important extensions of fertile, arable land are becoming more and more saline. This is due to management practices that do not aim at conserving the productive capacity of the soil, such as, for example, the non-existence of an efficient drainage system, the use of inadequate quality water in inadequate amounts, and also the incorrect and excessive use of chemical fertilizers.

Irrigation is an ancient agricultural practice, widely used throughout the world, principally in tropical regions where hot, dry climates prevail, such as, for example, the semi-arid region in northeastern Brazil, where the evapotranspiration rate exceeds the rainfall throughout the better part of the year. In these areas, where there is not sufficient water available to supply the hydric needs of the crops throughout the whole vegetative cycle, irrigation assumes a fundamentally important role in order to guarantee good agricultural harvests. Since all natural waters contain variable amounts of soluble salts, be they of meteoric (rain), surface (rivers, lakes, dams, etc.) or subterranean (aquifers) origin, the application of water to the soil by irrigation implies necessarily in the addition of salts to their profile.

Thus salinization of a soil depends on the quality of the water used for irrigation, on the existence and level of natural and/or artificial drainage of the soil, on the depth of the water-bearing stratum and on the original concentration of salt in the soil profile. The basic principle to avoid soil salinization is to maintain the equilibrium between the amount of salt provided to the soil by irrigation and the amount of salt removed by drainage. In arid or highly ventilated climes, evaporation of the water enriches the soil with solutes, increasing the danger of salinization. In the same way, soils with limited permeability tend to concentrate salts. In irrigated zones, the low rainfall, elevated evapotranspiration rates and the soil structure impede leaching of the salts, which accumulate in the surface layers.

Estimates by FAO indicate that of the 250 million hectares of irrigated land in the world, approximately 50% already show salinization and soil saturation problems, and 10 million hectares are abandoned annually due to these problems (CODEVASF, 2011)

The excessive amounts of salts provided by irrigation waters can have adverse effects on the chemical and physical properties of the soils and on their biological processes (Garcia & Hernandez, 1996; Rietz & Haynes, 2003; Tejada &Gonzalez, 2005). These effects include mineralization of the carbon and nitrogen and the enzymatic activity, which is crucial for the decomposition of organic matter and liberation of the nutrients necessary for sustainability of the production (Azam & Ifzal, 2006; Wong et al., 2008). In addition, the agricultural practices can increase or reduce the microbial population, thus altering the activity, source and persistence of the enzymes in the soil (Parham et al., 2003).

Organic fertilizers are considered useful for crops due to their nutritive value, principally of nitrogen (N), and for their merits in improving the physical properties of the soil (Jackson & Bertsch, 2001; Garbarino et al, 2003), but their salt content is usually ignored, which could prejudice plant growth and soil quality after continued application. The flow of nitrogen (N) and phosphorus (P) from the application of animal manure is considered to contribute to non-precise pollution (Parker, 2000; Anderson & Xia; 2001; Ekholm et al, 2005; Allen et al, 2006). Salinity is considered as a non-precise source of pollution, but the secondary salinity induced by the application of organic fertilizers has not been considered as particularly worrying. Li-Xian et al. (2007) evaluated the effect of applying poultry manure and its ionic composition on soil salinization. The authors showed that the increase or decrease in the concentration of a determined ion in the soil depended on its concentration in the manure, the application rate, the removal of the crops and to leaching. The ionic composition of the soil salinity changed according to the types and doses of the fertilizers used and to their applications. The results also showed that, even in humid regions, the potential for the risk of secondary soil salinization exists with the successive application of animal manure.

THE EFFECT OF SALINITY ON THE SOIL MICROORGANISMS

The microbial communities of the soil perform a fundamental role in cycling nutrients, in the volume of organic matter in the soil and in maintaining plant productivity. Thus it is important to understand the microbial response to environmental stress, such as high concentrations of heavy metals of salts, fire and the water content of the soil. Stress can be detrimental for sensitive microorganisms and decrease the activity of surviving cells, due to the metabolic load imposed by the need for stress tolerance mechanisms (Schimel et al, 2007; Yuan et al., 2007, Ibekwe et al., 2010; Chowdhury, 2011). In a dry hot climate, the low humidity and soil salinity are the most stressful factors for the soil microbial flora, and frequently occur simultaneously.

Saline stress can gain importance, especially in agricultural soils where the high salinity may be a result of irrigation practices and the application of chemical fertilizers. Research has been carried out on naturally saline soils, and the detrimental influence of salinity on the microbial soil communities and their activities reported in the majority of studies (Batra & Manna, 1997; Zahran, 1997; Rietz & Haynes, 2003; Sardinha et al., 2003).

The effect is always more pronounced in the rhizosphere according to the increase in water absorption by the plants due to transpiration. The simple explanation for this is that life in high salt concentrations has a high bio-energetic taxation, since the microorganisms need to maintain osmotic equilibrium between the cytoplasm and the surrounding medium, excluding sodium ions from inside the cell. As a result, energy sufficient for osmoadaptation is required (Oren, 2002; Jiang et al, 2007).

Fungi

The composition of the microbial community may be affected by salinity (Pankhurst et al., 2001; Gros et al., 2003; Gennari et al., 2007; Llamas et al., 2008; Chowdhury et al., 2011) since the microbial genotypes differ in their tolerance of a low osmotic potential (Mandeel, 2006; Llamas et al., 2008). In fungi, a low osmotic potential decreases spore germination and the growth of hyphae and changes the morphology (Juniper & Abbott, 2006) and gene expression (Liang et al., 2007), resulting in the formation of spores with thick walls (Mandeel, 2006).

Fungi have been reported to be more sensitive to osmotic stress than bacteria (Pankhurst et al., 2001; Sardinha et al., 2003; Wichern et al., 2006). There is a significant reduction in the total fungal count in soils salinized with different concentrations of sodium chloride. Similarly, with an increase in the salinity level to above 5%, the total count of bacteria and actinobacteria was drastically reduced (Omar et al., 1994). Van Bruggen & Semenov (2000) reported that on a long-term basis there is a decrease in the genetic diversity of fungi as a result of stress. On the other hand, Killham (1994) mentioned that the filamentous fungi are highly tolerant of hydric stress. However they have to deal with the increase in osmotic pressure and may therefore change their physiology (Killham, 1994) and morphology in response to this (Zahran, 1997). Two strategies used by microorganisms to adapt to osmotic stress were described by Killham (1994), both of which result in an accumulation of solutes in the cell to counteract the increase in osmotic pressure. One is the selective exclusion of the solute incorporated (for example, Na+, Cl-), thus accumulating the ions necessary for metabolism (for example, NH4+). The other cell adaptation mechanism is the production of organic compounds that will antagonize the concentration gradient between the soil solution and the cell cytoplasm. This adaptation finally results in a physiologically more active microbial community, and, in consequence, reduced substrate use efficiency. However these mechanisms are known for single microorganisms, but little has been studied at the community level.

According to Oren (2001) and Hagemann (2011), while sensitive cells are damaged by the low osmotic potential, some microorganisms can adapt by accumulating osmolytes (including amino acids in bacteria and polyols in fungi), that help retain water (Beales, 2004). Nevertheless the synthesis of osmolytes requires large amounts of energy: 30 to 110 ATPs, when compared to the 30 ATPs required to synthesize the cell wall (Oren, 1999), representing a significant metabolic responsibility for the microorganisms, and reduces the energy available for growth.

In order to better understand what happens to the microbial biomass and its activity in saline soils, one must also consider the water potential (osmotic potential + matrix potential), especially the low water content when the salt concentration in the soil solution increases. Since the water content changes, the microorganisms will be subject to different osmotic and water potentials, even though the modifications in the electrical conductivity (EC) measurement are small. Thus the EC is an indicator of little importance with respect to microbial stress in saline environments. According to Chowdhury et al. (2011), microorganisms have two strategies to respond to the water potential. A decrease in this potential to up to -2MPa damages a proportion of the microbial population, but the remaining microorganisms will adapt themselves and be active. For lower water potentials, the adaption mechanisms are not sufficient and, although the microorganisms survive, they do so with reduced activity per unit of biomass. However more studies are required in different soils and, in particular, in saline soils, in order to discern which effects can be generalized.

Considering the forecast for an increase in saline and sodic areas, an understanding of the effects of salinity and sodicity on the soil carbon (C) stock and flow is fundamental for environmental management. Wong et al. (2008) evaluated the effects of salinity and sodicity on the microbial biomass and on soil respiration, under controlled conditions, submitting perturbed soil samples to leaching after receiving different salts concentrations. The highest soil respiration rates were observed in soils with low salinity, and the lowest in soils with medium salinity, whilst the microbial biomass was greater in the treatments with high salinity and lower in those with low salinity. According to the authors, the results can be attributed to a greater availability of substrate in high salt concentrations, or by an increase in the dispersion of the aggregates of soil or from the dissolution or hydrolysis of the organic material in the soil, which can compensate, at least in part, the stress to which the microbial population is submitted in high salt concentrations. The apparent disparity between the evolution of respiration and that of the biomass could be due to a change induced in the microbial population from one dominated by more active microorganisms to one dominated by less active microorganisms.

The microbial biomass is an important labile fraction of the soil organic matter, functioning both as an agent of transformation and recycling of the organic matter and soil nutrients, as also of a source of nutrients for the plants. It is also a potential source of enzymes in the soil. High salinity reduces the microbial biomass (Tripathi et al., 2006; Wichern et al., 2006), affects amino acid capture and protein synthesis (Norbek & Blomberg, 1998) and respiration (Laura, 1974; Pathak & Rao, 1998; Gennari et al., 2007) and causes increases and decreases in C and N mineralization (Pathak & Rao, 1998; Wichern et al., 2006).

Since the soil organic matter, and consequently, the biomass and microbial activity, are generally more relevant in the first few centimeters at the surface of the soil, salinization close to the surface can significantly affect a series of microbiologically mediated processes. This is a considerable problem, since the microbial processes of the soil control its ecological functions and fertility.

The availability of nutrients for plants is regulated by the rhizospheric microbial activity. Thus any factor affecting this community and its functions influences the availability of nutrients and growth of the plants. One of the microbial responses playing a significant role in plant growth is the internal recycling of nitrogen (N) by way of immobilization and re-mineralization. In the majority of studies, the immobilization of NH_4^+-N is reported as being quicker than that of NO_3^--N, whilst the re-mineralization of the N immobilized in NH_4^+ is slower than that immobilized in NO_3^- (Herrmann et al., 2005). However, little has been reported about immobilization/re-mineralization in the two forms of N under conditions of salinity. Since nitrification is more or less inhibited in the presence of salts (Laura, 1977; Sethi et al., 1993) resulting in an accumulation of NH_4^+-N, the cycling of the two forms of N will have a significant impact on the dynamics and availability of N for the plants. According to Azam & Ifzal (2006) the presence of NaCl retards the N immobilization process. Both re-mineralization and nitrification were significantly retarded in the presence of NaCl, maximum inhibition occurring with 4000 mg NaCl kg^{-1} of soil. The inhibitory effect of NaCl on N re-mineralization was relatively higher in soils treated with NH_4^+. The results of this study suggest greater sensitivity to NaCl by microorganisms that have assimilated NO_3^-. However, N re-mineralization in the population that had assimilated NO_3^- was less affected by salinity when compared to the population that had assimilated NH_4^+.

Effect on enzymatic activities

Since the greater part of soil biochemical transformations are dependent on or related to the presence of enzymes, an evaluation of their activities could be useful to indicate if a soil is adequately carrying out the processes closely connected to its quality.

Soil enzymes carry out a fundamental role in the ecosystems, acting as catalysts of various reactions that result in the decomposition of organic residues, cycling of nutrients and the formation of organic matter in the soil, in addition to taking part in intercellular metabolic reactions responsible for the functioning and maintenance of living beings, quite apart from their biotechnological potential, with various applications in the industrial and environmental areas. They generally originate from microorganisms, but can also have animal and vegetable origins.

Amongst the diverse soil enzymes, dehydrogenase, β-glucosidase, urease and the phosphatases are important in the transformation of different nutrients for plants. The activity of dehydrogenase reflects the total oxidative capacity of the microbial biomass (Nannipieri et al., 1990) and is involved in the central aspect of metabolism. β-glucosidase is an important enzyme in the land carbon cycle, in the production of glucose, which constitutes an important energy source for the microbial mass (Tabatabai, 1994). Thus, the determination of β-glucosidase activity, amongst other hydrolytic enzyme activities, has been suggested as a good indicator of soil quality (Dick et al., 1996). The phosphatases play an important role in the transformation of organic phosphorus into inorganic forms more appropriate for plants. Phosphorus (P) is one of the essential nutrients for a plant, and the greater part of soil phosphorus occurs in the organic form.

Urease predominates amongst the enzymes involved in the N cycle of the soil (Tabatabai & Bremner, 1972; Cookson, 1999). It catalyzes the hydrolysis of urea into ammonia or the ammonium ion, depending on the pH of the soil and carbon dioxide. Urease and catalase are the enzymes responsible for the decomposition of vegetable residues. The activity of these enzymes transforms the residue into humus, which is then completely decomposed into the free nutrients (Ahmad & Khan, 1988). On the other hand, amylase hydrolyzes the polysaccharides, converting them into simpler constituents. The activity of this enzyme is associated with high productivity of the crops (Ahmad & Khan, 1988).

Under laboratory conditions, salinity influenced soil enzyme activity negatively, although the degree of inhibition varied according to the enzyme analyzed and the nature and amount of soil added (Frankenberger & Bingham, 1982). Dehydrogenase activity was severely inhibited whereas the hydrolases showed a milder degree of inhibition. The reduction of enzyme activity in saline soils could be due to the osmotic dehydration of the microbial cells that liberate intracellular enzymes, which become vulnerable to the attack by soil proteases, with a consequent decrease in enzyme activity. The salting-out effect modifies the ionic conformation of the protein-enzyme active site, and specific ionic toxicity causes a nutritional imbalance for microbial growth and subsequent enzyme synthesis (Frankenberger & Bingham, 1982). Ahmad & Khan (1988) and Rietz & Haynes (2003) obtained similar results. According to Rietz & Haynes (2003) the increase in salinity due to an influx of salty water under controlled conditions, decreased the carbon content of the soil microbial biomass and enzymes. Other researchers, for example Omar et al. (1994) and Jialiang (2008) also indicated the effects of soil salinity on the carbon of microbial biomass and on enzyme activity. Garcia & Hernandez (1996) and Ghollarata & Raiesi (2007) showed that an increase in soil salinity inhibited the enzyme activities of benzoyl argininamide alkaline phosphatase and β-glucosidase, and also microbial respiration. Invertase and urease activities were also severely reduced by an increasing concentration of sodium chloride (NaCl) during incubation. In addition, the effect was inhibitory of nitrate reductase in the majority of the treatments (Omar et al., 1994). On comparing the enzyme activities of saline soil with those of normal soil, Ahmed & Khan (1988) also observed a decline in amylase, catalase, phosphatase and urease activities with increasing salinity.

Controlled conditions (laboratory) do not usually reflect the natural situation prevailing in coastal region soils, where the salinity varies temporally. Tripathi et al. (2006; 2007) studied the influence of the salinity of arable soils in Indian coastal regions on the microbial biomass and the following enzyme activities: dehydrogenase, β-glucosidase, urease, and acid and alkaline phosphatases, in three different seasons of the year. The microbial and biochemical parameters were adversely affected by the salinity, and the most extreme situation occurred in the summer. Of the enzymes studied, the activity of dehydrogenase was the most affected.

Another particular ecosystem is the mangrove swamps, areas restricted to zones between coastal seas and islands in tropical regions, associated with estuaries, bays and lagoons in places protected from the impact of waves, where the salinity is between 5 and 30%, but can reach 90% (Museu do Una, 2010). This is a highly degraded natural environment for a variety of reasons, amongst which the discharge of domestic and industrial effluent. Variations in the salinity of this environment can affect the retention of the pollutants and the microbiological responses as a function of the discharge of effluent. On investigating such areas, Tam (1998) observed that the addition of effluent to mangrove swamps, independent of their salinity, stimulated microbial growth and increased the activities of the enzymes dehydrogenase and alkaline phosphatase. According to the author these effects were due to supplementation with additional carbon sources and other nutrients, provided by the effluent.

RECOVERY OF SALINE SOILS

The low productivity of saline soils can be attributed not only to their toxicity due to the salt or to the damage caused by excessive amounts of soluble salts, but also to low soil fertility. The fertility problems are usually evidenced by a lack of organic matter and of available mineral nutrients, especially N and P (Shi et al., 1994). These soils are also usually characterized by a reduction in the activities of some key soil enzymes, such as urease and phosphatase (Shi et al., 1994, Yuan et al., 1997), which are associated with biological transformations and the bioavailability of N and P. The adverse effect of soil salinity on crops depends both on their tolerance and on other factors with important roles in the selection of the natural soil microbial flora during salinization, such as: soil composition, organic matter, pH, heavy metals, water and oxygen availability (Ross et al., 2000).

Organic Amendents

Recently various organic supplements, such as ground coverings, manures and compounds, have been investigated for their efficiency in recovering saline soils. It has been shown that the application of organic matter can accelerate the leaching of NaCl, decrease the percentage of exchangeable sodium and the electrical conductivity, and increase water filtration, the water holding capacity and aggregate stability (El-Shakweer et al., 1998). In soils affected by salts and showing low productivity, the adoption of adequate agricultural practices is of fundamental importance for the success of their exploitation, including modifications in the organic fertilization (Garcia et al., 2000). For example, supplementation with organic matter improves the quality of saline soil and neutralizes the negative effects of the salt, since the microorganisms profit from the greater availability of substrate and can thus deal better with the high salinity. Supplementation also leads to a differentiation in the soil microbial community, with bacteria dominating the surface of the substrate (Wichern et al., 2006).

The application of decomposing cow manure, straw or decomposing stable manure significantly increased the productivities of rice, wheat, barley and sorghum, cultivated in saline soils (Swarup, 1985; Gaffar et al., 1992; Aich et al., 1997). The incorporation of sewage sludge and the epicarp-mesocarp of the almond tree fruit into saline soil increased the N, P and K concentrations in the soil and in tomato fruits (Gomez et al., 1992), and the iron and manganese concentrations in rice (Swarup, 1985). In contrast, the addition of stable manure reduced the sodium adsorption ratio (SAR) (Gaffar et al., 1992). Tejada & Gonzalez (2005) showed that an increase in the organic matter content of saline soils increased the soil structural stability and density and, consequently, the microbial biomass.

The C/N ratio is an extremely important property in the decomposition of organic matter by microorganisms, and for this reason, the organic matter added to saline soils performs an important role in the positive effect on the microbial activity and enzyme activities. The incorporation of rice straw, swine excrement or rice straw plus swine excrement significantly increased the activities of urease and phosphatase and the rate of respiration of the soil (Liang et al., 2003), coinciding with previous reports on the incorporation of other organic matters in saline soils (Blagodatsky & Richter, 1998; Luo & Sun, 1994).

Other organic residues with differentiated chemical compositions were studied by Tejada et al. (2006): one a compound obtained from a cotton de-stoner and the other non-composted chicken manure, in two different doses.

The application of both in the doses studied, under dry climate conditions, improved the physical, chemical and biological properties of the saline soil. These organic treatments also favor the appearance of spontaneous vegetation, which protects the soil and contributes to its correction. The alterations tested improved the soil structure, reducing the percentage of exchangeable sodium and promoting an increase in various enzyme activities. However, whereas the cotton compound had a greater effect on the physical properties of the soil and the percentage of exchangeable sodium, the chicken manure mainly increased the soil enzyme activity.

However, the excessive use of organic manure should be avoided, especially in areas flooded for long periods, in order to reduce the risk of toxic effects from reduced intermediates, which accumulate from the anaerobic decomposition of organic manure (Liang et al., 2003).

The use of residues as a soil corrective or conditioner is an economically and environmentally interesting practice, and coconut powder stands out amongst the organic materials that could be used to recover saline soils, since it is abundant in the Northeastern region of Brazil due to the great consumption of coconuts. It represents a solution for the use of discarded coconut shells, which are constituted of one fraction of fibers and another known as powder, which is aggregated to the fibers. Silva Junior et al. (2009) evaluated the basal respiration of soil incubated with different concentrations of coconut powder, submitted to different levels of salinity. The incorporation of organic matter increased the amount of C-CO_2 mineralization, even at high salinity levels. It also caused a reduction in the negative effect of salinity on microbial activity.

Plant Remediation Strategies

Another solution used for the recovery of saline soils in agricultural systems or salinized, abandoned areas is the use of plants (Hatton & Nulsen, 1999). The use of plants to remediate saline and sodic soils is a low-cost, emergent method, but with little acceptance due to its low profitability. However, some farmers have improved the salinity condition of their soils by planting salt-tolerant trees (Marcar et al., 1995) or forage shrubs (Barrett-Lennard & Malcolm, 1995; Porto et al., 2006). Various plant species (halophytic plants) grow naturally at the coast and in salinized areas, and can survive in salt concentrations equal or greater to that of sea water. The compartmenting of the ions in the vacuoles, the accumulation of compatible solutes in the cytoplasm and the presence of genes for salt tolerance, confer salt resistance on these plants (Gorham, 1995). The re-vegetation of salinized areas with halophytic plants is an example of pro-active phyto-remediation (Porto et al., 2001; 2006).

The introduction of the halophyte Glycyrrhiza glabra in the recovery of saline soils and restoration of the subsequent crop systems in irrigated agriculture has been demonstrated in various studies (Mihailova, 1966; Kerbabaev, 1971; Pauzner, 1971). From results presented by Ravindran et al., (2007), the authors concluded that of the six vegetable species evaluated, Suaeda maritima and Sesuvium portulacastrum L. exhibited a greater accumulation of salt in their tissues and greater salt reduction in the soil. Rabhi et al. (2009) compared Sesuvium portulacastrum L. (Aizoaceae) with two other native halophytes: A. indicum and S. fruticosa with respect to their abilities to desalinate saline soils. The authors showed that of the three species studied, Sesuvium portulacastrum L. was the most convenient for use in the leaching of salts from the rhizospere in arid and semi-arid regions, where the rainfall is low.

In the same way, the creation of high productivity plant forage systems by establishing palatable halophytic plants showed that it was possible to remediate a saline/sodic soil and provide extra income for the farmers at the same time (Hyder, 1981; Helalia et al., 1992; Dagar et al., 2004). Species of the saltbushes Atriplex have been used both as forage and to rehabilitate degraded areas (dunes, salt-mines, saline soils). They are dominant in many arid and semi-arid regions of the world, particularly in environments combining relatively high soil salinity with aridity (Ortiz-Dorda et al., 2005).

Quantifying the recovery of the biological activity of soils remediated with plants has been the focus of few studies. Silva et al. (2008) evaluated the effect of irrigation with the desalination waste from pink tilapia production tanks, on the chemical and microbiological properties of soils cultivated with Atriplex nummularia Lindl. The authors found that although the irrigation with saline waste affected the physical and chemical properties of the soil, cultivation of the halophyte favored microbial activity. Similar results were obtained by Pereira et al. (2004) studying soil cultivated for three years with Atriplex nummularia Lindl., and irrigated with saline waste. In the dry periods, the values for pH, electrical conductivity, fluorescein diacetate hydrolytic activity and alkaline phosphatase activity were higher than in other areas. However, a negative correlation was observed between the values for microbial carbon and the metabolic quotient. Carvava et al. (2005) studied the influence of the following eight halophytes: Asteriscus maritimus (L.) Less, Arthrocnemum macrostachyum (Moric.) Moris, Frankenia corymbosa Desf., Halimione portulacoides (L.) Aellen, Limonium cossonianum O. Kuntze, Limonium caesium (Girard) O. Kuntze, Lygeum spartum L., and Suaeda vera Forsskål ex J.F. Gmelin, on the microbiological and biochemical properties of the rhizospere and aggregate stability of a saline soil. There was good correlation between the enzyme activities, the C of the microbial biomass, colonization of the roots of the eight halophytes and the levels of stable aggregates. The results also showed that the microbial activity and the soil properties related to the microbial activity, as also the aggregate stability, were determined by the type of halophytic species. The modifications in microbial activity caused by the vegetation were also related to the variation in the activities of protease, phosphatase, urease and β-glucosidase (Ceccanti & Garcia, 1994). In the case of the halophytes Arthrocnemum macrostachyum and Sarcocornia fruticosa, when grown in a salty swampy area contaminated with metals, whose roots were colonized by arbuscular mycorrhizal fungi, it was found that the salinity and heavy metals negatively affected the degree of colonization by fungi and some of the parameters indicating microbial activity, such as dehydrogenase, urease, protease, phosphatase and β- glucosidase (Carrasco et al., 2006).

Consequences of remediation activities

The principal objective of the recovery of soils affected by salts is to reduce the concentration of soluble salts and of exchangeable sodium in the soil profile, to a level that does not prejudice the development of crops. A decrease in the degree of salinity involves the process of dissolution and consequent removal by percolation water, whereas a decrease in the exchangeable sodium content involves its displacement from the exchange complex by calcium before the leaching process. Since it is of low cost and relatively abundant in many parts of the world, plaster is the corrective most used to recover sodic and saline-sodic soils (Oad et al., 2002; Barros et al., 2004; Gharaibeh et al., 2009). The substituted sodium is leached from the radicular zone by way of excess irrigation, a process that demands an adequate flow of water through the soil (Qadir & Oster, 2004; Qadir et al, 2006).

There are reports in the literature that the efficiency of washing the radicular zone was higher when irrigation was carried out by dripping rather than by other methods (Bresler et al., 1982). The key question in the recovery of soils affected by salts using irrigation by dripping, is that a reasonable irrigation regime must be carried out to guarantee not only normal growth of the crops, but also an excess of water to leach out the salts. Recently, Kang et al. (2008) reported the recovery of heavy-textured saline soils using irrigation by dripping. However, the alterations in the soil properties during recovery are still not well defined. The physical, chemical and biological alterations occurring in a saline soil during the recovery process with a corn crop and irrigation by dripping were reported by Tan & Kang (2009). The results showed that the soil density in the first 0-20 cm decreased from 1.71 g cm^{-3} to 1.44 g cm^{-3} after three years of rehabilitation. The water content in the saturated soil of the 0- 10 cm layer increased from 20.3 to 30.2%. Both the soil salinity and pH value decreased significantly after three years of recovery. The organic matter contents reduced, whereas the total nitrogen, total phosphorus and total potassium tended to increase after cultivation and irrigation. The amount of bacteria, actinobacteria and fungi increased according to the number of years of rehabilitation, with a tendency for a homogenous distribution in the soil profile. The activities of urease and alkaline phosphatase also increased, but the activity of invertase altered little. Lin et al. (2006) observed that the bacteria, actinobacteria and fungi increased 2.3, 4.3 and 71 times, respectively, by planting Suaeda salsa L. and irrigating by dripping in coastal saline soil. There was a reduction in soil salinity and improvement in fertility.

The low solubility of Ca^{2+} during remediation could limit its efficiency, and thus the possibility of using it with microorganisms is being explored so as to provide more active Ca^{2+} from plaster. Experiments carried out with blue-green algae and plaster resulted in greater solubility of the plaster, thus providing recovery of the sodic soils (Subhashini & Kaushik, 1981). However, Syed et al. (2003) reported successful experiments in the recovery of saline-sodic soils when a mixture of different microorganisms was applied, without the prior application of plaster.

Sahina et al. (2011) studied the effect of microbial application in four different saline-sodic soils with saturated hydraulic conductivity, and treated with plaster. Suspensions of three fungal isolates (Aspergillus spp. FS 9, 11 and Alternaria spp. FS 8) and two bacterial strains (Bacillus subtilis OSU 142 and M3 Bacillus megaterium) were mixed with the leaching water of the soil treated with plaster, and subsequently applied to the soil columns. The measurement of the saturated hydraulic conductivity of the soil columns after treatment, indicated that it increased significantly (P Carter (1986) suggested that the addition of plaster caused a decrease in microbial activity, but tended to increase the microbial biomass in the soil. The effect of salinity on the carbon dynamics, with respect to the accumulation or loss of C is not well documented. The rate of C accumulation or loss depends on the balance between the amount entering and leaving. The entrance of C depends on the plant and the accumulation of biomass, when the organic carbon levels of the soil are dominated by the deposition of vegetable and root residues. The entrance of C into saline soils decreases with the decline in growth of the vegetation, due to the direct effect of the toxic ions and of the increase in osmotic potential, and the indirect effect of the structural decline of the sodic soil.

Wong et al. (2009) investigated the flow of C in saline soils to which plaster and organic matter were added. The microbial biomass was lower in the untreated saline soil, but the effect of adding plaster was insignificant. The accumulated respiration was greater in the soils receiving the organic supplement, whereas the \addition of plaster decreased the accumulated respiration rate when compared to the addition of organic matter and of organic matter and plaster. The lowest respiration rate and microbial biomass was attributed to the soil with the lowest rate of organic carbon, resulting from the little or absent entrance of C into the soil due to the high salinity, responsible for the lack or absence of vegetation. There was an increase in respiration and in microbial biomass with the addition of organic matter, independent of the adverse environmental conditions of the soil. The results suggest that the microbial biomass of a hibernating population of salt tolerant microorganisms was present, and multiplied quickly when substrate became available.

CONCLUSIONS

According to estimates made by the United Nations, the population projected for 2050 is one of 8.9 billion inhabitants, which would exacerbate the challenges of agriculture to meet the food demands of this population. In the past the main directive was to increase the potential for food cultivation and its productivity in the field. These days the paradigm has changed, and now demands that the increase in productivity must be accompanied by sustainable management. Sustainable agriculture involves the management of agricultural resources respecting human needs, the maintenance of environmental quality, and the conservation of natural resources for the future. Soil salinity is widely reported as the main agricultural problem, particularly in irrigated agriculture, and approximately 20% of arable land and 50% of agricultural land in the world are under saline stress. According to statistics from UNESCO and FAO, the area of saline soils in the world is of 9.5×10^7 km^2. Salinity causes, directly or indirectly, a harmful influence on the maintenance of soil quality, since it affects the physical, chemical and biological properties of the soil. Research has reported the detrimental influence of salinity on the microbial communities of the soil and their activities. Thus the recovery of soils affected by salt has an important role in the sense of mitigating the pressure, especially in agricultural areas. Saline soil is an important land resource for agriculture.

REFERENCES

1. Agricultura sustentável e conservação dos solos. Processos de degradação do solo. Salinização e sodificação. Ficha informativa no 4. Disponível em < http://soco.jrc.ec.europa.eu/documents/PTFactSheet-04.pdf> Acesso em 05/06/2011.
2. Ahmad, I. & Khan, K.M. (1988). Studies on enzymes activity in normal and saline soils. Pakistan Journal Agricultural Research, Vol. 9, No.4, pp. 506-508, ISSN 0251-0480.

3. Ahmad, Z.; Yahiro, Y.; Kai, H. & Harada, T. (1973). Transformation of the organic nitrogen becoming decomposable due to the drying of soil. Soil Science Plant Nutrition, Vol. 19, No 4, (December, 1973), pp. 287–298, ISSN: 0038-0768.
4. Aich, A.C.; Ahmed, A.H.M. & Mandal, R. (1997). Impact of organic matter, lime and gypsum on grain yield of wheat in salt affected soils irrigated with different grades of brackish water. Research, Vol. 10, No (1– 2), (December, 1997), pp. 79– 84, ISSN: 0970-5767.
5. Allen, S.C.; Fair, V.D.; Graetz, D.A.; SHIBU, J. & Ramachandran N.P.K. (2006). Phosphorus loss from organic versus inorganic fertilizers used in alley cropping on a Florida Ultisol. Agriculture, Ecosystems & Environment, Vol. 117, No 4, (December, 2006), pp. 290–298, ISSN: 0167-8809.
6. Almeida Neto, O.B. de. (2007). Dispersão da argila e condutividade hidráulica em solos com diferentes mineralogías, lixiviados com soluções salino-sódicas. Tese de Doutorado. Universidade Federal de Viçosa. Viçosa, MG,
7. Brasil. Anderson, R. & Xia, L.Z. (2001). Agronomic measures of P, Q/I parameters and lysimetercollectable P in subsurface soil horizons of a long-term slurry experiment. Chemosphere, Vol. 42, No 2, (January, 2001), pp. 171–178, ISSN: 0045-6535.
8. Azam, F. & Ifzal, M. (2006). Microbial populations immobilizing NH4 +-N and NO3 --N differ in their sensitivity to sodium chloride salinity in soil. Soil Biology & Biochemistry, Vol. 38, No 8, (August, 2006), pp. 2491–2494, ISSN: 0038-0717.
9. Barrett-Lennard, E.G. & Malcolm, C.V. (1995). Saltland Pastures in Australia—A Practical Guide. Department of Agriculture Western Australia ISBN: 1 920860 07 X, Perth, Western Australia.
10. Barros, M. de F.C.; Fontes, M.P.F.; Alvarez, V.H. & Ruiz, H.A. (2004). Recuperação de solos afetados por sais pela aplicação de gesso de jazida e calcário no Nordeste do Brasil. Revista Brasileira de Engenharia Agrícola e Ambiental, Vol. 8, No 1, (Janeiro-Abril, 2004), pp.59-64, ISSN: 1415-4366.
11. Batra, L. & Manna, M.C. (1997). Dehydrogenase activity and microbial biomass carbon in salt affected soils of semiarid and arid regions. Arid Soil Research and Rehabilitation, Vol. 11, No 3, (Available online: January, 2009), pp. 295–303, ISSN 0890-3069.
12. Beales, N. (2004). Adaptation of microorganisms to cold temperatures, weak acid preservatives, low pH, and osmotic stress: a review.

Comprehensive Reviews in Food Science and Food Safety, Vol. 3, No 1, (January, 2004), pp. 1-20, ISSN 1541-4337.

13. Blagodatsky, S.A. & Richter, O. (1998). Microbial growth in soil and nitrogen turnover: a theoretical model considering the activity state of microorganisms. Soil Biology & Biochemistry, Vol. 30, No 13, (November, 1998), pp. 1743–1755, ISSN 0038-0717.

14. Bot, A.; Nachtergaele, F. & Young, A. (2000). Land resource potential and constraints at regional and country levels. Rome: FAO, 2000. (FAO. World Soil Resources Report, 90).

15. Bresler E.; McNeal, B. L. & Carter, D L. (1982). Saline and Sodic Soils: Principles-DynamicsModeling. Springer-Verlag, ISBN: 3-540-11120-4, Berlin Heidelberg.

16. Caravaca, F.; Alguacil, M.M.; Torres, P. & Roldán, A. (2005). Plant type mediates rhizospheric microbial activities and soil aggregation in a semiarid Mediterranean salt marsh. Geoderma, Vol. 124, No 3-4, (February 2005), pp. 375-382, ISSN: 0016- 7061.

17. Carrasco, L.; Caravaca, F.; Alvarez-Rogel, J. & Roldan, A. (2006). Microbial processes in the rhizosphere soil of a heavy metals-contaminated Mediterranean salt marsh:A facilitating role of AM fungi. Chemosphere, Vol. 64, No 1, (June, 2006), pp. 104–111, ISSN: 0045-6535.

18. Carter, M.R. (1986). Microbial biomass and mineralizable nitrogen in Solonetzic soils: influence of gypsum and lime amendments. Soil Biology & Biochemistry, Vol. 18, No 5, (1986) pp. 531–537, ISSN: 0038-0717.

19. Ceccanti, B. & Garcia, C. (1994). Coupled chemical and biochemical methodologies to characterize a composting process and the humic substances. In: Humic Substances in the Global Environment and its Implication on Human Health, Senesi, N. & Miano, T. (Eds.), pp. 1279–1285, Elsevier, ISBN:10- 0444895930, New York, USA.

20. Chowdhury, N.; Marschner, P. & Burns, R.G. (2011). Soil microbial activity and community composition: impact of changes in matric and osmotic potential. Siol Biology and Biochemistry, Vol. 43, No 6, (June, 2011), pp. 1229-1236, ISSN: 0038-0717. CODEVASF - Salinização do solo. Disponível em < http://www.codevasf.gov.br/programas_acoes/irrigacao/salinizacao-do-solo> Acesso em 10/03/2011.

21. Cookson, P. 1999. Special variation in soil urease activity around irrigated date palms. Arid Soil Research and Rehabilitation, Vol. 13, No 2, (Available online: November, 2010), pp. 155–169, ISSN: 0890-3069.

22. Dagar, J.C.; Tomar, O.S.; Kumar, Y. & Yadav, R.K. (2004). Growing three aromatic grasses in different alkali soils in semi-arid regions of northern

India. Land Degradation and Development, Vol. 15, No 2, (March/April, 2004), pp. 143–151, ISSN: 1085-3278.

23. Dick, R.P., Breakwell, D.P. & Turco, R.F. (1996). Soil enzyme activities and biodiversity measurements as integrative microbiological indicators, In: Methods for Assessing Soil Quality Special Publication No. 49, Doran, J.W. & Jones, A.J. (Eds.), pp. 247–271, Soil Science Society America, ISBN-10: 0891188266, Madison, WI, USA.

24. Ekholm, P.; Turtola, E.; Gronroos, J.; Sauri, P. & Ylivainio, K. (2005). Phosphorus loss from different farming systems estimated from soil surface phosphorus balance, Agriculture, Ecosystems & Environment, Vol. 110, No 3-4, (November, 2005), pp. 266–278, ISSN: 0167-8809.

25. El-Shakweer, M.H.A.; El-Sayad, E.A. & Ejes, M.S.A. (1998). Soil and plant analysis as a guide for interpretation of the improvement efficiency of organic conditioners added to different soils in Egypt. Communications in Soil Science and Plant Analysis, Vol. 29, No 11-14, (Available online: November, 2008), pp. 2067–2088, ISSN: 0010-3624.

26. Frankenberger, W.T., Bingham, F.T. (1982). Influence of salinity on soil enzyme activities. Soil Science Society of America Journal, Vol. 46, No 6, (November-December, 1982), pp. 1173–1177, ISSN 0361-5995.

27. Freire, E. de A.; Laime, E.M.M.; Navilta, V. do N.; Lima, V. L. de & Santos, J.S. dos. (2009). Análise dos riscos de salinidade do solo do perímetro irrigado de Forquilha, Ceará. Revista Educação Agrícola Superior, Vol. 24, No 2, (December, 2009), pp.62-66, ISSN: 0101-756 X.

28. Gaffar, M.O.; Ibrahim, Y.M. & Wahab, D.A.A. (1992). Effect of farmyard manure and sand on the performance of sorghum and sodicity of soils. Journal of the Indian Society of Soil Science, Vol. 40, No 3, (September, 1992), pp. 540– 543, ISSN : 0019-638X.

29. Garbarino Jr.; Bednar A.J.; Rutherford, D. W.; Beyer, R.S. & Wershaw, R.L. (2003). Environmental fate of roxarsone in poultry litter. I. Degradation of roxarsone during composting. Environmental Science & Technology, Vol. 37, No 8, (Available online: March, 2003), pp. 1509–14, ISSN 0013-936X.

30. Garcıa, C. & Hernandez, T. (1996). Influence of salinity on the biological and biochemical activity of a calciorthird soil. Plant and Soil, Vol. 178, No. 2, (1996), pp. 255–263, ISSN: 0032-079X.

31. Garcia, C.; Hernandez, T.; Pascual, J.A.; Moreno, J.L. & Ros, M. (2000). Microbial activity in soils of SE Spain exposed to degradation and desertification processes. Strategies for their rehabilitation. In Research and Perspectives of Soil Enzymology in Spain.

32. Garcia, C., Hernandez, M.T. (Eds.), pp. 93–143, CEBAS-CSIC, ISBN: 84-605-9821-7, Murcia, Spain. Gennari, M., Abbate, C., La Porta, V., Baglieri, A. & Cignetti, A. (2007). Microbial response to Na2SO4 additions in a volcanic soil. Arid Land Research and Management, Vol. 21, No 3, (Available online: June, 2007), pp. 211-227, ISSN: 1532-4982.

33. Gharaibeh, M.A.; Eltaif, N.I. & Shunnar, O.F. (2009). Leaching and reclamation of calcareous saline–sodic soil by moderately saline and moderate-SAR water using gypsum and calcium chloride. Journal of Plant Nutrition Soil Science, Vol. 172, No 5, (Available online: May, 2009), pp. 713–719, ISSN: 1532-4982.

34. Gomez, I..; Navarro, P.J. & Mataix, J. (1992). The influence of saline irrigation and organic waste fertilization on the mineral content (N, P, K, Na, Ca and Mg) of tomatoes. Journal of the Science of Food and Agriculture, Vol. 59, No 4, (published online: September, 2006) pp. 483–487, ISSN: 0022-5142.

35. Gorham, J. (1995). Mechanism of salt tolerance of halophytes. In: Halophytes and biosaline agriculture, Choukr-Allah, R.; Malcolm, C.V. & Hamdy, A. (Eds.), pp. 207-233, Marcel Dekker, ISBN-10: 0824796640, New York, USA.

36. Gros, R.; Poly, F.; Jocteur-Monrozier, L. & Faivre, P. (2003). Plant and soil microbial community responses to solid waste leachates diffusion on grassland. Plant and Soil, Vol. 255, No 2, (March, 2003), pp. 445-455, ISSN: 0032-079X.

37. Hagemann, M. (2011). Molecular biology of cyanobacterial salt acclimation. Fems Microbiology Reviews, Vol. 35, No 1, (January, 2011), pp. 87-123, ISSN: 0168-6445. Hatton, T.J. & Nulsen, R.A. (1999). Towards achieving functional ecosystem mimicry with respect to water cycling in southern Australian agriculture. Agroforestry Systems, Vol. 45, No 1-3, (1999), pp.203–214, ISSN: 1572-9680.

38. Helalia, A.M.; El-Amir, S.; Abou-Zeid, S.T. & Zaghloul, K.F. (1992). Bio-reclamation of salinesodic soil by Amshot grass in northern Egypt. Soil and Tillage Research, Vol. 22, No 1-2, (January, 1992), pp.109–115, ISSN: 01671987.

39. Herrmann, A.; Witter, E. & Katterer, T. (2005). A method to assess whether 'preferential use' occurs after 15N ammonium addition; implication for the 15N isotope dilution technique. Soil Biology & Biochemistry, Vol. 37, No 1, (January, 2005), pp. 183–186, ISSN: 0038-0717.

40. Iannetta, M. & Colonna, N. Salinização. Lucinda, Série B, No. 3. Disponível em < http://geografia.fcsh.unl.pt/lucinda/Leaflets/B3_Leaflet_PT.pdf> Acesso em 05 de junho de 2011.

41. Ibekwe, A.M.; Poss, J.A.; Grattan, S.R.; Grieve, C.M. & Suarez, D. (2010). Bacterial diversity in cucumber (Cucumis sativus) rhizosphere in response to salinity, soil pH, and boron. Soil Biology & Biochemistry, Vol. 42, No 4, (April, 2010), pp. 567-575, ISSN 0038-0717.

42. Informativo CAPEBE 06/12/2010 - Um patrimônio chamado solo. Disponível em >http://www.capebe.org.br/informativo.php?id=355> Acesso em 13 de junho de 2011.

43. Jackson, B.P. & Bertsch, P.M. (2001). Determination of arsenic speciation in poultry wastes by IC-ICP-MS. Environmental Science & Technology, Vol. 35, No 24, (November, 2001), pp. 4868–4873, ISSN 0013-936X.

44. Jialiang, L: (2008).Research on the effect of saline wetland from pulp wastewater irrigation in Yellow River Dleta, Doctoral dissertation, Ocean University of China, Chine.

45. Jiang, H.; Dong, H.; Yu, B.; Liu, X.; Li, Y.; Ji, S. & Zhang, C.L. (2007). Microbial response to salinity change in Lake Chaka, a hypersaline lake on Tibetan plateau. Environmental Microbiology, Vol. 9, No 10, (July, 2007), pp. 2603-2621, ISSN: 1462-2920.

46. Juniper, S., Abbott, L.K. (2006). Soil salinity delays germination and limits growth of hyphae from propagules of arbuscular mycorrhizal fungi. Mycorrhiza, Vol. 16, No 5, (July, 2006), pp. 371-379, ISSN: 1432-1890

47. Kang, Y.; Wan, S.; Jiao, Y.; Tan, J. & Sun, Z. (2008). Saline soil salinity and water management with tensiometer under drip irrigation. In: Symposia on the Fifth Annual Meeting of Agricultural Land and Water Engineering of Chinese Society of Agricultural Engineering, pp. 124-131, Beijing

48. Keren, R. (2000). Salinity. In: Handbook of Soil Science, Sumner, M.E. (Ed.), pp. G3–G25, CRC Press, ISBN: 9780849331367, Boca Raton.

49. Killham, K. (1994). Soil Ecology (1), Cambridge University Press, ISBN: 0 521 43521 8, United Kingdom. Laura, R.D. (1974). Effects of neutral salts on carbon and nitrogen mineralization of organic matter in soil. Plant and Soil, Vol. 41, No 1, (1974), pp. 113-127, ISSN: 0032-079X.

50. Laura, R.D. (1977). Salinity and nitrogen mineralization in soil. Soil Biology & Biochemistry, Vol. 9, No 5, pp. 333–336, ISSN: 0038-0717.

51. Liang, Y.; Chen, H.; Tang, M.J. & Shen, S.H. (2007). Proteome analysis of an ectomycorrhizal fungus Boletus edulis under salt shock. Mycological

Research, Vol. 111, No 8, (August, 2007), pp. 939-946, ISSN: 0953-7562.

52. Liang, Y.C.; Yang, Y.F.; Yang, C.G.; Shen, Q.Q.; Zhou, J.M. & Yang, L.Z. (2003). Soil enzymatic activity and growth of rice and barley as influenced by organic matter in an anthropogenic soil. Geoderma, Vol. 115, No 1-2, (July, 2003), pp. 149–160, ISSN: 0016-7061.

53. Lin, X. Z.; Chen, K. S.; He, P.Q.; Shen, J.H. & Huang X. H. (2006). The effects of Suaeda salsa L. planting on the soil microflora in coastal saline soil. Acta Ecologica Sinica, Vol. 26, No 3, (March, 2006), pp. 801-807 (in Chinese), ISSN: 1872-2032.

54. Li-Xian, Y.; Guo-Liang, L.; Shi-Hua, T.; Gavin, S. & Zhao-Huan, H. (2007). Salinity of animal manure and potential risk of secondary soil salinization through successive manure application. Science of the Total Environment, Vol. 383, No 1-3, (June, 2007), pp. 106– 114, ISSN 0048-9697.

55. Llamas, D.P., Gonzales, M.D., Gonzales, C.I., Lopez, G.R., Marquina, J.C. (2008). Effects of water potential on spore germination and viability of Fusarium species. Journal of Industrial Microbiology & Biotechnology, Vol. 35, No 11, (November, 2008), pp. 1411- 1418, ISSN: 1367-5435.

56. Luo, A. & SUN, X. (1994). Effect of organic manure on the biological activities associated with insoluble phosphorus release in a blue purple paddy soil. Communications in Soil Science and Plant Analisys, Vol. 25, No 13-14, (Available online: Nov 2008), pp. 2513– 2522, ISSN: 0010-3624.

57. Machado, R. E. (2002). Simulação de escoamento e de produção de sedimentos em uma microbacia hidrográfica utilizando técnicas de modelagem e geoprocessamento. 154p. Tese (Doutorado em Agronomia) – Escola Superior de Agricultura Luiz de Queiroz, Piracicaba, Brasil. Ghollaratta, M. & Raiesi, F. (2007). The adverse effects of soil salinization on the growth of Trifolium alexandrinum

58. L. and associated microbial and biochemical properties in a soil from Iran. Soil Biology & Biochemistry, Vol. 39, No 7, (July, 2007), pp. 1699-1702, ISSN: 0038-0717.

59. Mandeel, Q.A. (2006). Biodiversity of the genus Fusarium in saline soil habitats. Journal of Basic Microbiology, Vol. 46, No 6, (December, 2006), pp. 480-494, ISSN: 0233-111X.

60. Marcar, N.; Crawford, D.; Leppert, P.; Jovanovic, T.; Floyd, R. & Farrow, R. (1995). Trees for Saltland: In: A Guide to Selecting Native Species for Australia. CSIRO Press, ISBN: 0 643 05819 2, Melbourne,Australia. Museu do Una. Manguezais e estuário. Disponível em < http://www.

museudouna.com.br/mangue.htm> Acesso em 05/06/2011.

61. Nannipieri, P.; Gregos, S. & Ceccanti, B. (1990). Ecological significance of the biological activity in soil. In: Soil Biochemistry, vol. 6, Bollag, J.M. & Stotzy, G. (Eds.), pp. 293– 355, ISBN: 0-8247-8232-1,Marcel Dekker, New York, USA.

62. Oad, F.C.; Samo, M.A.; Soomro, A.; Oad, D.L.; Oad, N.L. & Siyal, A.G. (2002). Amelioration of salt affected soils. Pakistan Journal of Applied Sciences, Vol. 2, No 1, (January, 2002), pp.1–9, ISSN: 1607-8926.

63. Oldeman, I. R. (1994). The global extent of soil degradation. In: Soil resilience and sustainable land use. D. J. Greenland & I. Szabolcs, (Ed.), pp. 99-118, CAB International, ISBN- 10: 0851988717, Wallingford, England. Omar, S.A.; Abdel-Sater M.A.; Khallil, A.M.; Abdalla, M.H. (1994). Growth and enzyme activities of fungi and bacteria in soil salinized with sodium chloride. Folia Microbiologica, Vol.39, No 1, (1994), pp. 23-28, ISSN: 0015-5632 Oren, A. (1999). Bioenergetic aspects of halophilism. Microbiology and Molecular Biology Reviews, Vol. 63, No 2, (June, 1999), pp. 334-348, ISSN: 1098-5557.

64. Oren, A. (2001). The bioenergetic basis for the decrease in metabolic diversity at increasing salt concentrations: implication of the functioning of salt lake ecosystems. Hidrobiología, Vol. 466, No 1-3, pp. 61-72, ISSN: 0073-2087.

65. Oren, A. (2002). Molecular ecology of extremely halophilic archaea and bacteria. FEMS Microbiology Ecology, Vol. 39, No 1, (January, 2002), pp. 1–7, ISSN: 0095-3628.

66. Ortiz-Dorda, J.; Martinez-Mora, C.; Correal, E.; Simon, B. & Cenis, J.L. (2005). Genetic structure of Atriplex halimus populations in the mediterranean basin. Annals of Botany, Vol. 95, No 5, pp. 827–834, ISSN: 0305-7364.

67. Pankhurst, C.E., Yu, S., Hawke, B.G. & Harch, B.D. (2001). Capacity of fatty acid profiles and substrate utilization patters to describe differences in soil microbial communities associated with increased salinity or alkalinity at three locations in South Australia. Biology and Fertility of Soils, Vol. 33, No 3, (March, 2001), pp. 204–217, ISSN: 0178- 2762.

68. Parker, D. (2000). Controlling agricultural nonpoint water pollution: costs of implementing the Maryland Water Quality Improvement Act of 1998. Agricultural Economics, Vol. 24, No 1, (August, 2005), pp. 23–31, ISSN: 1574-0862.

69. Pathak, H., Rao, D.L.N. (1998). Carbon and nitrogen mineralisation from added organic matter in saline and alkali soils. Soil Biology &

Biochemistry, Vol. 30, No 6, (June, 1998), pp. 695-702, ISSN: 0038-0717.

70. Pereira, S.V.; Martinez, C.R.; Porto, E.R.; Oliveira, B.R.B. & Maia, L.C. (2004). Atividade microbiana em solo do Semi-Árido sob cultivo de Atriplex nummularia. Pesquisa Agropecuária Brasileira, Vol. 39, No 8, (Agosto, 2004), pp. 757-762, ISSN: 0100- 204X.

71. Porto, E.R.; Amorim, M.C.C. de; Dutra, M.T.; Paulino, R.V.; Brito, L.T. de L. & Matos, A.N.B. (2006). Rendimento da Atriplex nummularia irrigada com efluentes da criação de tilápia em rejeito da dessalinização de água. Revista Brasileira de Engenharia Agrícola e Ambiental, Vol. 10, No 1, (Janeiro-março, 2006), pp. 97-103, ISSN: 1415-4366.

72. Porto, E.R.; Amorim, M.C.C. de & Silva Junior, L.G. de A. Uso de rejeito da dessalinização de água salobra para irrigação da erva-sal (Atriplex nummularia). (2001). Revista Brasileira de Engenharia Agrícola e Ambiental, Vol. 5, No 1, (Janeiro-Abril, 2001), pp.111-114, ISSN: 1415-4366.

73. Qadir, M., Ghafoor, A., Murtaza, G. (2000). Amelioration strategies for saline soils: a review. Land Degradation & Development, Vol. 11, No 6, (January, 2001), pp. 501–521, ISSN: 1099-145X.

74. Qadir, M.; Noble, A.D., Schubert, S.; Thomas, R.J. & Arslan, A. (2006). Sodicity-induced land degradation and its sustainable management: problems and prospects. Land Degradation & Development, Vol. 17, No 6, (May, 2006), pp. 661–676, ISSN: 1099- 145X.

75. Qadir, M. & Oster, J.D. (2004). Crop and irrigation management strategies for saline–sodic soils and waters aimed at environmentally sustainable agriculture. The Science of the total Environment, Vol. 323, No 1-3, (May, 2004), pp. 1–19, ISSN: 0048-9697.

76. Rabhi, M.; Hafsi, C.; Lakhdar, A.; Hajji, S.; Barhoumi, Z.; Hamrouni, M.H.; Abdelly, C. & Smaoui, A. (2009). Evaluation of the capacity of three halophytes to desalinize their rhizosphere as grown on saline soils under nonleaching conditions. African Journal of Ecology, Vol. 47, No 4, (August, 2009), pp. 463–468, ISSN: 0141-6707.

77. Ravindran, K.C.; Venkatesan, K.; Balakrishnan, V.; Chellappan, K.P. & Balasubramanian, T. (2007). Restoration of saline land by halophytes for Indian soils. Soil Biology & Biochemistry, Vol. 39, No 10, (October, 2007), pp. 2661–2664, ISSN 0038-0717.

78. Rietz, D.N., Haynes, R.J., 2003. Effects of irrigation induced salinity and sodicity on soil microbial activity. Soil Biology & Biochemistry, Vol. 35, No 6, (June, 2003), pp. 845– 854, ISSN 0038-0717.

79. Ross, I.L.; Alami, Y.; Harvey, P.R.; Achouak, W. & Ryder, M.H. (2000). Genetic diversity and biological control activity of novel species of closely related pseudomonads isolated from wheat field soils in South Australia. Applied and Environmental Microbiology, Vol. 66, No 4, (April, 2000), pp. 1609-1616, ISSN: 0099-2240.

80. Sahin, U.; Eroglu, S. & Sahin, F. Microbial application with gypsum increases the saturated hydraulic conductivity of saline–sodic soils. (2011). Applied Soil Ecology, Vol. 48, No 2, (June, 2011), pp. 247–250, ISSN: 0929-1393.

81. Sardinha, M.; Muller, T.; Schmeisky, H. & Joergensen, R.G. (2003). Microbial performance in soils along a salinity gradient under acidic conditions. Applied Soil Ecology, Vol. 23, No 3, (July, 2003), pp. 237–244, ISSN: 0929-1393.

82. Schimel, J.P.; Balser, T.C. & Wallenstein, M. (2007). Microbial stress response physiology and its implications for ecosystem function. Ecology, Vol. 88, No 6, (June, 2007), pp. 1386-1394, ISSN: 0012-9658.

83. Sethi, V.; Kaushik, A. & Kaushik, C.P. (1993). Salt stress effects on soil urease activity, nitrifier and Azotobacter populations in gram rhizosphere. Tropical Ecology, Vol. 34, No 2, (December, 1993), pp. 189–198, ISSN: 0564-3295.

84. Shi, W.; Cheng, M.; Li, C. & Ma, G. (1994). Effect of Cl_ 1 on behavior of fertilizer nitrogen, number of microorganisms and enzyme activities in soils. Pedosphere, Vol. 4, No 4, (December, 1994), pp. 357– 364, ISSN: 1002-0160.

85. Silva Júnior, J.M.T.da; Tavares, R. de C; Mendes Filho, P.F. & Gomes, V.F.F. (2009). Efeitos de níveis de salinidade sobre a atividade microbiana de um Argissolo Amarelo incubado com diferentes adubos orgânicos. Revista Brasileira de Ciências Agrárias, Vol. 4, No 4, (out-dez, 2009), pp. 378-382, ISSN: 1981-1160

86. Silva, C.M.M.S.; Vieira, R.F. & Oliveira, P.R. de. (2008). Salinidade, sodicidade e propriedades microbiológicas de Argissolo cultivado com erva-sal e irrigado com rejeito salino. Pesquisa Agropecuária Brasileira, Vol. 43, No 10, (Out, 2008), pp. 1389-1396, ISSN 0100-204X .

87. Sterr, A. Making development sustainable. (1998). Advanced Geo-Ecology, Vol. 31, pp. 857- 865, SSN: 0145 8752.

88. Subhashini, D. & Kaushik, B.D. (1981). Amelioration of sodic soils with blue–green algae. Australian Journal Soil Research, Vol. 19, No 3, (1981), pp. 361–366, ISSN: 0004-9573.

89. Swarup, A. (1985). Yield and nutrition of rice as influenced by pre-

submergence and amendments in a highly sodic soil. Journal of the Indian Society of Soil Science, Vol. 33, No 2, (April-June, 1985), pp. 352– 357, ISSN: 0019-638X.

90. Syed, A., Satou, N. & Higa, T. (2003). Mechanisms of effective microorganisms (EM) in removing salt from saline soils. In: 13th Annual West Coast Conference on Contaminated Soils, Sediments and Water , Mission Valley Marriott, San Diego, CA, USA, March 2003.

91. Tabatabai, M.A., (1994). Soil enzymes. In Methods of Soil Analysis. Part 2: Microbial and Biochemical Properties, Weaver, R.W., Angel, J.S., Bottomley, P.S. (Eds.), pp. 775–833, Soil Science Society America, ISBN 10 089118810X, Madison, WI, USA.

92. Tabatabai, M.A. & Bremner, J.M. (1972). Assay of urease activity in soils. Soil Biology & Biochemistry, Vol. 4, No 4, (November,1972), pp. 479–487, ISSN: 0038-0717.

93. Tam, N.F.Y. (1998). Effects of wastewater discharge on microbial populations and enzyme activities in mangrove soils. Environmental Pollution, Vol. 102, No 2-3, (August, 1998), pp. 233-242, ISSN: 02697491.

94. Tan, J. & Kang, Y. (2009). Changes in Soil Properties Under the Influences of Cropping and Drip Irrigation During the Reclamation of Severe Salt-Affected Soils. Agricultural Sciences in China, 2009, Vol. 8, No 10, (October, 2009), pp. 1228-1237, ISSN: 1671- 2927.

95. Tejada, M. & Gonzalez, J.L. (2005). Beet vinasse applied to wheat under dry land conditions affects soil properties and yield. European Journal of Agronomy, Vol. 23, No 4, (December, 2005), pp. 336-347, ISSN: 1161-0301.

96. Tejada, M.; Garcia, C.; Gonzalez, J.L. & Hernandez, M.T. (2006). Use of organic amendment as a strategy for saline soil remediation: Influence on the physical, chemical and biological properties of soil. Soil Biology & Biochemistry, Vol. 38, No 6, (June, 2006), pp. 1413–1421, ISSN 0038-0717.

97. Tripathi, S.; Kumari, S.; Chakraborty, A.; Gupta, A.; Chakrabarti, K. & Bandyapadhyay, B.K. 2006. Microbial biomass and its activities in salt-affected soils. Biology and Fertility of Soils, Vol. 42, No 3, (February, 2006), pp. 273–277, ISSN: 0178-2762.

98. Tripathi, S.; Chakrabarty, A.; Chakrabarti, K. & Bandyopadhyay, B.K. (2007). Enzyme activities and microbial biomass in coastal soils of India. Soil Biology &Biochemistry, Vol. 39, No 11, (November, 2007), pp. 2840–2848, ISSN 0038-0717.

99. Van Bruggen, A.H.C. & Semenov, A.M. (2000). In search of biological indicators for soil health and disease suppression. Applied Soil Ecology, Vol. 15, No 1, (August, 2000), pp. 13–24, ISSN: 0929-1393.
100. Wichern, J.; Wichern, F. & Joergensen, R.G. (2006). Impacto of salinity on soil microbial communities and the decomposition of maize in acidic soils. Geoderma, Vol. 137, No 1-2, (December, 2006), pp. 100-108, ISSN: 0016-7061.
101. Wong, V.N.L.; Dalal, R.C. & Greene, R. S. B. (2008). Salinity and sodicity effects on respiration and microbial biomass of soil. Biology and Fertility of Soils, Vol. 44, No 7, (August, 2008), pp. 943-953, ISSN: 0178-2762.
102. Wong, V.N.L.; Dalal, R.C. & Greene, R. S. B. (2009). Carbon dynamics of sodic and saline soils following gypsum and organic material additions: A laboratory incubation. Applied Soil Ecology, Vol. 41, No 1, (January, 2009), pp. 29– 40, ISSN: 0929-1393.
103. Yuan, L.; Huang, J. & Yu, S. (1997). Responses of nitrogen and related enzyme activities to fertilization in rhizosphere of wheat. Pedosphere, Vol. 7, No 2, (June, 1997), pp. 141– 148, ISSN: 1002-0160.
104. Yuan, B-C; Li, Z-Z; Liu, H; Gao, M. & Zhang, Y-Y. (2007). Microbial biomass and activity in salt affected soils under arid conditions. Applied Soil Ecology, Vol. 35, No 2, (February, 2007), pp. 319-328, ISSN: 0929-1393.
105. Zahran, Z. (1997). Diversity, adaptation and activity of the bacterial flora in saline environments. Biology and Fertility of Soils, Vol. 25, No 3, (September, 2007), pp. 211– 223, ISSN: 0178-2762.

Chapter 9

SOIL FERTILITY OF TROPICAL INTENSIVELY MANAGED FORAGE SYSTEM FOR GRAZING CATTLE IN BRAZIL

Alberto C. de Campos Bernardi, Patrícia P. A. Oliveira, and Odo Primavesi
East region, Sao Carlos - SP, Brazil

INTRODUCTION

In Brazil the predominant beef and dairy cattle production systems are based mostly on grazing and rely on native and cultivated pastures, which are grazed by continuous stocking all year round and are the main source of animal feed. About 90% of the nutrients required by the ruminants are obtained directly through grazing and supplemental forage feeding is only utilized in intensive dairy production systems and feed-lot systems (Euclides et al., 2010).

Most of the Brazilian cattle are maintained on pastures grown on acidic and low fertility soils that do not receive any lime or fertilizer. This lack of lime and nutrient input in the establishment and the maintenance phases, and the inadequate management of grasses are the main causes of pasture degradation in Brazil (Cantarella et al., 2002). Estimates indicate that approximately 80% of the 50 million hectares of pastures are degraded or under degradation process in the Cerrado (Savannah) region. Pasture degradation is the main cause of low productivity in the Brazilian livestock sector. In these low-input forage systems, lack of regular nitrogen (N) supply plus declining soil P levels are expected to eventually necessitate a reduction in stocking rate and animal production (Macedo, 2002).

The average stocking rate of Brazilian pasture is less than 1.0 animal per ha per year (Carvalho, 2002) with an annual productivity of meat and of milk around 50 and 2,000 kg per hectare, respectively. On the other hand, using appropriate technology in cultivated pastures it is possible to improve animal production significantly. Well established pastures that are properly managed and fertilized are the main source of food for cattle and most practical as the least costly source of feeding (Camargo et al., 2002). In intensive cattle production, well managed pastures allow for increased rates of stocking and productivity (Corsi and Nussio, 1993; Primavesi et al, 1999; Lugão et al., 2003) based on replanting with better grass varieties, grazing rotation, forage availability in the dry season, regulation of stocking densities, genetic improvement of cattle herds and building up soil fertility by balanced fertilization supply and improving soil organic matter (SOM) content.

In the intensive cattle production system, pastures are mainly composed of Brachiaria and Panicum species (C4 grasses with high herbage dry matter yield potential). These grass forages are used in the summer season, and in the winter, feed is based on chopped sugarcane (corrected with crude protein or nitrogen) or silage (corn or sorghum), due the decrease in forage yield. This feeding system leads to increasing stocking rate up to 6 to 12 cow ha^{-1} in pasture for about 200 days $year^{1}$ without irrigation, with higher stocking rates possible when pastures are irrigated during the dry season (Corsi et al., 2001; Santos et al., 2003). Crop-pasture rotation and the tropical grass pasture management intensification increases animal production to more than 25,000 kg of milk per ha per year and 900 kg of liveweight gain per ha per year (Corsi et al., 2001).

Carvalho and Batello (2009) pointed out that as stocking rate is increased, individual animal performance decreases, while production/unit-area increases to some maximum and then declines as a result of concurrent process controlling plant production and utilization by the grazing animal. Increasing grazing intensity decreases plant solar energy capture because of the negative impact on leaf area index. The harvest efficiency is increased with increasing grazing intensity because the forage intake per unit area is increased. Conversely, with increasing the number of animals, the competition for forage decreases the individual animal intake, which diminishes the assimilation efficiency.

Of the controllable factors determining forage yield and quality, soil fertility including fertilizer application is one of the most important. On these tropical acid soils naturally poor in plant nutrients, soil liming and a balanced nutrient supply of nitrogen (N), phosphorus (P), potassium (K), calcium (Ca), magnesium (Mg), sulphur (S), boron (B), copper (Cu), molybdenum (Mo), manganese (Mn) and zinc (Zn) are therefore essential to ensure high yielding and high quality forage (Corsi and Nussio, 1993; Primavesi et al, 1999; Camargo et al., 2002). In order to intensify cattle production and reach high animal productivity, the main point to be considered is the correction of soil acidity and the balanced supply of mineral nutrients.

Maintenance of soil fertility depends on the nutrient recycling and inputs to the system. In extensive systems, natural rates of nutrient cycling may be sufficient, since it works with very low stocking rates, requiring plenty of waste returned to the soil, but it presupposes the need for large areas. In the case of intensive production, which is carried out in small areas, requiring high stocking rate, and thus higher biomass productivity of the grass, is essential to correct soil acidity with lime and fertilizer use.

However, the used rates of fertilizer in Brazilian pastures are still extremely low, around 5 to 6 kg per ha of NPK. In intensively managed pastures, the maximum doses of nutrients for technical and economical response are quite high. Research results point to economic doses around 800 kg per ha per year of N and K_2O in irrigated pastures, applied as 8 to 9 split applications, and 500 kg per ha per year of these nutrients in non-irrigated pastures (Martha Júnior et al., 2004; Sousa et al., 2004; Primavesi et al., 2004; Primavesi et al., 2003; Oliveira et al., 2003).

Intensive management also contributes to turn to a more profitable and competitive livestock production because it reduces the pasture land area, reduces the potential deforestation of natural forests, increases the possibility of environmental conservation, and release areas for another agricultural land use (Kaimowitz and Angelsen, 2008). Moreover, it contributes to increase carbon sequestration, lower energy loss by animals that otherwise would take long walks in search of food and water, and generates less methane per unit of product. So, intensification is a sustainable practice, since the recent development of Brazilian agriculture has been strongly based on productivity gains, and, to a minor extent, on land area expansion (Contini and Martha Jr., 2010).

This chapter provides information regarding the management of liming and fertilization of intensively managed pastures (based on soil analysis and requirement of the grass). Sources of information are scientific articles (mostly in Portuguese) and personal experience of experts.

FERTILITY OF BRAZILIAN SOILS

Brazilian tropical soils usually produce low yields due to the high Al saturation, low concentrations of most mineral nutrients that are essential for plant development, low organic matter content, leading to low CEC and high P fixation (Bernardi et al., 2002). A summary of the extent of soil-related limitations, both physical and chemical, in the acid infertile soils of the tropical Latin America region was given by Sanchez and Salinas (1981) and is presented in Table 1. Deficiency of N and P was shown to be the most severe limitation to crop growth. The list of major chemical constraints is completed by the toxicity of Al, deficiency of K, high P fixation, and low cation exchange capacity. Other physical hindrances are shown but they are of minor relevance.

Table 1: Geographical extent of the major soil constraints to crop production in tropical America.

Soil constraints	Tropical America		Acid soils	
	1.000.000 ha	% total	1.000.000 ha	% total
Nitrogen deficiency	1332	89	969	93
Phosphorus deficiency	1217	82	1002	96
Potassium deficiency	799	54	799	77
Calcium deficiency	732	49	732	70
Magnesium deficiency	731	49	739	70
Sulphur deficiency	756	51	745	71
Cu deficiency	310	21	310	30
Zn deficiency	741	50	645	62
High P fixation	788	53	672	64
Low CEC	620	41	577	55
Aluminum toxicity	756	51	756	72
Low water availability	626	42	583	56
High erosion risk	543	36	304	29
Flooding	306	20	123	12
Compaction	169	11	169	16
Laterization	126	8	81	8
Water stress (> 3 month)	634	42	299	29

Source: Adapted from Sanchez and Salinas (1981).

The most important function of organic matter in soil is as a reserve of nutrients required by plants (Craswell and Lefroy, 2001). Nevertheless, SOM also plays an extremely important role in tropical soils, since it affects soil properties such as electrical charge and nutrient supply (Sanchez, 1976). The main factor responsible for negative charges, and therefore the SOM which contributes 60 - 80% of total soil CEC (Raij, 1969). The organic matter content is affected by vegetation type, as well as the parent material from geological formations and increases with soil clay content and rainfall (Tognon et al., 1998).

SOM can be increased by addition of crop residues, cover crops, green manure crops, compost, animal manure, by reduced or no-tillage and by avoiding residue burning. Enhanced SOM increases soil aggregation, water holding capacity and P availability; reduces P fixation, toxicity of Al and Mn, leaching nutrients by adsorbing exchangeable Ca, Mg and K (Baligar and Fageria, 1997). SOM also provides a source of nutrients, as was shown by Pereira et al. (2000) who evaluated changes in chemical properties of a Xanthic Hapludox managed under pasture, using two rotational systems with Brachiaria brizantha and Panicum maximum. The organic material incorporated into the soil through vegetable and animal residues, influenced the chemical characteristics, increasing the levels of Ca, Mg, K, P, N, C, OM and pH, and decreasing the Al levels, indicating that SOM has a buffering effect and a complexing effect on Al.

SOIL TESTING

Liming and fertilizer recommendations for pasture should be based mainly on soil analysis and expected yield. Potassium rates are recommended based on soil exchangeable K values. However P recommendation is based on two analytical methods: ion exchange resinextractable P and Mehlich-1 P. Due to the differences between the analytical protocols used for P determination, there are differences in the interpretation levels as shown in Table 2 and 3. Soil analysis for micronutrients, extracted with hot water (B) and DTPA-TEA or Mehlich-1 (Fe, Cu, Mn, and Zn) is also used as criterion for fertilizer recommendation (Tables 4 and 5).

Table 2: Soil fertility classes and limits for interpretation of soil-P availability by the P resin method

Fertility class	P resin			
	Forestry	Perennial	Annual	Horticultural
		mg kg^{-1}		
Very low	0 - 2	0 - 5	0 - 6	0 - 10
Low	3 - 5	6 - 12	7 - 15	11 - 25
Medium	6 - 8	13 - 30	16 - 40	26 - 60
High	9 - 16	31 - 60	41 - 80	61 - 120
Very high	>16	>60	>80	>120

Source: adapted from Raij et al. (1996).

Table 3: Soil fertility classes and limits for interpretation of soil-P availability by the Mehlich-1 P method, considering soil clay content.

Clay content	Soil fertility class				
	Very low	Low	Medium	Adequate	High
%			mg kg^{-1}		
60-100	≤ 2.7	2.8 – 5.4	5.5 – 8.0	8.1 – 12.0	> 12.0
35-60	≤ 4.0	4.1 – 8.0	8.1 – 12.0	12.1 – 18.0	> 18.0
15-35	≤ 6.6	6.7 – 12.0	12.1 – 20.0	20.1 – 30.0	> 30.0
0-15	≤ 10.0	10.1 – 20.0	20.1 – 30.0	30.1 – 45.0	> 45.0

Source: adapted from Alvarez et al. (1999).

Table 4: Soil fertility classes and limits for interpretation of micronutrient availability by the hot water (B) and DTPA (Cu, Fe, Mn and Zn) methods.

Soil fertility class	B (hot water)	Cu	Fe	Mn	Zn
			DTPA-extractable		
			mg kg^{-1}		
Low	0 - 0.2	0 - 0,2	0 - 4	0 - 1,2	0 - 0,5
Medium	0,21 - 0,6	0,3 - 0,8	5 - 12	1,3 - 5,0	0,6 - 1,2
High	>0,6	>0,8	>12	>5,0	>1,2

Source: adapted from Raij et al. (1996).

Table 5: Soil fertility classes and limits for interpretation of micronutrient (Cu, Fe, Mn and Zn) availability by the Mehlich-1 extraction method

Micronutrient	Soil fertility class				
	Very low	Low	Medium	Adequate	High
			mg kg^{-1}		
Cu	≤ 0.3	0.4 -0.7	0.8 – 1.2	1.3 – 1.8	> 1.8
Fe	≤ 8	9 – 18	19 – 30	31 – 45	> 45
Mn	≤ 2	3 – 5	6 -8	9 -12	> 12
Zn	≤ 0.4	0.5 – 0.9	1.0 – 1.5	1.6 – 2.2	> 2.2

Source: adapted from Alvarez et al. (1999).

FOLIAR DIAGNOSIS

Leaves are the first and the principal part grazed by animals (Moraes and Palhano, 2002) so their chemical composition reveals their nutritional value. The principle of foliar diagnosis is based on comparing nutrient concentrations in leaves with standard values. Crops are considered to integrate factors such as presence and availability of soil nutrients, weather variables, and crop management. So plant tissue analyses are the best reflection of what the plant has taken up. Traditionally, whole plant shoots are sampled for forage nutritional diagnosis (Monteiro, 2004). The range of levels considered adequate for forages in Brazil by Werner et al. (1996) are shown in Table 6.

Usually, forage value is based on its protein content however tissue mineral composition also plays an important role in animal nutrition. Gerdes et al. (2000) mentioned that the low nutritional value of tropical forage, often mentioned in literature, is associated with reduced crude protein and minerals, the high fiber content and low dry matter digestibility.

Table 6: Adequate shoot macronutrient concentrations for tropical forages and alfalfa.

Forage	N	P	K	Ca	Mg	S
				g kg^{-1}		
Panicum maximum cv Colonião	15 - 25	1.0 - 3.0	15 - 30	3 - 8	1.5 - 5.0	1.0 - 3.0
Pennisetum purpureum cv Napier	15 - 25	1.0 - 3.0	15 - 30	3 - 8	1.5 - 4.0	1.0 - 3.0
Cynodon dactylon cv. Coast-cross	15 - 25	1.5 - 3.0	15 - 30	3 - 8	2.0 - 4.0	1.0 - 3.0
Cynodon spp cv Tifton	20 - 26	1.5 - 3.0	15 - 30	3 - 8	1.5 - 4.0	1.5 - 3.0
Brachiaria brizantha	13 - 20	0.8 - 3.0	12 - 30	3 - 6	1.5 - 4.0	0.8 - 2.5
Andropogon gaianus	12 - 25	1.1 - 3.0	12 - 25	3 - 6	1.5 - 4.0	0.8 - 2.5
Brachiaria decumbens	12 - 20	0.8 - 3.0	12 - 25	2 - 6	1.5 - 4.0	0.8 - 2.5
Medicago sativa	34 - 56	2.5 - 5.0	20 - 35	10 - 25	3.0 - 8.0	2.0 - 4.0

Source: Adapted from Werner et al. (1996).

PASTURE FERTILIZATION

The pasture productivity is determined by many factors as species, climatic and soil conditions and management practices. Research in tropical and subtropical regions has highlighted the need to supply the pasture system with macro- and micro-nutrients, as well as soil amendments, since fertilization is one of the factors that most contribute to increase forage dry matter (DM) productivity and quality (Primavesi et al., 1999; Cantarella et al., 2002).

Pastures fertilization consists of two phases: fertilization during the establishment period, which aims to provide nutrients for the development of fresh established pasture and correcting deficiencies in the soil nutrient supply; and maintenance fertilization, which aims to provide or restore nutrients extracted or lost during grazing. Table 7 based on Macedo (2004) summarizes the N, P and K recommendations for pastures from three sources: Werner et al. (1997), Cantarutti et al. (1999), and Vilela et al. (2002). These actual lime and fertilizer recommendations for forage in the published literature are adequate for semiintensive forage management systems, but do not capture the whole production potential of the tropical forage.

Bernardi et al (2008) carried out a survey with 232 farmers and rural extension workers who adopted the intensive managed forage system. Questions about the adoption of technical soil conservation and fertility, leaf and soil analysis, irrigation, use of fertilizers and limestone and costs of these techniques on production were made. The results pointed to an usual use of soil analysis, lime, micronutrients and soil conservation practices by 96%, 97%, 78% and 99% of interviewees, respectively. Regardless of the soil test, most used a N, P_2O_5 and K_2O formulation (NPK) for pasture fertilization of

400 kg ha^{-1} of 8-28-16 at planting or seeding, and 700 kg ha^{-1} of 20-5-20 at topdressing. However, 93.8% of the producers do not perform leaf analysis. The approximate relationship between fertilizer use and animal production for this group of interviewees was 1 ton fertilizer to 1000 liters milk and 1 ton fertilizer to 195 kg meat.

Table 7: Nitrogen, P and K fertilizer recommendations (levels and fertilization criteria) for establishment and maintenance of grass-based forage for grazing cattle in tropical pastures of Brazil.

Recommendation	Grass-based forage phases			
	Establishment		Maintenance	
	N (kg ha^{-1})	Criteria	N (kg ha^{-1})	Criteria
Werner et al. (1996)	40	20-40 days after germination	40 to 80 50	Nutritional need each grazing cycles
Cantarutti et al. (1999)	0 to 150	60% of soil coverage by pasture	50 100-150 200-300	Technological level: Extensive Medium Intensive
Vilela et al. (2002)	40 to 50	75% of soil coverage by pasture	40	Extensive
	P_2O_5 (kg ha^{-1})	Criteria	P_2O_5 (kg ha^{-1})	Criteria
Werner et al. (1996)	20 to 150	Nutritional need	20 to 50	Nutritional need (annual application)
Cantarutti et al. (1999)	15 to 120	Technological level	15 to 60	Technological level (annual application)
Vilela et al. (2002)	20 to 180	Nutritional need	20	Nutritional need (biannual application)
	K_2O (kg ha^{-1})	Criteria	K_2O (kg ha^{-1})	Criteria
Werner et al. (1996)	20 to 80	Nutritional need	20 to 60	Nutritional need (annual application)
Cantarutti et al. (1999)	20 to 60	Technological level	40 to 200	Technological level (annual application)
Vilela et al. (2002)	20 to 60	Nutritional need	50	K < 30 mg dm^{-3}

Source: adapted from Macedo (2004).

NITROGEN

In general, N is the nutrient that is the most limiting for plant growth and affects the productivity of pastures (Jarvis et al., 1995). Many authors pointed out that an increment in N fertilization increases grass dry matter yield. Vicente-Chandler et al. (1959) found a positive response to the application of up to 1,800 kg N per ha per year. Growth response of 40 to 70 kg DM kg^{-1} of applied N can be expected at fertilizer rates of as high as 400 to 600 kg N ha^{-1} during the growing season (Vicente-Chandler et al., 1974; Minson et al., 1993).

Other authors, however, demonstrated that higher responses are obtained with doses of 300 to 400 kg N per ha per year (Werner et al. 1977; Olsen, 1972,

Gomes et al., 1987). According to Cantarella et al. (2002) N use efficiency (expressed in kg of dry matter produced per kg fertilizer N applied), decreases with increasing dose of N applied. Additions of N through the mineralization of SOM, atmospheric (wet and dry) deposition and biological-fixing activities in soils are uncertain and are generally inadequate to sustain high pasture productivity (Martha Jr. et al., 2004).

Positive results of N fertilization on dry mass production of species of Brachiaria were achieved by Bonfim-da-Silva and Monteiro (2006), Primavesi et al. (2006), Rodrigues et al. (2006), and Benett et al. (2008), for Cynodon dactylon cv. coast-cross by Corrêa et al. (2007), for P. maximum by Lugão et al. (2003), Volpe et al. (2008) and for Pennisetum by Martha Jr. et al. (2004).

From a forage quality standpoint, N fertilization increased whole plant crude protein concentration (Alvim et al. 1998) but effects on other forage quality variables were less consistent. Results on the effect of N on in vitro dry matter digestibility (IVDMD) are conflicting (Monson and Burton, 1982; Gomide and Costa, 1984; Cáceres et al., 1989).

Urea has been the most used N-source in Brazil (ANDA, 2008), due to lower cost per unit of N. But N use efficiency of urea may be reduced because of losses from agricultural system by volatilization of ammonia to atmosphere. This is one of the main factors responsible for the low efficiency of urea, and may reach extreme values, with losses close to 80% of N applied (Lara-Cabezas et al., 1997, Cantarella et al., 1999; Martha Junior et al., 2004). Such results have been reported even in acid soils, since the liming increases soil pH and favors volatilization. Mulch present in no-tillage or pasture systems may also increase the amount of N lost by volatilization, especially when urea is applied on soil surface. Anjos and Tedesco (1976) compared the losses by N volatilization from N sources and reported values of 30% for urea and less than 1% for ammonium sulfate. In urea fertilized pastures, the volatilization of ammonia (NH3) into the atmosphere is the most significant N loss (Whitehead, 1995). These losses are enhanced when the urea is applied in the topdressing and at the end of the rainy season (Martha Júnior et al., 2004, Primavesi et al., 2004), specifically with greatest losses (60%) when applied on wet soil (field capacity) followed by no rain or irrigation, and lowest losses (1%) when used on dry soil followed by around 10 mm rain or irrigation (Primavesi et al. 2001). These losses after N fertilization reduce pasture growth and consequently, the stocking rate and weight gain (Whitehead, 1995). In a Marandu grass (Brachaiaria brizantha) pasture in Brazil, Oliveira et al. (2007) found gaseous losses of N fertilized with urea ranging from 14 to 38%, but the losses were much lower when urea was incorporated in the soil by a no-till double-disc soil cultivator.

Ammonia losses from N fertilizer can be reduced through the use of sources less susceptible to volatilization (nitrate-based fertilizers), or by soil incorporation of the urea (a process hindered by direct fertilization) (Primavesi et al., 2003). Slow urea liberation has also be observed with the addition of acids, salts of K, Ca and Mg, and by the choice of specific urea's grain size distribution (Allaire and Parent 2004). The urea-N losses can also be reduced using zeolites as additives in the fertilizers to control the retention and release of NH_4^+ (Bernardi et al., 2009).

Teitzel et al. (1991) associated the use of N fertilizers in intensively managed tropical pastures with positive economic responses. Results from Euclides et al. (2007) showed that fertilizing P. maximum with N had significant responses on animal yield as weight gain and production per hectare.

Results from Primavesi et al. (2004) and Primavesi et al. (2006) indicated that the N fertilization doses of 500 kg ha^{-1} $year^{-1}$ of N can be reduced in approximately 10% in subsequent years until stabilizing in the sixth or seventh year with maintenance dose, due the increased SOM levels.

Efficient pasture use in intensive production systems depends on the balanced mineral nutrition of the forage plant (Hopkins et al., 1994). Nevertheless physiological processes of plants are affected specially by high doses of N fertilizer. Data presented by Primavesi et al. (2001 and 2003) suggest that the maximum level of N in tropical forage grasses is 24 g kg^{-1}, from which starts the accumulation of nitrate in forage losses occur more intense and nitrate leaching. The forage protein content should be at least 70 g kg^{-1} or higher to stimulate an animal intake and digestibility.

When N fertilizers are supplied to grasses, there may be increasing in levels of provided nutrient, but there may be side effects of this application, resulting in increases or reductions in levels of other nutrients. Primavesi et al. (2005) determined the uptake of cations and anions of coastcross grass fertilized with N from 0 to 1,000 kg ha^{-1} (split in 5 applications during the rainy season). High doses of N fertilizer as urea or ammonium nitrate applied on coastcross grass favored absorption of cations and anions, although increasing rates of N caused higher K^+ uptake in relation to other cations and in Cl^- among the anions. Batista and Monteiro (2010) evaluated changes in K, Ca and Mg concentrations of B. brizantha, cv. Marandu (Marandu palisadegrass) due N and S fertilizer inputs. N fertilization influenced Ca and Mg concentrations as well as the proportions of K, Ca and Mg in the above-ground part of Marandu palisadegrass.

PHOSPHORUS

Acid tropical soils normally contain a limited P reserve and often have a high sorption capacity (Novais and Smith, 1999). According to Sanchez (1976), there are two main processes responsible for P fixation in acid soils: (i) precipitation by exchangeable Al and; (ii) adsorption on the surface of sesquioxides. Phosphorus fixation tends to be high in acid soils where the Fe and Al–oxyhydroxides are ubiquitous. The reversibility of P sorption is important since desorption often is a limiting step in the uptake of P by crops. Hence, P is considered to be the most limiting nutrient in uncultivated tropical soils and frequently found only as a trace (below 1 mg per kg of soil).

Phosphorus Acid tropical soils normally contain a limited P reserve and often have a high sorption capacity (Novais and Smith, 1999). According to Sanchez (1976), there are two main processes responsible for P fixation in acid soils: (i) precipitation by exchangeable Al and; (ii) adsorption on the surface of sesquioxides. Phosphorus fixation tends to be high in acid soils where the Fe and Al–oxyhydroxides are ubiquitous. The reversibility of P sorption is important since desorption often is a limiting step in the uptake of P by crops. Hence, P is considered to be the most limiting nutrient in uncultivated tropical soils and frequently found only as a trace (below 1 mg per kg of soil).

Haag (1993), Hoffmann et al. (1995) and Belarmino et al. (2003) pointed out that P fertilization significantly increased root and tiller growth. Once P is available in sufficient supply, N availability drives pasture production. Cultivars of P. maximum generally show high response to P fertilization (Gheri et al., 2000).

POTASSIUM

Potassium is the cation in higher concentration in forage plants and has relevant physiological and metabolic functions such as enzymes activation, photosynthesis, photoassimilates translocation, stabilization of internal pH, stomatal function, turgor-related processes, N absorption and protein synthesis. The addition of K increases its levels in plant tissue and reduces the Ca and Mg levels in equivalent quantities (Mattos et al., 2002).

Providing an adequate supply of K is important for plant production and is essential to maintain high quality and profitable yields. In order to determine the best time and way of supplying of a nutrient source, their dynamics in soil and role on plant metabolism should also be considered (Benites et al., 2010). Band application of K in the furrow or topdressing application on soil surface is possible due its uptake by mass flux and its high mobility within the plant (Benites et al., 2010). Proper management of K fertilizer in relation to doses and application methods (banding, broadcast and split applications) can minimize losses, avoid depletion of soil K, increase the soil available K pool for a beneficial residual effect of infrequent K fertilization and increase crop yields per unit of K applied to soil (Vilela et al., 2002). The most appropriate time and manner of application of K and of any other nutrient are determined according to plant requirement and element dynamics in soil. The strategy for K fertilization must be accomplished in two steps: first corrective fertilization and then maintenance fertilization.

LIMING

Soil acidity is one of the most yield-limiting factors for crop production, since is a complex of numerous factors involving nutrient/element deficiencies and toxicities, low activity of beneficial microorganisms and reduced plant root growth that limit nutrient and water uptake (Fageria and Baligar, 2003). High amounts of Al, and sometimes Mn, and the low contents of Ca, Mg, and other nutrients frequently account for the low productivity of crops grown on the acid soils. High concentrations of Al inhibit root development and tend to limit absorption of other nutrients, especially of Ca and Mg since their uptake is directly related to root growth and plant development (Lathwell and Grove, 1986).

The clay fraction of Oxisols and Ultisols s usually dominated by sesquioxides, gibbisite, kaolinite and intergrade minerals. These compounds have low intrinsic amounts of negative charges and, therefore, most of the CEC of these soils depends on organic matter (see below) and depends on the soil solution pH. As a consequence, such soils exhibit a strong relationship between charge and pH. In some cases the soils may show net positive charge at low pH, which affects the availability of some nutrients (Sanchez, 1976). Cation exchange capacity is responsible for the equilibrium of ions in the solid/liquid interface in soils. So the usually low values of CEC combined with low pH lead to leaching of K, Ca, and Mg. Low concentrations of K, Ca and Mg, and the low CEC associated with high Al contents are serious fertility constraints in acid tropical soils. Evaluation of these parameters in subsurface layers (below 0.2 m) should be undertaken.

Liming is a low-cost and effective way to neutralize soil acidity and to improve crop yields. Liming reduces Al and Mn toxicity, improves P, Ca and Mg availability, increases CEC, promotes N_2 fixation, and improves soil structure. Overall, liming improves soil capacity to supply needed nutrients and the ability of plants to absorb nutrients and water due to better root growth and activities of beneficial microorganisms. Also an increase in exchangeable bases and pH can stimulate decomposition and mineralization of organic matter by creating a more favorable environment for microbial populations (Sanchez, 1976; Havlin et al., 1999; Fageria and Baligar, 2008). The quantity of lime required depends on the soil type, quality of liming material, costs and crop species or cultivars (Fageria and Baligar, 2008).

Information on appropriate liming rates for tropical forages grown on acid soils in Brazil requires further studies, because the forage response to this practice have been differentiated (Cruz et al., 1994; Oliveira et al., 2003; Paulino et al., 2006), probably due to differences in soil properties and the variability of tolerance to soil acidity of tropical forage grasses. Macedo (2005) believes that the controversy lies in the Cerrado, since clays in these Oxisols affect their response to liming in a way that is quite different from other regions of Brazil. Besides, forages commonly used in this region have high tolerance to soil acidity. Limestone rates are calculated to raise soil base saturation as a percentage of the cation exchange capacity (CEC) of the soil at pH 7.0, to levels which vary with forage species. Werner et al. (1996) recommended a soil base saturation (BS) of 70% for planting and 60% for pasture maintenance, for Pennisetum purpureum, P. maximum, Cynodon dactylon, Digitaria decumbens, Hyparrhenia rufa and Chloris gayana pastures; 60% of BS for the establishment period and 50% of BS for the maintenance of B. brizantha, Andropogon gayanus e Cynodon plectostachyus pastures; and 40% of BS for the establishment and maintenance of Brachiaria decumbens, Brachiaria humidicola, Melinis minutiflora, Paspalum notatum and Setaria anceps pastures. B. decumbens pasture, which had the adequate basis saturation of 40%, after four years of intensive use of N and K fertilizer was observed high depletion in Ca and Mg levels in soil that could lead to plants death (Primavesi et al., 2004 and 2008).

Nevertheless, some intensive managed pastures appear to require more lime, with 80% of basis saturation being the optimum level. This apparent controversy can be explained by the fact that commonly used N, P and S fertilizers are acid-forming. The acidifying effect varies with the forms of these elements in the specific fertilizer used. For example, urea, ammonium nitrate and ammonium sulphate are soil-acidifying and require respectively 1.8, 1.8 and 5.4 kg of lime per kg of fertilizer to neutralize the produced soil acidity (Havlin et al., 1999).

Disking for lime incorporation in soil cultivated with pastures is a controversial practice (Arruda et al., 1987; Soares Filho et al., 1992; Luz et al., 1998). A slight increase in forage yield (130 kg ha^{-1}) was observed when lime was incorporated into the soil of a degraded P. maximum pasture by disking (Luz et al., 1998). On the other hand, disking decreased the dry aboveground matter of B. decumbens pasture (Soares Filho et al., 1992). In the recovery of degraded pastures, Primavesi et al. (2004 e 2008) showed that lime can be applied superficially especially when high N dose are used.

SULPHUR

Sulphur (S) is an important macronutrient for plant metabolism and growth, being a component of essential amino acids (methionine, cysteine) and other organic compounds. The extraction of sulfur by forage plants may be around 50 kg ha^{-1} yr^{-1}, considering yields of 20 t ha^{-1} year-1 of dry matter and S concentration in shoots of 2.5 g kg^{-1} (Werner et al., 1996). Since plants have a lower demand for S than N, it has been neglected in Brazilian pasture fertilization (Monteiro et al., 2004). In Brazil, as a result of the constant use of concentrated NPK fertilizers, besides some edaphic and climatic factors, S became a limiting nutrient for plant development. Moreover in intensive forage management system low response to N fertilizer may be associated to low levels of S in the soil (Cunha et al. 2001; Mattos and Monteiro, 2003, Oliveira et al. 2005; Bomfim and Monteiro-Silva, 2006).

Monteiro et al. (2004) suggested that S fertilization of pasture grasses should be recommended when these forages are well fertilized with N. Stevens (1985) emphasized that both N and S supply are directly related and they must be in plant tissues in adequate proportions and amounts for the optimal synthesis of protein. The N:S ratio is an important nutritional status index since it remains constant at different stages of grasses development (Vitti & Novaes, 1986). According with Scott (1983) to ensure proper development to forage plants, the optimum N:S ratio must be around 16.5:1. The uses of ammonium sulphate or simple superphosphate or gypsum are adequate sulphur sources.

MICRONUTRIENTS

Micronutrients play important roles in plant metabolism, acting as a constituent of organic compounds or as regulators of the functioning of enzyme systems. Micronutrients, also known as trace minerals, which chiefly include boron (B), molybdenum (Mo), copper (Cu), zinc (Zn), manganese (Mn) and iron (Fe), are required in extremely small quantities by crops and cattle. Review articles related to micronutrients on tropical forages prepared by Gupta et al. (2001) and Monteiro et al. (2004) provide information about forage yield responses to these minerals.

Soil acidity is one of the primary factors affecting the availability of micronutrients to crops (Sanchez, 1976). With the exception of Mo, the plant availability of other micronutrients, e.g., Zn, Mn, B and Fe decreases with liming (Gupta et al., 2001; Monteiro et al., 2004). Micronutrients needs are assessed by soil testing and eventually supplied by fertilizers. In more intensive agricultural systems, such as intensively managed grassland with high productivity and high stocking rates, the tendency is a greater need for micronutrients. So an adequate supply is important to avoid a reduction in forage production. Positive response of grasses to zinc addition has been reported in Cerrado low fertility soil and for intensive managed pasture system with high N fertilization (Monteiro et al., 2004).

"ADUBAPASTO" SOFTWARE

As presented in this chapter, the criteria for lime and fertilization recommendations for intensively managed pastures are not organized in a specific publication. Thus, there was a clear need to gather, organize, and make available this existing information to producers and agricultural extension agents. So, Embrapa Pecuária Sudeste made free, online software Adubapasto 1.0 with remote access software by Web service. The structure of this software is based on: 1) architecture of the environment: CLIENT / SERVER, 2) Server Operating System: LINUX, 3) Web Server: APACHE, 4) Application Server: Zope/Plone; 5) Database Server: FIREBIRD and 6) Language development: PYTHON / JAVA SCRIPT.

Software is available at the site: http://www.cppse.embrapa.br/adubapasto. Based on the results of soil analysis, characteristics of a property and cattle schedule (stocking rates, number of days in pasture, supplemental feeds, etc), algorithms for lime and fertilizer recommendation were established, based on results of studies published in scientific and technical literature and experience of experts in soil fertility, fertilizer use, plant nutrition, animal nutrition and forage production. The calculation routines include recommendations for liming, gypsum, N, P and K fertilizer for pasture establishment and maintenance periods, depending on the forage species, cattle management and stocking rate.

As a result, the software generates reports of recommendations for correction and fertilization, stocking rate expected and achieved. It is also possible to assess the historical evolution of soil fertility, since the data is stored in the database. This software operates as a management tool for agricultural technicians, extension workers, producers and researchers who can organize their information in the database.

FINAL REMARKS

The chapter showed that intensive pasture management practices can make cattle more profitable and competitive, and also conserve soil and water, thereby reducing the potential for deforestation and increasing the possibility of environmental preservation. So, High pasture productivity, leading to improve and high cattle production (milk and meat) is a sustainable practice that can meet societal demands for development without environmental degradation, better quality of life and improved resource availability with the opportunity to progressively combat social inequalities in all sectors and especially in the agricultural sector. Soil chemical analysis is an important tool to know the soil fertility and make appropriate lime and fertilizer recommendations. There are differences between analytical protocols used for soil testing in Brazil, therefore, attention should be paid to the results to avoid any ambiguity in the interpretation and recommended doses.

Pastures need N to accumulate carbon, and also K, Ca, Mg, P, S and micronutrients for high yield and quality and profitability of cattle. Nutrient deficiencies and some soil chemical constraints can be avoided by regular monitoring of soil fertility. Careful attention to stocking rates to prevent overgrazing is also important. The practices of soil amendment and fertilizer depend on the production system that farmer adopts and will always want to obtain satisfactory economic return with low environmental impacts. Table 8 summarizes the suggestions for liming and fertilization of intensive managed pasture systems on tropical acid soils in Brazil. Soil testing must be carried out every year

Table 8: Suggestions for liming and fertilization of intensive managed pasture systems in Brazil.

Liming	Increase basis saturation to 70 or 80% (Ca: 55 to 60% of CEC*; Mg: 15 to 20% of CEC)
Nitrogen	Doses established as a function of forage species, animal stocking rate and soil organic matter: 40 to 50 kg ha^{-1} of N per AU** considering 3 to 7 AU** per ha. Reducing doses in 10% each year, from 6 or 7th year just N fertilizer for replace the exportation. Foliar diagnosis for evaluation keeping N in shoots approximately 24 - 25 g kg^{-1}.
Phosphorus	Begin with 10 mg dm^{-3} and increase until 30 mg dm^{-3}
Potassium	Begin with 3% of CEC* and increase until 6% of CEC*
Sulphur	60 - 90 kg ha^{-1}
Micronutrients	B = 0.5 to 1.0 kg ha^{-1} Cu = 1.0 to 2.0 kg ha^{-1} Zn = 2.0 to 4.0 kg ha^{-1}, or FTE*** = 30 to 40 kg ha^{-1}

*CEC = cation exchange capacity; **A.U. = animal unit = 450 kg of live weight; ***FTE = fritted trace elements (composition: Ca = 7.1%; S = 5.7%; B = 1.8%; Cu = 0.8%; Mn = 2.0%; Mo = 0.1% and Zn = 9.0%).

ACKNOWLEDGMENTS

The authors wish to thank IPI (International Potash Institute) for the financial support to soil fertility studies of pasture systems.

REFERENCES

1. Allaire, S.E.; Parent, L.E. 2004. Physical properties of granular organic-based fertilizers. Part 1 Static properties. Biosystems Engineering, 87: 79-87.
2. Alvarez V., V.H.; Novais, R.F.; Barros, N.F.; Cantarutti, R.B. & Lopes, A.S. 1999. Interpretação dos resultados das análises de solo. In: Riberio, A.C.; Guimarães, P.T.G. & Alvarez V., V.H. Recomendações para o uso de corretivos e fertilizantes em Minas Gerais (5a Aproximação). Viçosa, CFSEMG, p.25-32
3. Alvim, M.J.; Xavier, D.F.; Botrel, M.A. et al. 1998. Resposta do Coastcross (Cynodon dactylon) a diferentes doses de nitrogênio e freqüências de corte. Revista Brasileira de Zootecnia, 27: 833-840.
4. Anda. Associação Nacional para Difusão de Adubos. 2008. Anuário estatístico do setor de fertilizantes. São Paulo. Anjos, J.T.; Tedesco, M. J. 1976. Volatilização de amônia proveniente de dois fertilizantes nitrogenados aplicados em solos cultivados. Científica, 4:49-55.
5. Arruda, N.G.; Cantarutti, R.B.; Moreira, E.M. 1987. Tratamentos físico-mecânicos e fertilização na recuperação de pastagens de Brachiaria decumbens em solos de Tabuleiro. Pasturas Tropicales, 9: 36-39.
6. Baligar, V.C.; Fageria, N.K., 1997. Nutrient use efficiency in acid soils: nutrient management and plant use efficiency. In: Moniz, A.C.; Furlani, A.M.C.; Schaeffert, R.E.; Fageria, N.K.; Rosolem, C.A.; Cantarella, H. (eds.). Plant-soil interactions at low pH. Campinas, SP, Viçosa, MG: Brazilian Soil Sci. Soc., p.75-96.
7. Batista, K.; Monteiro, F.A. 2010. Variações nos teores de potássio, cálcio e magnésio em capim-marandu adubado com doses de nitrogênio e de enxofre. Revista Brasileira de Ciência do Solo, 34: 151-161.
8. Belarmino, M.C.J.; Pinto, J.C.; Rocha, G.P. et al. 2003. Altura de perfilho e rendimento de matéria seca de capim Tanzânia em função de diferentes doses de superfosfato simples e sulfato de amônio. Ciência Agrotecnologia, 27: 879-885.
9. Benett, C. G. S.; Buzetti, S.; Silva, K. S.; Bergamaschine, A. F.; Fabrício, J. A. 2008. Produtividade e composição bromatológica do capim marandu a fontes e doses de nitrogênio. Ciência e Agrotecnologia, 32: 1629-1636.

10. Benites, V.M; Polidoro, J.C.; Carvalho, M.C.S.; Resende, A.V.; Bernardi, A.C.C. Álvares, F.A. O potássio, o cálcio e o magnésio na agricultura brasileira. In: Prochnow, L.I. Boas práticas para uso eficiente de fertilizantes. Piracicaba: IPNI. 2010.
11. Bernardi A.C.C., Machado, P.L.O.A.; Silva, C.A. 2002. Fertilidade do solo e demanda por nutrientes no Brasil. In: Manzatto; C.M., Freitas Júnior, E.; Peres, J.R.R. Uso agrícola dos solos brasileiros. Rio de Janeiro: Embrapa Solos, p 61-77.
12. Bernardi, A.C.C.; Monte, M.B.M. 2009. O uso de zeólitas na agricultura. p. 493-508. In: Lapido-Loureiro, F.E.V., Melamed, R. Figueiredo Neto, J. eds. Fertilizantes, agroindústria e sustentabilidade. CETEM/MCT, Rio de Janeiro, RJ, Brazil.
13. Bomfim-Silva, E. M.; Monteiro, F. A. 2006. Nitrogênio e enxofre em características produtivas do capim-braquiária proveniente de área de pastagem em degradação. Revista Brasileira de Zootecnia, 35: 1289-1297.
14. Cáceres, O.; Santana, H.; Delgado, R. 1989. Influencia de la fertilización nitrogenada sobre el valor nutritivo y rendimiento de nutrimentos. Pastos y Forrages, 12: 189-195.
15. Camargo, A.C., Novo, A.L., Novaes, N.J., Esteves, S.N., Manzano, A.; Machado, R. 2002. Produção de leite a pasto. In: 'Simpósio sobre o manejo da pastagem', 18., Piracicaba. Anais... Piracicaba: Fealq. p. 285-319.
16. Cantarella, H.; Correa, L.A.; Primavesi, O.; Primavesi, A.C. 2002. Fertilidade do solo em sistemas intensivos de manejo de pastagens. p.99-131. In: Anais do Simpósio sobre Manejo de Pastagens. 2002. FEALQ, Piracicaba, SP, Brazil.
17. Carvalho, P.C.F. 2002. Country Pasture/Forage Resources Profile: Brazil, FAO, Rome,
18. Carvalho, P.C.F; Batello, C. 2009. Access to land, livestock production and ecosystem conservation in the Brazilian Campos biome: the natural grasslands dilemma. Livestock Science, 120: 158-162.
19. Contini, E.; Martha Jr., G.B. 2010. Brazilian agriculture, its productivity and change. Bertebos Conference on "Food security and the futures of farms: 2020 and toward 2050". Falkenberg: Royal Swedish Academy of Agriculture and Forestry, August 29-31,
20. Correa, J.C.; Reichardt, K. 1995. Efeito do tempo de uso das pastagens sobre as propriedades de um latossolo amarelo da Amazônia Central. Pesquisa Agropecuária Brasileira, 30:107-114,

21. Corrêa, L.A.; Cantarella, H.; Primavesi, A.C.; Primavesi, O.; Freitas, A.R.; Silva, A.G. 2007. Efeito de fontes e doses de nitrogênio na produção e qualidade da forragem de capim-coastcross. Revista Brasileira de Zootecnia, 36: 763-772.

22. Corsi, M.; Martha Jr., G.B.; Nascimento Jr., D.; Balsalobre, M.A.A. 2001. Impact of grazing management on productivity of tropical grasslands. In: International Grassland CongresS, 19., São Pedro, 2001. Proceedings. São Pedro, SBZ, p.801-805.

23. Corsi, M.; Nussio, L.G. 1993. Manejo do capim elefante: correção e adubação do solo. In: 'Simpósio sobre o manejo da pastagem', 10., Piracicaba. Anais... Piracicaba: Fealq, pp.87-115.

24. Craswell, E.T., Lefroy, R.D.B. 2001. The role and function of organic matter nin tropical soils. In: Martius, C. Tiessen, H., Vlek, P.L.G. (Eds.). managing organic matter in tropical soils: scope and limitations. Kluwer, Dordrecht, The Netherlands, p.7-18.

25. Cruz, M.C.P.; Ferreira, M.E.; Lucheta, S. 1994. Efeito da calagem sobre a produção de matéria seca de três gramíneas forrageiras. Pesquisa Agropecuária Brasileira, 29: 303-312.

26. Cunha, M. K.; Siewerdt, L.; Silveira Jr., P; Siewerdt, F. 2001. Doses de nitrogênio e enxofre na produção e qualidade a forragem de campo natural de planossolo no Rio Grande do Sul. Revista Brasileira de Zootecnia, 30: 651-658,

27. Euclides, V.P.B. Valle, C.B.; Macedo, M.C.M.; Almeida, R.G., Montagner, D.B., Barbosa, R.A. 2010. Brazilian scientific progress in pasture research during the first decade of XXI century. Revista Brasileira de Zootecnia, 39: 151-168.

28. Euclides, V.P.B.; Costa, F.P.; Macedo, M.C.M. 2007. Eficiência biológica e econômica de pasto de capim-tanzânia adubado com nitrogênio no final do verão. Pesquisa Agropecuária Brasileira, 42: 1345-1355.

29. Fageria, N.K. And Baligar, V.C. 2003. Fertility management of tropical acid soil for sustainable crop production. In: Rengel, Z., ed. Handbook of soil acidity. New York, Marcel Dekker, p.359-385.

30. Fageria, N.K. and Baligar, V.C. 2008. Ameliorating soil acidity of tropical Oxisols by liming for sustainable crop production. Advances Agronomy, 99:345-431.

31. Gerdes, L.; Werner, J.C.; Colozza, M.T.; Possenti, R.A.; Schammass, E.A. 2000. Avaliação de características de valor nutritivo das gramíneas forrageiras marandu, setária e tanzânia nas estações do ano. Revista Brasileira de Zootecnia, 29: 955-963,

32. Gheri, E.O.; Cruz, M.C.P.; Ferreira, M.E. et al. 2000. Nível crítico de fósforo no solo para Panicum maximum Jacq. cv. Tanzânia. Pesquisa Agropecuária Brasileira, 35: 1809-1816.

33. Gomes, J.F.; Siewerdt, L.; Silveira Jr., P. 1987. Avaliação da produtividade e economicidade do feno de capim pangola (Digitaria decumbens Stent) fertilizado com nitrogênio. Revista da Sociedade Brasileira de Zootecnia, 16: 491-499.

34. Gomide, J.A.; Costa, G.G. 1984. Adubação nitrogenada e consorciação de capim-colonião e capim-jaraguá. III. Efeitos de níveis de nitrogênio sobre a composição mineral e digestibilidade da matéria seca das gramíneas. Revista da Sociedade Brasileira de Zootecnia, 13: 215-224.

35. Gupta, U.C.; Monteiro, F.A.; Werner, J.C. 2001. Micronutrients in grassland production. In: INTERNATIONAL GRASSLAND CONGRESS, 19., São Pedro, 2001. Proceedings. São Pedro, SBZ, p.149-156.

36. Guss, A.; Gomide, J.A.; Novais, R.F. 1990.Exigência de fósforo para o estabelecimento de quatro espécies de Brachiaria em solos com características físico-químicas distintas. R. Bras. Zootec, 19: 278-289.

37. Havlin, J.; Beaton, J.D.; Tisdale, S.L. & Nelson, W.L. 1999. Soil fertility and fertilizers: an introduction nutrient management. Upper Saddle River: Prentice Hall. 499p.

38. Hopkins, A.; Adamson, A.H. and BOWLING, P.J. 1994. Response of permanent and reseeded grassland to fertilizer nitrogen: 2. Effects on concentrations of Ca, Mg, K, Na, S, P, Mn, Zn, Cu, Co and Mo in herbage at a range of sites. Grass Forage Science, 49:9-20.

39. Jarvis, S.C.; Scholefield, D.; Pain, B. 1995. Nitrogen cycling in grazing systems. p.381-419. In: Bacon, P.E. ed. Nitrogen fertilization in the environment. Marcel Dekker, New York, EUA.

40. Kaimowitz, D.; Angelsen, A. 2008. Will livestock intensification help save Latin America´s forests? Journal of Sustainable Forestry, 27: 6-24.

41. Lara-Cabezas, W.A.R; Korndörfer, G.H.; Motta, A.S. 1997. Volatilização de N-NH3 na cultura de milho: II. Avaliação de fontes sólidas e fluidas em sistema de plantio direto e convencional. Revista Brasileira Ciência do Solo 21:489-496.

42. Lathwell, D.J.; Grove, T.L., 1986. Soil-plant relationship in the tropics. Annu. Rev. Ecol. Syst., 17:1-16.

43. Lugão, S.M.B.; Rodrigues, L.R.A.; Abrahão, J.J.S.; Malheiros, E.B.; Morais, A. 2003. Acúmulo de forragem e eficiência de utilização do

nitrogênio em pastagens de Panicum maximum Jacq. (acesso BRA-006998) adubadas com nitrogênio. Acta Scientiarum. Animal Sciences, 25: 371-379.

44. Luz, P.H.C.; Braga, G.J.; Herling, V.R.; Lima, C.G.L. 1998. Efeitos de tipos, doses e incorporação de calcário sobre as características agronômicas do Panicum maximum Jacq. cv. Tobiatã. In: REUNIÃO DA SOCIEDADE BRASILEIRA DE ZOOTECNIA, 35., Botucatu, 1998. Anais. Brasília: SBZ, p.248-250.

45. Macedo, M.C.M. 2002. Degradação,renovação e recuperação de pastagens cultivadas:ênfase sobre a região dos Cerrados. In: Simpósio Sobre Manejo Estratégico da Pastagem, 2002, Viçosa. Anais do Simpósio Sobre Manejo Estratégico da Pastagem. Viçosa, MG : Departamento de Zootecnia, UFV, p.85-108.

46. Macedo, M.C.M. 2004.Analise comparativa de recomendações de adubação em pastagens. In: 21° Simposio sobre Manejo da Pastagem, 2004, Piracicaba, SP. Fertilidade do Solo para Pastagens Produtivas. Piracicaba, SP: FEALQ, v. 1. p. 317-355.

47. Macedo, M.C.M. 2005. Pastagens no ecossistema Cerrados: evolução das pesquisas para o desenvolvimento sustentável. In: REUNIÃO ANUAL DA SOCIEDADE BRASILEIRA DE ZOOTECNIA, 42., 2005, Goiânia. Anais... Goiânia: Sociedade Brasileira de Zootecnia, p.56-84.

48. Martha Jr., G.B.; Corsi, M.; Trivelin, P.C.O. 2004. Nitrogen recovery and loss in a fertilized elephantgrass pasture. Grass and Forage Science, 59: 80-90.

49. Martha Junior, G.B.; Corsi, M.; Trivelin, P.C.O.; Alves, M.C. 2004a. Nitrogen recovery and loss in a fertilized elephant grass pasture. Grass and Forage Science 59:80-90.

50. Martha Júnior, G.B.; Vilela, L.; Barioni, L.G.; Sousa, D.M.G. 2004b. Manejo da adubação nitrogenada em pastagens. In: Pedreira, C.G.S.; Moura, J.C.; Faria, V.P. Fertilidade do solos para pastagens produtivas. Piracicaba: FEALQ, p. 101-138.

51. Mattos, W. T.; Monteiro, F. A. Produção e nutrição do capim braquiária em função de doses de nitrogênio e enxofre. Boletim de Indústria Animal, v.60, p.1-10, 2003.

52. Mattos, W. T.; Santos, A. R.; Almeida, A. A. S.; Carreiro, B. D. C.; Monteiro, F. A. 2002. Aspectos produtivos e diagnose nutricional do capim-Tanzânia submetido a doses de potássio. Magistra, 14: 37-44.

53. Minson D.J., Cowan T. And Havilah E. 1993. Summer pastures and crops. Tropical Grasslands, 27, 131–149. Monson, W.G.; Burton, G.W. 1982.

Harvest frequency and fertilizer effects on yield, quality and persistence of eight bermuda grasses. Agronomy Journal, 74: 371-374.

54. Monteiro, F.A. 2004. Concentração e distribuição de nutrientes em gramíneas e leguminosas forrageiras. In: Simpósio Sobre Manejo Estratégico Das Pastagem, 2., Viçosa, 2004. Anais. Viçosa: DZO,UFV, p.71-107.

55. Monteiro, F.A.; Colozza, M.T.; Werner, J.C. 2004. Enxofre e micronutrientes em pastagens. In: Simpósio Sobre Manejo Da Pastagem, 21., Piracicaba, 2004. Anais. Piracicaba: FEALQ,. p.279-301.

56. Moraes, A.; Palhano, A.L. 2002. Fisiologia de produção de plantas forrageiras. In: Wachowicz, C.M; Carvalho, R.I.N. (Ed.). Fisiologia vegetal - Produção e póscolheita. Curitiba: Champagnat, p. 249-271.

57. Mott, G.O. 1981. Measuring forage quantity and quality in grazing trials. In: Southern Pasture And Forage Crop Improvement Conference, 37., 1981, Nashville. Proceedings... Nashville: United States Department of Agriculture, p.3-9.

58. Novais, R.F.; Smyth, T.J., 1999. Fósforo em solo e planta em condições tropicais. Viçosa, MG: Universidade Federal de Viçosa. UFV, Departamento de Solos. DPS. 399p.

59. Oliveira, P. P. A.; Trivelin, P. C. O.; Oliveira, W. S.; Corsi, M. 2005. Fertilização com N e S na recuperação de pastagem de Brachiaria brizantha cv. Marandu em um neossolo quartzarênico. Revista Brasileira de Zootecnia, 34: 1121-1129.

60. Oliveira, P.P.A.; Boaretto, A.E.; Trivelin, P.C.O. et al. 2003. Liming and fertilization to restore degraded Brachiaria decumbens pastures grown on an entisol. Scientia Agrícola, 60: 125-131.

61. Oliveira, P.P.A.; Trivelin, P.C.O.; Oliveira, W.S. 2003. Eficiência de fertilização nitrogenada com uréia (15N) em Brachiaria brizantha cv. Marandu associada ao parcelamento de superfosfato simples e cloreto de potássio. Revista Brasileira de Ciência do Solo, 27: 613-620.

62. Oliveira, P.P.A.; Trivelin, P.C.O.; Oliveira, W.S. 2007. Balanço do nitrogênio (15N) da úreia nos componentes de uma pastagem de capim-marandu sob recuperação em diferentes épocas de calagem. Revista Brasileira de Zootecnia 36: 1982-1989.

63. Olsen, F.J. 1972. Effect of large application of nitrogen fertilizer on the productivity and protein content of four tropical grasses in Uganda. Tropical Agricultural, 49: 251-260.

64. Paulino, V.T.; Costa, N.L.; Rodrigues, A.N.A. et al. 2006. Resposta de Panicum maximum cv Massai à níveis de calagem. In: Reunião Anual Da Sociedade Brasileira De Zootecnia, 43., 2006, João Pessoa.. Anais... João Pessoa: Sociedade Brasileira de Zootecnia (CD-ROM).

65. Pereira, W.L.M.; Veloso, C.A.C.; Gama, J.R.N.F., 2000. Propriedades químicas de um Latossolo Amarelo cultivado com pastagens na Amazônia Oriental. Scientia Agricola, 57:531-537.

66. Primavesi, A.C.; Primavesi, O.; Corrêa, L.A. Cantarella, H.; Silva, A.G.; Freitas, A.R.; Vivaldi, L.J. 2004. Adubação nitrogenada em capim-coastcross: efeitos na extração de nutrientes e recuperação aparente do nitrogênio. Revista Brasileira de Zootecnia, 33: 68-78.

67. Primavesi, A.C.; Primavesi, O.; Corrêa, L.A.; Cantarella, H. and Silva, A.G. 2005. Absorção de cátions e ânions pelo capim-coastcross adubado com ureia e nitrato de amônio. Pesq. Agropec. Bras., 40: 247-253.

68. Primavesi, A.C.; Primavesi, O.; Corrêa, L.A.; Silva, A.G.; Cantarella, H. 2006. Nutrientes na fitomassa de capim-marandu em função de fontes e doses de nitrogênio. Ciência e Agrotecnologia, 30: 562-568.

69. Primavesi, O.; Corrêa, L. A.; Primavesi, A. C.; Cantarella, H.; Silva, A. G. Adubação com uréia em pastagem de Brachiaria brizantha sob manejo rotacionado: eficiência e perdas. São Carlos: Embrapa Pecuária Sudeste, (nov) 2003. 6p. (Embrapa Pecuária Sudeste, Comunicado Técnico, 41). Available in: http://www.cppse.embrapa.br/080servicos/070publicacaogratuita/comunicadote cnico/ComuTecnico41.pdf

70. Primavesi, O.; Corrêa, L. A.; Primavesi, A.C.; Cantarella, H.; Armelin, M. J. A.; Silva, A. G.; Freitas, A. R. de. Adubação com uréia em pastagem de Cynodon dactylon cv. Coastcross: eficiência e perdas. São Carlos: Embrapa Pecuária Sudeste, (jun) 2001. 42p. (Embrapa Pecuária Sudeste, Circular Técnica, 30). Available in: http://www.cppse.embrapa.br/080servicos/070publicacaogratuita/circulartecnica/Circular30.pdf

71. Primavesi, O.; Primavesi, A.C.; Camargo, A.C. 1999. Conhecimento e controle, no uso de corretivos e fertilizantes, para manejo sustentável de sistemas intensivos de produção de leite de bovinos a pasto. Revista de Agricultura 74:249-266.

72. Primavesi, O.; Primavesi, A.C.; Corrêa, L.A.; Armelin, M.J.A.; Freitas, A.R. Calagem em pastagem de Brachiaria decumbens recuperada com adubação nitrogenada em cobertura. São Carlos: Embrapa Pecuária Sudeste, (dez) 2004. 32p. (Embrapa Pecuária Sudeste, Circular Técnica, 37). Available in: http://www.cppse.embrapa.br/080servicos/070publicacaogratuita/circulartecnica/Circular37.pdf

73. Primavesi, O.; Primavesi, A.C.; Correa, L.A.; Silva, A.G.; Cantarella, Heitor. Lixiviação de nitrato em pastagem de coastcross adubada com nitrogênio. Revista Brasileira de Zootecnia, 35: 683-690, 2006.

74. Raij, B. van, 1969. Capacidade de troca de frações orgânicas e minerais dos solos. Bragantia, 28:85-112.

75. Raij, B. Van; Cantarella, H.; Quaggio, J.A.; Furlani, A.M.C. Recomendações de adubação e calagem para o Estado de São Paulo. Campinas, Instituto Agronômico/Fundação IAC, 1996. 285p. (Boletim Técnico, 100).

76. Rodrigues, R.C.; Alves, A.C.; Brennecke, K.; Plese, L.P.M.; Luz, P.H.C. 2006. Densidade populacional de perfilhos, produção de massa seca e área foliar do capimxaraés cultivado sob doses de nitrogênio e potássio. Boletim de Indústria Animal, 63: 27-33.

77. Sanchez, P. A.; Salinas, J. G. 1981. Low-input technology for managing Oxisols and Ultisols in tropical America. Advances in Agronomy, 34: 280-407. Sanchez, P.A., 1976. Properties and management of soil in tropics. New York: John Wiley. 619p.

78. Santos, F.A.P.; Martinez, J.C.; Voltolini, T.V.; Nussio, C.M.B. 2003. Associação de plantas forrageiras de clima temperado e tropical em sistemas de produção animal de regiões sub-tropicais. In: Simpósio Sobre Manejo Da Pastagem; Produção Animal Em Pastagem, Situação Atual E Perspectivas, 20., Piracicaba, 2003. Anais. Piracicaba: FEALQ, p.215.

79. Scott, N.M.; Watson, M.E.; Caldwell, K.S. 1983. Response of grassland to the application of sulphur at two sites in Northeast Scotland. Journal of the Food and Agriculture, 34: 357-361.

80. Soares Filho, C.V.; Monteiro, F.A.; Corsi, M. 1992. Recuperação de pastagens degradadas: 1. Efeito de diferentes tratamentos de fertilização e manejo. Pasturas Tropicales, 14: 2-6.

81. Sousa, D.M.G.; Lobato, E.; Rein, T.A. 2004. Adubação com fósforo. In: Sousa, D.M.G.; Lobato, E. (Eds.) Cerrado correção do solo e adubação. 2.ed. Brasília: EMBRAPA, p.147-168.

82. Sousa, D.M.G.; Martha Júnior, G.B.; Vilela, L. 2004. Manejo da adubação fosfatada em pastagens. In: Pedreira, C.G.S.; Moura, J.C.; Faria, V.P. Fertilidade do solos para pastagens produtivas. Piracicaba: FEALQ, p. 155-215.

83. Stevens, R.J. 1985. Evaluation of the sulphur status of some grasses for silage in Northern Ireland. Journal of Agricultural Science, 105: 581-585.

84. Teitzel J.K., Gilbert M.A. And Cowan R.T. 1991. Sustaining productive

pastures in the tropics. 6. Nitrogen fertilized grass pastures. Tropical Grasslands, 25, 111–118.

85. Tognon, A.A., Demattê, J.L.I.; Demattê, J.A.M. 1998. Teor e distribuição da matéria orgânica em latossolos das regiões da floresta amazônica e dos cerrados do Brasil Central. Scientia Agricola, 55(3): 343-354.

86. Vicente-Chandler J., Abruña F., Caro-Costas R., Figarella J., Silva S. And Pearson R.W. 1974. Intensive grassland management in the humid tropics of Puerto Rico. Bulletin 23, University of Puerto Rico, Rio Piedras, Puerto Rico: University of Puerto.

87. Vicente-Chandler, J.; Silva, S.; Figarella, J. 1959. The effect of nitrogen fertilization and frequency of cutting on the yield and composition of three tropical grasses. Agronomy Journal, 51: 202-206.

88. Vilela, L.; Soares, W.V.; Sousa, D.M.G.; Macedo, M.C.M. 2002. Calagem E Adubação Para Pastagens. Sousa, D.M.G.; Lobato, E. Cerrado: correção do solo e adubação. Planaltina, Embrapa Cerrados. p.367-382.

89. Vilela, L.; Sousa, D. M. G.; Silva, J. E. 2002. Adubação Potássica. In: Souza. D. M. G.; Lobato, E. Cerrado: correção do solo e adubação. Planaltina: Embrapa Cerrados, p.169-183.

90. Vitti, G.C.; Novaes, N.J. 1986. Adubação Com Enxofre. In: Simpósio Sobre Calagem E Adubação De Pastagens, 1., Nova Odessa. Anais. Piracicaba: POTAFOS, 1986. p.191-231.

91. Volpe, E.; Marchetti, M.E.; Macedo, M.C.M. et al. 2008. Acúmulo de forragem e características do solo e da planta no estabelecimento de capim-massai com diferentes níveis de saturação por bases, fósforo e nitrogênio. Revista Brasileira de Zootecnia, 37: 228-237.

92. Werner, J.C.; Paulino, V.T.; Cantarella, H.; Andrade, N.O.; Quaggio, J.A. 1996. Forrageiras. In: Raij, B. Van; Cantarella, H.; Quaggio, J.A.; Furlani, A.M.C. Recomendações de adubação e calagem para o Estado de São Paulo. Campinas, Instituto Agronômico/Fundação IAC, p.261-273 (Boletim Técnico, 100).

93. Werner, J.C.; Pedreira, J.V.S.; Caiele, E.L. 1977. Estudo de parcelamento e níveis de adubação nitrogenada com capim-pangola (Digitaria decumbens Stent). Boletim da Indústria Animal, 24: 147-151.

94. Whitehead, D.C. 1995. Volatilization of ammonia. p.152-179. In: Whitehead, D.C. ed. Grassland nitrogen. CAB International, Wallingford, England.

Chapter 10

EFFECTS ON SOIL FERTILITY AND MICROBIAL POPULATIONS OF BROADCAST-TRANSPLANTING RICE SEEDLINGS IN HIGH STANDING-STUBBLE UNDER NO-TILLAGE IN PADDY FIELDS

Ren Wan-Jun, Huang Yun, and Yang Wen-Yu

Sichuan Agricultural University, Wenjiang, Sichuan, China

INTRODUCTION

No-tillage broadcast-transplanting of rice is a new cultivation technology developed in recent years (Xia, 2003; Liu et al., 2002). Our system involves no-tillage of rice paddies, which leaves high standing stubble, and raising seedlings on dryland beds. Then, we developed the technology of broadcast-transplanting seedlings in the field with high standing-stubble under no-tillage condition (Fig.1~4), which has several advantages such as lower fuel costs, savings in labour and time (yang et al., 2000; Liu, 2006; Ren et al., 2008). The first experiments of broadcast-transplanting of rice began in 1950s, reported by Peiris (1956) in Sri Lanka. In the 1980s, broadcast-transplanting of rice was successfully used in China (Zhang et al., 1993). From the middle and late 1990s, Guangdong province (Liu et al., 2002) and Guangxi province (Jiang et al., 2005) experimented with no-tillage broadcasttransplanting of rice under two rice crops and Sichuan province (Liu and Li, 2002) under single indica rice crop. Compared with conventional tillage broadcast-transplanting, the notillage broadcast-transplanting rice grew slower and had fewer tillers at the early stage (Liu et al., 2002) , as well, root biomass and length were reduced during the seedling standing period (Jiang et al., 2005), but it brought to a higher spike-bearing rate and more grains.

Residue management is an important component of this new rice production system (broadcast-transplanting seedlings into high standing-stubble under no-tillage condition). In addition to the 20 to 50 cm of high stubble left upon harvest of the previous crops (wheat or rape) harvested, the farmer is encouraged to return the harvested stubble from previous crops to the field (yang et al., 2000; Ren et al., 2003). This practice increases the soil fertility: soil organic matter, total N and K, and available N and K concentrations were higher than those of conventional tillage paddy fields, either with conventional rice transplanting or broadcast-transplanting rice in one study (Ren et al., 2007). This promising result needs to be verified in other paddy fields, to determine the effects of broadcast-transplanting seedlings into no-tillage fields with high standing-stubble on soil fertility and microbial populations in paddy field. The objectives of this study were to clarify the influence of the new technology on soil fertility and microbial populations of paddy field, and to provide a scientific basis and give some evidence in the introduction – what type of yields can be achieved with the broadcast-transplanting method in a no-tillage field?

Figure 1: Field covered with high standing-stubble from the previous rape crop

Figure 2: Broadcasting rice seedlings into high standing-stubble

Figure 3: Early growth stage of rice seedlings

Figure 4: Straw decay and soil surface conditions after rice harvest

MATERIALS AND METHODS

Plant Material and Condition

Ya' an experiment: The hybrid rice combinations Gangyou22, IIyou162 and Kyou047 were the cultivars used in this study.The experiments were conducted in the paddy at Daxing town of Ya' an city, Sichuan province, China (lat. 29°59 N, long. 102°59 E) in 2003. The experiment was on a sandy-loam soil with the following chemical properties: organic matter 23.66 g kg^{-1}; total N 1.422 g kg^{-1}, alkali hydrolysable N 109.36 mg kg^{-1}; total P 0.458 g kg^{-1}, available P 12.01 mg kg^{-1}; total K 11.66 g kg^{-1}, available K 30.22 mg kg^{-1}. Previous crop was rape.

Pixian experiment: The hybrid rice combination IIyou21 were used as plant materials in this study. The experiment was conducted in the paddy at Gucheng town of Chengdu city, Sichuan province, China (lat. 30°93 N, long. 103°91 E) in 2005. The experiment was on a fluviatile loamsandy soil, and its chemical properties were as follows: organic matter, 18.73 g kg^{-1}; total N, 1.18 g kg^{-1}, alkali hydrolysable N, 78.26 mg kg^{-1}; total P 0.21 g kg^{-1}, available P, 43.53 mg kg^{-1}; total K, 22.73 g kg^{-1}, available K, 42.21 mg kg^{-1}. Previous crop was wheat.

Experimental Treatment and Cultivation Management

Ya' an experiment: The experiment was laid out in a split plot design with tillage and cultivation methods in the main plots and hybrid combination in the sub-plots with three replicates. Plot area was 15.84 m2. In the tillage and cultivation methods, there were three levels, including as conventional tillage and transplanting (CTT), conventional tillage and broadcast-transplanting of seedlings (CTB) and broadcast-transplanting of seedlings in the field with high standing-stubble under no-tillage condition (BSNT). The hybrid combination tested were Gangyou22, IIyou162 and Kyou047. The seedlings of CTT were raised on wet land and seeded at a rate of 30 g m^{-2}; those of CTB and BSNT were raised on plastic trays with 428 holes and each hole was seeded about 2 seeds. Seeds were sown on March 30, then transplanted on May 9. When previous rape was harvested, 50 cm of stubble was left in the field. The soil was plowed and plugged after removing the rape stubble in tillage treatments; for no-tillage treatments, a herbicide was used 5 days after harvesting rape. Then paddies were irrigated to reach soil water saturation prior to broadcast-transplanting. Fertilizer inputs as a basal dressing were at the following rates: 90 kg P_2O_5 ha^{-1} as superphosphate, 45 kg K_2O ha^{-1} as KCl and 105 kg N ha^{-1} as urea. Additional 45 kg N ha^{-1} was broadcasted at the tillering stage and 45 kg K_2O ha^{-1} at the booting stage. The transplanting density was 30×10^4 hills ha^{-1}. Irrigation water was kept at 1-5 cm depth for 1 week following transplanting to aid in seedling establishment. Pests were controlled according to the standard recommendation and other rice management was similar as that in the paddy field (Ren et al., 2003).

Pixian experiment: The experiment was carried out in a randomized plot design with four treatments: No-tillage+Straw (NT+S), No-tillage (NT), Tillage+Straw (T+S) and Tillage (T). There were three replicates of each treatment, in plots of 30 m^2 . Ridges were built between plots to avoid fertilizer and water movement to adjacent plots. The +S treatment involved returning straw to soil that was removed when the previous wheat crop was harvested (leaving 25 cm of stubble in the field). After rice seedlings was survived, the straw was returned and scattered evenly in the plots. Returned straw provided an organic residue input of about 5-6 t ha^{-1}. In the NT+S treatment, straw was left on the soil surface, but the T+S treatment involved cultivating and mixing straw with soil. For tillage treatments, we plowed and plugged the soil after removing the wheat stubble. In the no-tillage treatments, a herbicide was used 5 days after harvesting wheat. Then we irrigated the paddy until it reached soil water saturation prior to broadcast-transplanting. Seeds were sown on March 20, then transplanted on May 23. Fertilizer inputs as a basal dressing were at a rate of 75 kg P_2O_5 ha^{-1} as superphosphate, 75 kg K_2O ha-1 as KCl and 105 kg N ha^{-1} as urea. Additional 30 kg N ha-1 was broadcasted at the tillering stage and 75 kg K_2O ha-1 and 15 kg N ha-1at the booting stage. The transplanting density was 25.5×10^4 hills ha^{-1}. Irrigation water was kept at 1-5 cm depth for 1 week following transplanting to aid in seedling establishment. Pests were controlled according to the standard recommendation and other rice management was similar as that in the paddy field (Ren et al., 2003).

Soil and plant analysis

Soil fertility

Samples for nutrient measurement were sampled randomly with soil drill (25 mm in diameter).Five cores were taken after rice harvest from the 0-10 cm (upper layer) and 10- 20 cm (deep layer) from each plot and pooled together. Large pieces of plant and animal resideues, gravel, etc., were removed by sieving soil through a 2-mm mesh, mixing and a subsample was taken for analysis. Concentrations of soil organic matter, total N, P and K, alkali hydrolysable N, available P and K were tested by the methods described by Lu (2000).

Soil microorganism enumeration

Soil samples for microbiological assessment and cellulose decomposition intensity were collected from the Pixian experiment only at the tillering stage, elongation stage, booting stage and maturity stage of rice from the 0-10 cm (upper layer) and 10-20 cm (deep layer) from each plot and pooled together. Samples were cooled with ice packs in the field immediately after collection, and were stored at 4°C for later analysis. Microbial communities were enumerated using the dilution plate method. Bacteria was isolated by BF medium, fungi was cultured by Martin agar medium and actinomyces were determined by modified Goss I medium (Department of Microbiology, 1985). Cellulose decomposition intensity was tested by the method of burying fabric into situ in the laboratory (Department of Microbiology, 1985).

Rice development and yield

Sixty (60) hill plants from each plot were investigated to study panicle development. Five (5) hills from each plot were harvested at maturity to estimate the yield components. The grain filling percentage was determined according to Zhu et al (1995). All plants from each plot were harvested at maturity for the determination of grain yield.

Statistical method

Statistical analyses were made by Office Excel 2003 and the SAS-Stat package (SAS Institute Inc. 1996). The Duncan test and 1east significant difference test at the 90% and 95% confidence levels were used to compare treatment means.

RESULTS

Effects on soil fertility of tillage and residue management in a system with broadcast-transplanting of rice seedlings

Soil fertility was increased when rice seedlings were broadcast-transplanted among high standing-stubble under no-tillage condition (BSNT), as the organic matter, total N and K, and available N and K concentrations were higher than those of conventional tillage systems with conventional transplanting or broadcast-transplanting (Table1).

Table 1: Effects of different tillage and transplanting methods on soil fertility (0-20 cm depth) in paddy field. The experiment was conducted near Ya' an city, Sichuan province, China in 2003.

Treatment	Organic matter ($g \cdot kg^{-1}$)	Total N ($g \cdot kg^{-1}$)	Total P ($g \cdot kg^{-1}$)	Total K ($g \cdot kg^{-1}$)	Alkali hydrolysable N ($mg \cdot kg^{-1}$)	Available P ($mg \cdot kg^{-1}$)	Available K ($mg \cdot kg^{-1}$)
CTT	23.42	1.444	0.604	14.24	109.89	15.88	33.45
CTB	23.31	1.506	0.583	12.55	105.29	13.66	36.59
BSNT	30.04	2.375	0.524	15.18	131.72	14.25	61.93
Initial soil fertility	23.66	1.422	0.458	11.66	109.36	12.01	30.22

CTT: conventional tillage and conventional transplanting; CTB: conventional tillage and broadcast-transplanting of seedlings; BSNT : broadcast-transplanting of seedlings in the field with high standing-stubble under no-tillage condition.

As shown in Table 2, in the upper 0-10 cm soil layer, the organic matter content for 'notillage + returning straw' treatment was 5.33, 2.79 and 5.37 $g \cdot kg^{-1}$ higher than that for 'notillage', 'tillage + returning straw' and 'tillage' treatment, respectively. However, in the deep 10-20 cm layer, content of organic matter in 'tillage + returning straw' and 'tillage' treatments were higher than others. Since residues are mostly left on the surface in the notillage treatment, there organic matter accumulation was found in the surface soil, whereas a tilled soil had organic matter incorporated into the deeper soil layer. The maximum difference of two soil layers was noted for the 'no-tillage + returning straw' treatment, and there was little difference in soil organic matter in the 'no-tillage' treatment, and 'tillage' treatment without the extra straw residue input.

In upper soil layer, total and available N, P and K concentrations were greatest in the 'notillage + returning straw' treatment, those in the 'no-tillage' treatment and 'tillage + returning straw' treatment followed, and those in the 'tillage' treatment were the lowest. In the deep layer, the soil fertility indicators for 'tillage + returning straw' treatment were greater than those for other treatments. At the same time, the difference of nutrient status between two soil layers was the maximum for 'no-tillage + returning straw' treatment, therefore, the 'no-tillage + returning straw' treatment enriched soil fertility in the surface soil layer.

Table 2: Effect of different tillage methods on soil fertility in paddy field with broadcasttransplanting of rice. The upper soil layer was 0-10 cm depth and the lower soil layer was 10-20 cm depth. The experiment was conducted at Pixian near Chengdu city, Sichuan province, China in 2005.

Soil layer	Treatment	Organic matter (g·kg^{-1})	Total N (g·kg^{-1})	Total P (g·kg^{-1})	Total K (g·kg^{-1})	Alkali hydrolysable N (mg·kg^{-1})	Available P (mg·kg^{-1})	Available K (mg·kg^{-1})
Upper layer	NT+S	30.02a	1.67a	0.260a	29.114a	120.250a	82.593a	81.063a
	NT	24.69b	1.62ab	0.255a	26.556a	113.436b	70.739b	71.095ab
	T+S	27.23ab	1.59b	0.254a	22.900b	110.966c	69.013bc	64.718ab
	T	24.65b	1.56b	0.244b	21.920c	101.778d	64.001c	60.497b
Deep layer	NT+S	20.69a	1.35b	0.223a	11.255b	85.162a	40.305a	48.983a
	NT	18.12b	1.34b	0.221a	10.241b	84.572a	40.622a	46.781a
	T+S	22.20a	1.42a	0.224a	18.784a	89.686a	42.890a	54.674a
	T	22.20a	1.40ab	0.226a	16.345a	87.918a	40.755a	51.640a

NT+S: No-tillage+Straw; NT: No-tillage; T+S: Tillage+Straw; T: Tillage. Values followed by different small letters meant significant difference at 0.05 level, respectively (LSD test).

Effects on soil microorganisms of tillage and residue management in a system with broadcast-transplanting of rice seedlings

Bacterial Numbers

The average numbers of soil bacteria under different treatments are given in Table 3. The results showed that soil bacterial numbers for 'returning straw to soil' treatments were higher than that for 'no returning straw' treatments, and those in upper soil layer were higher than in deep layers too. In upper soil layer, the highest soil bacteria' numbers appeared in the 'no-tillage + returning straw' treatment at five growth stages, and the lowest bacteria' numbers were in the soil under 'tillage'. Bacterial numbers for 'no-tillage + returning straw' treatment were 15.86%, 14.0%, 2.53% and 40.44% higher than that for 'notillage', and 22.62%, 13.25%, 6.32% and 29.05% higher than that for 'tillage + returning straw' treatment at the tillering stage, elongation stage, booting stage and maturity stage of rice, respectively. In the deep soil layer, bacterial numbers appeared lower at tillering stage, and increased during elongation to booting stage, reaching a maximum value at rice booting and declined thereafter. In this soil layer, the highest bacterial numbers appeared in the 'tillage + returning straw' treatment, and the lowest bacteria' numbers were in the soil treated with 'tillage', too.

Fungal numbers

The data in Table 3 showed that fungal populations assessed by direct counting were lower than the other microbial groups. Fungal numbers increased during the rice growing season. Soil fungi were more abundant in the upper soil layer than in the deep layer, and greater in 'returning straw to soil' treatments than that for 'no returning straw' treatments. In the upper soil layer, fungal numbers were the highest in the 'no-tillage + returning straw' treatment, followed by the 'no-tillage' treatment, and the lowest fungal numbers were in the soil treated with 'tillage'. In deep soil layer, the order of fungal numbers in all treatments was 'tillage + returning straw'>'no-tillage + returning straw' >'no-tillage'> 'tillage'.

Table 3: Effect of different tillage methods on soil microbial population in paddy field with broadcast-transplanting of rice. The upper soil layer was 0-10 cm depth and the lower soil layer was 10-20 cm depth. The experiment was conducted at Pixian near Chengdu city, Sichuan province, China in 2005

Soil layer	Treatment	Bacteria (×10^3 CFU·g^{-1})				Fungi (×10^3 CFU·g^{-1})				Actinomyce (×10^3 CFU·g^{-1})			
		Tillering	Elongation	Booting	Maturity	Tillering	Elongation	Booting	Maturity	Tillering	Elongation	Booting	Maturity
Upper layer	NT+S	168	171	202	191	2.6	2.89	5.16	6.23	130	108	155	189
	NT	145	150	197	136	1.6	2.64	4.35	5.29	135	90	120	134
	T+S	137	151	190	148	1.2	1.7	3.38	4.82	134	72	104	108
	T	85	103	157	94	0.8	1.44	2.56	4.45	104	44	96	97
Deep layer	NT+S	72	89	132	94	1	1.1	2.7	3.1	50	30	67	57
	NT	75	78	138	98	0.5	0.7	2.3	2.2	46	27	59	48
	T+S	83	99	143	100	1.1	1.5	2.9	3.9	64	41	77	66
	T	64	70	109	63	0.3	0.5	1.5	2.7	32	12	48	39

NT+S: No-tillage+Straw; NT: No-tillage; T+S: Tillage+Straw; T: Tillage.

Actinomycete numbers

Actinomycete numbers are affected by gas permeability of soil because that this microbial group are facultative aerobes. The data in Table 3 showed that actinomycete numbers were the lowest at elongation stage, increasing thereafter and reaching a peak at booting stage (deep soil layer) or maturity stage (upper soil layer) of rice. In upper soil layer, 'no-tillage + returning straw' treatment appeared to possessaerobic microsites, therefore, it had the highest actinomycete numbers except at tillering stage. The lowest actinomycete numbers were in the 'tillage' treatment, presumably because conditions were not favorable for aerobic microorganisms, particularly in the deep soil layer, which had fewer actinomycetes than in the upper soil layer. In deep soil layer, actinomycetes abundance in four treatments followed the order 'tillage + returning straw'>'no-tillage + returning straw' >'no-tillage'> 'tillage'.

Effects on cellulose decomposition intensity of tillage and residue management in a system with broadcast-transplanting of rice seedlings

The cellulose decomposition intensity under as affected by residue management and tillage at the Pixian experiment was shown in Fig. 5. The highest cellulose decomposition intensity appeared at the tillering stage and booting stage during different growth periods of rice. At the same time, cellulose decomposition intensity in 'returning straw to soil' treatments were higher than 'no returning straw' treatments. In upper soil layer, cellulose decomposition intensity in 'no-tillage + returning straw' treatment was always higher than that in other treatments. At the tillering stage of rice, cellulose decomposition intensity in 'no-tillage + returning straw' treatment was 28.20%, 39.76% and 52.93% higher than that in 'tillage + returning straw', 'no-tillage' and 'tillage' treatment; furthermore, 25.51%, 46.27% and 91.62% higher at elongation stage of rice; 38.76%, 68.60% and 79.71% higher at booting stage of rice; and 26.44%, 79.01% and 98.15% higher at maturity stage of rice, respectively. There was little difference between 'no-tillage' and 'tillage' treatment. In the deeper soil layer, the highest cellulose decomposition intensity was the treatment of 'tillage + returning straw', probably due to incorporation of straw into the deep soil layer, which enhanced the activity of cellulose decomposing bacteria.

Correlation between soil microorganisms and soil fertility in paddy field with broadcast-transplanting of rice seedlings in no-tillage and conventional tillage plots

The relationship between soil microorganisms and soil fertility at the Pixian experiment was analyzed (Table 4). There were significant positive correlations between soil bacterial numbers and the soil organic matter, total N, total P, total K, alkali hydrolysable N and available P concentrations. In contrast, soil fungi numbers were not correlated with soil fertility indicators.

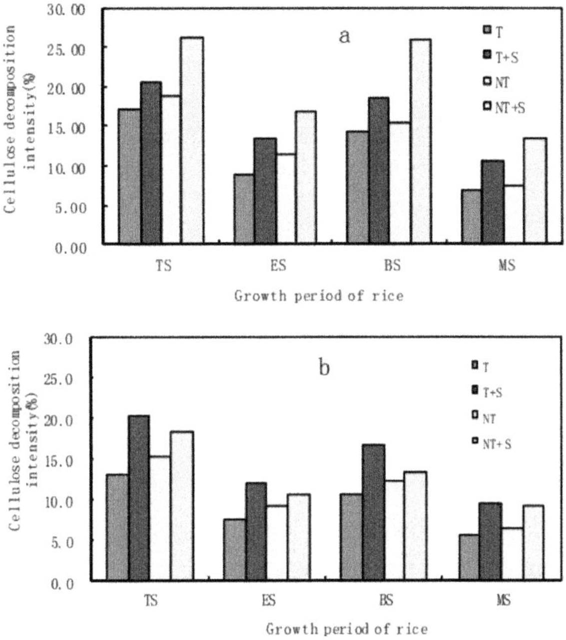

TS: Tillering stage; ES: Elongation stage; BS: Booting stage; MS: Maturity stage. NT+S: No-tillage+Straw; NT: No-tillage; T+S: Tillage+Straw; T: Tillage.

Figure 5: Effect of different tillage and residue management methods on cellulose decomposition intensity in paddy field with broadcast-transplanting of rice. Results are for (a) the upper soil layer, 0-10 cm depth and (b) the lower soil layer, 10-20 cm depth. The experiment was conducted at Pixian near Chengdu city, Sichuan province, China in 2005.

There were significant positive correlations between soil actinomycete numbers and the soil organic matter, total N, total P, total K, alkali hydrolysable N, available P and K concentrations. The cellulose decomposition intensity was significantly and positively correlated with the soil fertility indicators, which indicated that high cellulose decomposition intensity was related to improved soil nutrient status.

Table 4: Correlation (r values) between soil microorganisms and soil fertility in paddy field with broadcast-transplanting of rice seedlings in no-tillage and conventional tillage plots. The experiment was conducted at Pixian near Chengdu city, Sichuan province, China in 2005

Factor	Total K	Total N	Total P	Available K	Alkali hydrolysable N	Available P	Organic matter
Bacteria	0.5858*	0.6447*	0.7107**	0.5253	0.6610*	0.5625*	0.7139**
Fungi	0.3956	-0.0925	0.5613	-0.6073	-0.4090	-0.5324	0.0360
Actinomycete	0.8217**	0.8310**	0.8393**	0.7307**	0.8404**	0.7552**	0.6110*
Cellulose decomposition intensity	0.6781*	0.7854**	0.8848**	0.8535**	0.8339**	0.8623**	0.8818**

All the data was used for correlation analysis.*$P<0.05$; **$P<0.01$.

Effects on rice yield and its components in paddy field with contrasting tillage and transplanting methods, including broadcast-transplanting of rice seedlings

Analysis of variance showed that the main effect of yield by tillage and transplanting methods was significant at $P<0.1$ ($F_A=3.17$, $F_{0.05}=3.63$, $P=0.069$) and yield followed the order: CTT>BSNT>CTB (Table 1). There was no difference due to the hybrid grown ($F_B=0.36$, $F_{0.05}=3.63$) and interaction of the hybrid by tillage and transplanting method ($F_{A\times B}=1.99$, $F_{0.05}=3.01$) was not significant. Among three tillage and transplanting methods, effective panicles were affected as follows: CTB>CTT>BSNT, and there were significantly more effective panicles in CTB than in BSNT ($F_A=4.06$, $F_{0.05}=3.63$, $F_{0.01}=6.23$). Under three tillage and transplanting methods, effective panicles of Kyou047 were the highest among three combinations. There were significant differences in the number of spikelets among different tillage and transplanting methods and between hybrids ($F_A=12.30$, $F_B=50.68$, $F_{0.01}=6.23$), and BSNT had 8.74% and 15.27% more spikelets than CTT and CTB, respectively.

Table 5: Effect on rice yield and yield components of rice of different tillage and trans-plantingmethods, and hybrid types, in a paddy field. The experiment was conducted near Ya' an city, Sichuan province, China in 2003

Treatment		Effective panicle (No. m-2)	Spikelet (No. ear-1)	Seed-setting rate (%)	Filled grain percentage (%)	1 000-grain weight (g)	Yields (kg.m-2)
	Gangyou22	175.2 b	167.6 a	87.69 a	77.67 a	26.31 a	0.683 a
CTT	IIyou162	189.0 b	179.4 a	77.23 b	73.71 a	26.65 a	0.689 a
	Kyou047	214.7 a	140.1 b	89.20 a	77.93 a	26.46 a	0.697 a
Mean		193.0 ab	162.4 b	84.71 a	76.44 ab	26.47 a	0.690 a
	Gangyou22	177.4 b	155.5 a	82.16 a	73.56 b	25.57 a	0.650 ab
CTB	IIyou162	184.6 b	172.4 a	72.90 b	72.20 b	25.49 a	0.596 b
	Kyou047	227.0 a	131.8 b	80.90 a	78.47 a	25.68 a	0.676 a
Mean		196.3 a	153.2 b	78.65 b	74.74 b	25.58 b	0.641 b
	Gangyou22	172.7 b	187.6 a	82.35 b	77.58 ab	26.77 a	0.676 a
BSNT	IIyou162	172.0 b	201.0 a	79.48 b	75.29 b	26.98 a	0.677 a
	Kyou047	205.5 a	141.1 b	87.12 a	80.57 a	26.60 a	0.628 a
Mean		183.4 b	176.6 Aa	82.99 a	77.81 a	26.78 a	0.660 ab

CTT: conventional tillage and transplanting; CTB: conventional tillage and broadcasting seedlings; BSNT:broadcasting seedlings in the field with high standing-stubbles under no-tillage condition.

Among three combinations, IIyou162 had the greatest number of spikelets, and Kyou047 had the fewest. CTB had the lowest seed-setting rate, at 6.06% and 4.34% less seed-setting rate than CTT and BSNT, and this was significant ($P<0.05$). Among three hybrids, IIyou162 had the lowest seed-setting rate. BSNT had the highest percentage of filled grain, which was significantly ($P<0.05$) higher than that of CTB; filled grain percentage of Kyou047 was also significantly ($P<0.05$) higher than Gangyou22 and IIyou162. The 1000-grain-weight values were ranked BSNT>CTT>CTB, and those of BSNT and CTT were 1.20 g and 0.89 g higher than CTB, but no significant difference due to or hybrid combinations was detected for 1000- grain-weight.

DISCUSSION

Our results showed that the method of broadcast-transplanting seedlings in the field with high standing-stubble and extra straw residues under no-tillage condition ('no-tillage + returning straw' treatment) promoted soil microbial populations at the surface soil layer because it returned about 5-6 t ha^{-1} of straw to soil. Larger soil microbial populations are attributed to a more favorable soil ecological environment, with good conditions of water, nutrients, gas exchange and heat, that were beneficial for soil microorganisms. Therefore, the populations of bacteria, fungi, actinomycetes and cellulose decomposition intensity in the upper soil layer (0-10 cm depth) of the 'no-tillage + returning straw' treatment were the highest among four tillage and residue management systems. However, the cultivation method of 'tillage + returning straw' had a highest soil microbial communities in deep soil layer because it incorporated the straw input into the deeper soil layer (10-20 cm depth) and promoted soil microbial growth and activity.

Soil quality is a concept that generally refers to the soil's ability to sustain productivity, environmental quality, human and animal health (Doran and Parkin, 1994). Therefore, analysis of soil quality should consist of a minimum data set that includes measures of soil physical, chemical, and biological properties (Papendick et al., 1994). Soil microorganisms are important for maintaining soil quality due to their role in decomposition of organic matter and nutrient cycling and storage, and potentially represent a very sensitive biological marker (Turco et al., 1994). Morris and Boerner (1999) suggested that the spatial distributions of soil microorganisms and the factors affecting them should be further investigated since both soil chemistry and vegetation are affected by soil microbial communities. Different tillage systems have significantly effect on soil microorganism (Gao et al., 2004; Balesdent et al., 1990; Wei et al., 1993). Ridge no-tillage was advantageous to improve soil ecological environment and soil fertility in paddy field (Fan and Liu, 2002).

The straw return significantly improved soil structure and increased soil nutrient concentration (Fan and Liu, 2002), and increased soil organic matter and available N, P, and K (Luo and Zhang, 1999), furthermore, the straw's cellulose, hemicelluloses, lignin and other components were decomposed slowly to release nutrients in the field with no-tillage condition, so there was a higher soil fertility later in the rice growth period. In the upper soil layer, the organic matter content for 'no-tillage + returning straw' treatment was 5.33, 2.79 and 5.37 $g \cdot kg^{-1}$ higher than that for 'no-tillage', 'tillage + returning straw' and 'tillage' treatment, respectively. Also, the concentration of total and available N, P and K were greatest in the 'no-tillage + returning straw' treatment, followed by the 'no-tillage' treatment and 'tillage + returning straw' treatment, and the lowest concentrations were in the 'tillage' treatment. In the deep layer, the fertility indicators for 'tillage + returning straw' treatment were higher than those for other treatments. The populations of bacteria, actinomycetes and cellulose decomposition intensity were significantly and positively correlated with the soil fertility indicators, which indicated that soil bacteria and actinomycetes were especially important for improving soil nutrient availability. The method of broadcast-transplanting seedlings in the field with high standing-stubble under no-tillage condition increased soil nutrient concentrations and soil microbial communities, and delayed senescence of root and shoots in the later growth stage (Xiao et al., 2009), and prolonged the grain-filling time, therefore, the spikelets of per ear, seed-setting rate, and 1 000-grains weight were increased, in comparison with conventional tillage and broadcasting seedlings. The findings provide insight into the physiological and ecological mechanisms whereby stable and high yield of broadcast-transplanted rice can be achieved in paddy field with high standing-stubble under no-tillage condition.

ACKNOWLEDGEMENTS

This study was supports by Chinese Ministry of Science and Technology (2004BA520A05; 2006BAD02A05; 2011BAD16B05).

REFERENCES

1. Xia J Y. Development and countermeasure about no-tillage cast-transplanted technology of high quality rice. China Agricultura Technology Extension, 2003, (6): 9-11 (in Chinese)
2. Liu J, Huang H, Fu H, et al. Physiological mechanism of high and stable yield of notillage cast-transplanted rice. Sci Agric Sin, 2002, 35: 152-156 (in Chinese)
3. Yang W Y, Ren W J. The new technology of broadcasting in the field with high standingstubbles under no-tillage condition. Sichuan Agricultural Sciences and Technology, 2000, (4): 13-14 (in Chinese)
4. Liu D Y. The results and experience about extension of no-tillage cast-transplanted rice in Sichuan province. China Rice, 2006, (1): 54-55 (in Chinese)
5. Ren W J, Liu D Y, WU J X, et al. Effect of broadcasting rice seedlings in the field with high standing-stubbles under no-tillage condition on yield and some physiological characteristics. Acta Agronomica Sinica, 2008, 34(11): 1994-2002 (in Chinese)
6. Peiris M E. Broadseedling- a promising new technique in paddy cultivation. Trop Agriculrist, 1956, 112: 105-108
7. Zhang H C, Dai Q G, Zhong M X, et al. Studies on the yield formation and ecological characters of scattered-planting rice. Sci Agric Sin, 1993, 26(3): 39-49 (in Chinese)

8. Jiang L G, Li R P, Wei S Q, et al. Root growth and stanging characteristics of Jinyou253 seedlings under no-tillage with cast transplantation. J Guangxi Agric Biol Sci, 2005, 24(1): 30-34 (in Chinese)
9. Liu J Z, Li Y K. Studies on the no-tillage and seedling-throwing culture in hybrid rice. Hybrid rice , 1999, 14(3): 33-34 (in Chinese)
10. Ren W J, Yang W Y, Liu D Y, et al. The technology of broadcasting in the field with high standing-stubbles under no-tillage condition. China Rice , 2003, (2): 22-23 (in Chinese)
11. Ren W J, Yang W Y, Fan G Q, et al. Effect of different tillage and transplanting methods on soil fertility and root growth of rice. J Soil Water Conserv, 2007, 21: 108-110,162 (in Chinese)
12. Lu R-K. Analytical Methods of Soil and Agro-Chemistry. Beijing: China Agricultural Science and Technology Press, 2000 (in Chinese)
13. Department of Microbiology, Institute of Soil Science, Chinese Academy of Sciences. Analytical Methods of Soil Microbe. Beijing: Science Press, 1985 (in Chinese)
14. Zhu Q S, Wang Z Q, Zhang Z J, et al. Study on indicators of grain-filling of rice. J Jiangsu Agric coll, 1995, 16(2): 1-4 (in Chinese)
15. Doran J W, Parkin T B. Defining and assessing soil quality. In: Doran J W, Coleman D C, Bezdicek D F and Stewart B A ed. Defining Soil Quality for a Substainable Environment. Madison, W I, USA: SSSA Special Publication Number 35, 1994, 3- 21
16. Papendick R.I., Parr J.F., and J. van Schilfgaarde. Soil quality: New perspectives for a sustainable agriculture. In Proceedings for International SoilConservation Organization. New Delhi, India, December 4-8, 1994
17. Turco R.F., Kennedy A.C., Jawson M.D. Microbial indicators of soil quality. In J.W. Doranet al. (ed.) Defining soil qualityfor a sustainable environment. SSSA Spec. Publ. 35. SSSA, Madison, WI. 1994, 73-90
18. Morris S.J., and Boerner R.E.J. Spatial distribution of fungal and bacterial biomass in southern Ohio hardwood forest soils: Scale dependency and landscape patterns. SoilBiol. Biochem. 1999, 31:887-902
19. Gao M , Zhou B-T, Wei C-F , et al. Effect of tillage system on soil animal, microorganism and enzyme activity in paddy field. Chinese Journal of Applied Ecology, 2004, 15(7): 1177-1181 (in Chinese)
20. Balesdent J, Mariotti A, Boisgontier D. Effect of tillage on soil organic carbon mineralization estimated from abundance in maize fields. Soil Science, 1990, 41: 587- 896

21. Wei C-F, Gao M, Che F-C, et al. A study on infiltrated-ridged paddy soil ecosystem. Chinese Journal of Ecology, 1993, 12(3): 26-30 (in Chinese)
22. Fan J-H, Liu M. Effect of different utilization methods on microorganism and its activation. Journal of tarim universivty of abricultural reclamation, 2002, 4(1): 15-17 (in Chinese)
23. Luo A-C, Zhang Y-S. Effect of organic manure on the numbers of microbes and enzyme activity in rice rhizosphere. Plant nutrition and fertilizer science, 1999, 5(4): 321-327 (in Chinese)
24. Xiao Q-Y, Ren W-J, Yang W-Y, et al. Effect of cultivation method of broadcasting rice seedlings in the field with standing-stubbles under no-tillage condition on senescence characteristics of leaves during late stages of rice development. Acta Agronomica Sinica, 2009, 35(8): 1562-1567 (in Chinese)

Chapter 11

INDIGENOUS FERTILIZING MATERIALS TO ENHANCE SOIL PRODUCTIVITY IN GHANA

Roland Nuhu Issaka[1], Moro Mohammed Buri[1], Satoshi Tobita[2], Satoshi Nakamura[2], and Eric Owusu-Adjei[1]

[1]CSIR-Soil Research Institute, Academy Post Office, Kwadaso-Kumasi Ghana

[2]Japan International Research Center for Agricultural Sciences, Ohwashi, Tsukuba, Japan

INTRODUCTION

Ghana is divided into six ecological zones namely; Sudan Savannah, Guinea Savannah, Forest Savannah Transition, Semi-Decideous Rainforest, High Rainforest and Coastal Savannah (Figure 1). The Guinea Savannah zone covers the whole of Upper West and Northern regions. It also occupies parts of Uppr East region and the northern part of Brong Ahafo and Volta regions. This zone has a single rainfall season lasting from May to October. Annual rainfall is about 1000 mm. The Sudan Savannah occupies the north-eastern part of Upper East region with an annual rainfall of between 500 – 700 mm. The Forest Savannah Transition lies within the middle portion of Brong Ahafo region, the northern part of both Ashanti and Eastern regions and the western part of Volta region. This zone has a bimodal rainfall with an annual rainfall of about 1200 mm. The Semi-Decideous zone cut across the northern part of Western region through southern Brong Ahafo, Ashanti and Eastern regions. It also occupies the estern part of Volta region and most parts of th Central region. It also has a bimodal rainfall with an annual rainfall of 1400mm.

Most parts of Western region is within the High Rainfall zone. A small part of Central region also falls within this zone. Annual rainfall is over 2000 mm with a bimodal partern. The Coastal Savannah stretches from Central region through Greater Accra to the Votal region. It has only one rainy season of about 600 mm.

Agriculture Productivity

Table 1 shows major crops and the respective areas on which these crops were cultivated in 2007. Due to inappropriate farming practices actual yield per unit area is less than 30% of achievable yield. The trend is similar for most crops. After several interventions including the introduction of improved varieties yield per unit area for most crops is still very low. Soil fertility has been identified as a major factor militating against crop yield. Mineral fertilizers to boost crop production are expensive and sometimes unavailable.

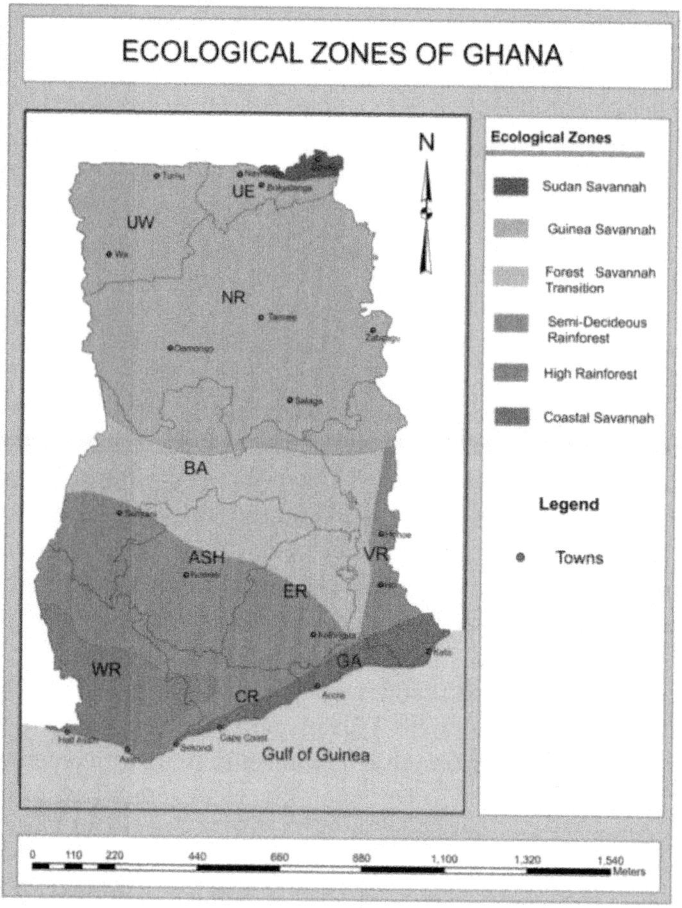

Figure 1: Ecological zones of Ghana

Table 1: Cultivated Area, Average Yield and Potential Yield of Selected Crops in Ghana

Crop	Area ('000 ha)	Average Yield (Mg/ha)	Achievable Yield (Mg/ha)	Potential Yield Gap (Mg/ha)
Cassava	886	13.8	48.7	34.9
Maize	954	1.7	6.0	4.3
Rice	162	2.4	6.5	4.1
Yam	379	15.3	49.0	33.7
Plantain	325	11.0	20.0	9.0
Cocoa	1,600	0.4	1.0	0.6

MoFA (2009).

STATUS OF SOIL FERTILITY IN GHANA

Major Soils

The Ghanaian classification (Brammer, 1962) was equated to the World Reference Base (WRB) classification, ISSS/ISRIC/FAO (1998) by Adjei-Gyapong and Asiamah (2000) as follows:

Savanna Ochrosols (WRB: Lixisols/Luvisols) – These soils occur in northern Ghana and parts of the coastal savanna. They are highly weathered and moderately to strongly acid in the surface soil. Organic matter is low (<15 gkg^{-1} soil). Soil fertility is generally low.

Forest Ochrosols (WRB: Acrisols/Alfisols/Lixisols/Ferralsols/Nitisols/Plinthosols) - These soils occur within the forest zone and parts of the forest-savanna transition. They are deep and highly weathered and are generally moderate to strongly acid in the surface soil. These soils have high organic matter content in the top horizon which may contribute significantly to the phosphorus pool, exchangeable bases and Nitrogen levels.

Forest Oxysols (WRB: Ferralsols/Acrisols) - These occur in the high rainfall zone (south-west of Western Region). These soils are deep and highly weathered but strongly acid (pH<5.0). Organic matter is very high with high potential in N and P supply. P fixation is very high due to the presence of large amounts of Al and Fe oxides.

Groundwater Laterites (WRB: Plinthosol/Planosol) - These soils occur mostly in northern Ghana. They are shallow to plinthite and low in organic matter. Soil fertility is generally poor. They have high P fixation due to the presence of abundant iron concretions.

Table 2a: Mean soil fertility characteristics of lowlands within the Guinea Savannah.

Parameter	Mean	Range	St. Dev.
pH (water)	4.6	3.7 - 7.4	0.5
Total C (g kg^{-1})	6.10	0.6 - 19	3.0
Total N (g kg^{-1})	0.65	0.1 - 1.6	0.3
C:N ratio	9.3	5.0-14.3	1.4
Available P (mg kg^{-1})	1.5	Tr - 5.4	0.9
Exchangeable K {cmol (+) kg^{-1}}	0.22	0.04 - 1.1	0.17
Exchangeable Ca {(+) kg^{-1}}	2.10	0.53 - 15	1.9
Exchangeable Mg {(+) kg^{-1}}	1.00	0.27 - 5.87	0.27
Exchangeable Na {(+) kg^{-1}}	0.12	0.1 - 0.72	0.11
Exchangeable. Acidity {(+) kg^{-1}}	1.00	0.05 - 1.80	0.48
Clay content (g kg^{-1})	66	40 - 241	39
Silt content (g kg^{-1})	607	347 - 810	107

Number of samples: 90; Source: Buri et al. 2000; Topsoil (0-20 cm)

Table 2b: Mean soil fertility characteristics of lowlands within the Semi-deciduous rainforest

Parameter	Mean	Range	St. Dev.
pH (water)	5.7	4.1 – 7.6	0.89
Organic C (g kg^{-1})	12	3.6 – 36.5	0.58
Total N (g kg^{-1})	1.1	0.30 – 3.20	0.05
C: N ratio	11	4.9 – 14.2	1.26
Available P (mg kg^{-1})	4.9	0.1 – 28.5	5.36
Exchangeable K {(+) kg^{-1}}	0.42	0.03 – 1.28	0.25
Exchangeable Ca {(+) kg^{-1}}	7.5	1.1 – 26.0	5.1
Exchangeable Mg {(+) kg^{-1}}	4.1	0.3 – 12.3	2.6
Exchangeable Na {(+) kg^{-1}}	0.32	0.04 – 1.74	0.26
Exchangeable. Acidity {(+) kg^{-1}}	0.31	0.04 – 1.15	0.29
Clay content (g kg^{-1})	127	41 - 301	8.2
Silt content (g kg^{-1})	502	187 - 770	45.8

Number of samples: 122; Source: Buri et al. 2009. Topsoil 0-20 cm

Characteristics of lowland soils: Most lowlands within the Guinea savannah and Semideciduous rainforest are mainly inland valleys and river flood plains. Rectilinear valleys occur within the Savannah agro-ecological zone while convex valleys are common within the Forest agro-ecological zone. Concave valleys, however, occur in both zones. Major soil types in these two zones are basically Gleysols and to a lesser extent, Fluvisols. Volta and Lima series are prominent within the savanna while Oda, Kakum and Temang series are prominent in the forest zone.

Soil fertility levels as observed for selected parameters are low across locations, particularly within the Savanna zone (Buri et al. 2009). Available phosphorus (P) is the most deficient nutrient in both zones. Soils of the Savanna were also observed to be quite acidic. Exchangeable Cations (K, Ca, Mg, Na) are quite moderate across locations within the Forest agro-ecology but relatively low for the Savannah, particularly Ca. Both total carbon and nitrogen levels, even though low, were comparatively higher for the forest than the savanna zone. To increase yield levels under these conditions the fertility levels of these soils must be improved.

Characteristics of upland soils: Nutrient levels for upland soils are characteristically very low (Tables 3 a, b and c). The soils are generally acidic. Soil pH values for the high rainforest zones are strongly acidic with very low exchangeable cations. Prolonged weathering and leaching under high rainfall regime has resulted in soils with very low pH regimes. In the Semi-deciduous rainforest the soils are relatively richer. Nutrient levels, however, suggest the need for improvement for any profitable production levels to be achieved. In the Savannah zone, nutrient levels are similar to the high rainfall zone. Soil pH values are, however, higher in the Savannah zone. While organic matter and nitrogen levels vary between ecologies, available P is a problem throughout the country. Low total P coupled with high fixation are major factors affecting P availability.

Table 3a: Mean soil fertility characteristics of lowlands within the High rainforest (Western region)

Parameter	Mean	Range
pH (water)	4.1	3.4-5.4
Total C (g kg^{-1})	13.0	3.9-24.3
Total N (g kg^{-1})	1.9	0.7-3.5
Available P (mg kg^{-1})	2.8	0.4-13.6
Exchangeable K {(+) kg^{-1}}	0.13	0.04-0.33
Exchangeable Mg {(+) kg^{-1}}	1.0	0.3-3.3
Exchangeable Ca {(+) kg^{-1}}	1.9	0.5-5.1
Exchangeable. Acidity {(+) kg^{-1}}	4.1	2.1-5.6
ECEC {(+) kg^{-1}}	3.1	0.9-8.1
Base saturation (%)	78	38-98
Sand (%)	-	-
Silt (%)	-	-
Clay (%)	6.1	2.1-18.1

Mean of 30 samples (0-20 cm)

Table 3b: Mean soil fertility characteristics of lowlands within the Semi-deciduous rainforest

Parameter	Mean	Range
pH (water)	5.5	4.6-6.6
Total C (g kg^{-1})	18.0	13.9-22.3
Total N (g kg^{-1})	1.3	0.8-2.5
Available P (mg kg^{-1})	6.8	2.4-18.6
Exchangeable K {(+) kg^{-1}}	0.18	0.05-033
Exchangeable Mg {(+) kg^{-1}}	2.5	1.5-4.6
Exchangeable Ca {(+) kg^{-1}}	4.9	2.0-10.1
Exchangeable. Acidity {(+) kg^{-1}}	1.5	0.2-3.6
ECEC {(+) kg^{-1}}	8.5	3.8-14.5
Base saturation (%)	85	75-95
Sand (%)	88	80-92
Silt (%)	15	5-21
Clay (%)	9	3-15

Mean of 40 samples (0-20 cm)

Table 3c: Mean soil fertility characteristics of lowlands within the Savannah agro-ecological zone

Parameter	Mean	Range
pH (water)	5.4	4.6-6.6
Total C (g kg^{-1})	12.0	2.9-18.3
Total N (g kg^{-1})	0.9	0.8-1.5
Available P (mg kg^{-1})	4.8	0.4-11.6
Exchangeable K {(+) kg^{-1}}	0.11	0.03-0.23
Exchangeable Mg {(+) kg^{-1}}	1.5	0.8-2.6
Exchangeable Ca {(+) kg^{-1}}	2.9	1.0-5.1
Exchangeable. Acidity {(+) kg^{-1}}	1.1	0.5-2.6
ECEC {(+) kg^{-1}}	5.1	2.8-8.5
Base saturation (%)	88	65-98
Sand (%)	92	82-92
Silt (%)	11	4-21
Clay (%)	7	2-12

Mean of 40 samples (0-20 cm)

The inability of farmers to buy adequate amounts of mineral fertilizers to improve their crop yields is a major factor affecting food security. Use of locally available materials for soil improvement is an option that must be fully exploited. Use of these materials (manures, dungs, crop residue, mineral deposits) will significantly improve on the soils ability to sustain higher crop yield.

Table 4: Effect of organic amendments and mineral fertilizer on paddy yield (t/ha)

Treatments	Potrikrom	Biemso No. 1	Biemso No. 2
Absolute Control	1.7a	2.6a	1.5a
90-60-60 (N-P_2O_5-K_2O) kg/ha (Urea as N source)	7.0d	7.5d	4.0ef
90-60-60 (N-P_2O_5-K_2O) kg/ha (SA as N source)	6.6d	7.3d	3.9def
Poultry Manure (7.0 t/ha)	6.0d	6.4c	3.3bc
PM (3.5 t/ha) +45-30-30 (N-P_2O_5-K_2O) kg/ha (Urea as N source)	6.3d	7.3d	3.7cde
Cattle Manure (7.0 t/ha)	4.5c	6.3c	3.4cd
CM (3.5 t/ha) +45-30-30 (N-P_2O_5-K_2O) kg/ha (Urea as N source)	4.9c	6.2c	3.7cde
Rice Husk (7.0 t/ha)	3.3b	5.5b	2.8b
RH (3.5 t/ha) +45-30-30 (N-P_2O_5-K_2O) kg/ha (Urea as N source)	4.4c	6.3c	3.3c

Source: Buri et al. 2004

Table 4 shows the effect of some of these materials (poultry manure, cow dung and rice husk) on rice yield. Sole application of these materials or in combination with mineral fertilizer increased rice yield over the control. In some instances sole application of some of these materials was as good as applying the recommended rate of mineral fertilizer. This clearly shows that the productivity of these poor soils can be improved through the use of locally available fertilizing materials.

FERTILIZING POTENTIAL OF INDIGENOUS MATERIALS

Organic waste is produced wherever there is human habitation. The main forms of organic waste are household food waste, agricultural waste, human and animal waste. The economies of most developing countries dictates that materials and resources must be used to their fullest potential, leading to a culture of reuse, repair and recycling. In many developing countries there exists a whole sector of recyclers, scavengers and collectors, whose business is to salvage 'waste' material and reclaim it for further use. Where large quantities of waste are created, usually in the major cities, there are inadequate facilities for dealing with it. Much of this waste is either left to rot in the streets, or is collected and dumped on open land near the city limits. There are few environmental controls in these countries to prevent such practices. In addition, mineral deposits can also be exploited for several purposes including agriculture, particularly soil fertility improvement.

Types of organic waste include: (a) Domestic or household waste (cooked or uncooked food scraps), (b) Agricultural residue (e.g. stover of crops, rice husks, etc.). (c) Commercially produced organic waste (waste generated from schools, hotels, restaurants etc.), (d) Human faecal residue and (e) Animal residue (dung and manures).

Plant Origin

Large amount of plant waste are annually generated in the form of plant materials. These include maize stover, rice straw, rice husk, millet/sorghum resulting from annual production of these crops. Large amounts of saw dust and wood shavings also come from the wood industry (mostly from timber firms and to a lesser extent from commercial carpenters). Maize stover is normally left on the field after harvest. On the other hand rice is generally brought to a particular spot on the farm for threshing. This results in huge amount of rice straw being put at particular spots on the farm. Rice husk is generated during milling. Large amount of rice husk (about 33 % of paddy weight) is seen in hips close to rice mills in villages or towns. These materials can be used in various forms (direct application, composting, charring or ash) to increase the productivity of the soil.

Rice straw and husk: An estimation of of rice straw and rice husk produced in Ghana in 2007 show that over 366,000 Mt of rice straw and 63,000 Mt of rice husk were produced as waste. Large amount of rice straw (> 120,000 Mt) was produced in the Northern region followed by the Volta (Figure 2A). Upper East region, Eastern and Western regions also produced substantial quantities of rice straw. The trend for rice husk is similar to rice straw but of lower quantities.

Nutrient equivalent of rice straw and husk is presented in Figure 2B. About 2528 Mt of N, 990 Mt of P_2O_5, 5,459 Mt of K_2O (Table 2) is potentially available in these materials. Large amount of calcium and magnesium are also potentially available. Over 50% of these nutrients are potentially available in the large amounts of rice straw and husk produced in the Northern region. These materials can be used as amendments to improve the productivity of lowland soils, for higher grain yields of rice. Particularly the very low organic matter and P status of lowland soils in Northern Ghana (savanna zone) can be improved through effective management of these materials. Total amount of rice straw and husk generated in 2007 is presented in Table 5. These materials can be used to improve both physical and chemical properties of the soil.

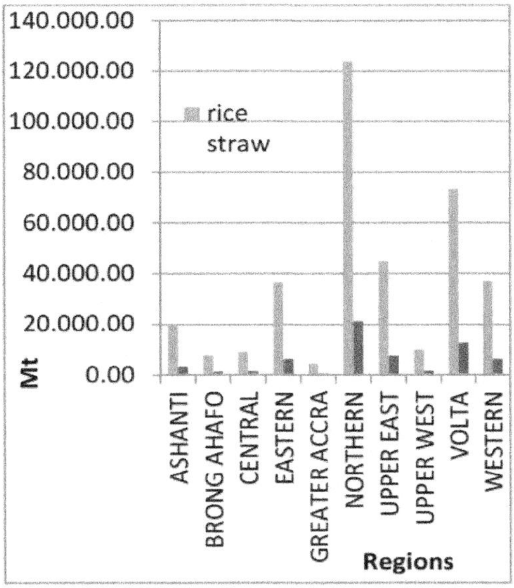

(a)

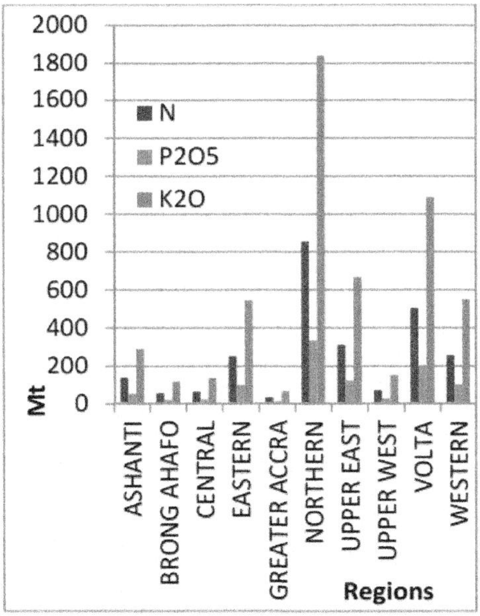

(b)

Figure 2: Quantity of rice straw and husk (A) and nutrient equivalent (B) for the various regions

Table 5: Estimated quantities of rice straw and husk

Organic source	Quantity (Mt)	N (Mt)	P_2O_5 (Mt)	K_2O (Mt)	CaO (Mt)	MgO (Mt)
Straw	366,975.2	1,834.9	587.2	5,137.7	-	-
Husk	63015.9	693.2	403.3	321.4	352.9	894.8
Total	429,991.1	2,528.1	990.5	5,459.1	352.9	894.8

Saw dust: Generally timber firms are located in the Brong Ahafo, Ashanti and Eastern regions. Some can also be found in the Central and Western regions. Large hips of saw dust can be located on the out skirts of towns or cities where these factories are located. Saw dust constitute a good source of fertilizing material when properly managed.

Animal Origin

Animal waste that can be classified as fertilizing materials include, poultry manure, cow dung, manures from sheep, goats and pigs. Human excreta is another source of fertilizing material. The amount of cow dung produce annually is far larger than poultry manure (Figure 3). Cattle are mostly reared in Northern Ghana (Northern, Upper East and Upper West regions) and hence large amounts of cow dung are obtained from these areas. The three regions together produce over 1.2 million Mt of cow dung annually. On the other hand poultry rearing is more concentrated in Greater Accra and Ashanti regions and like cow dung each of the regions produce some amount of it.

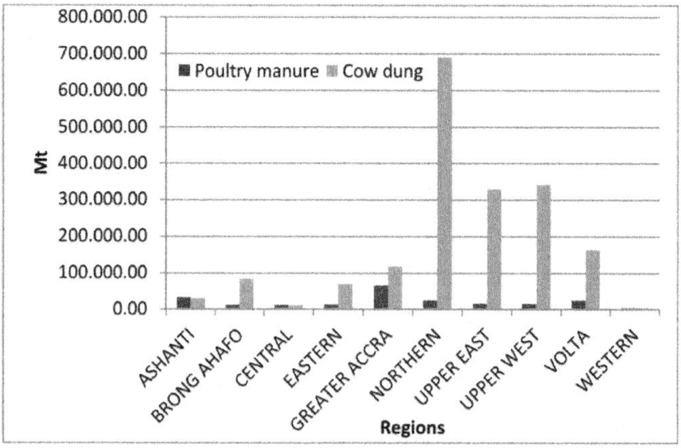

Figure 3: Quantities of poultry manure and cow dung for the various regions

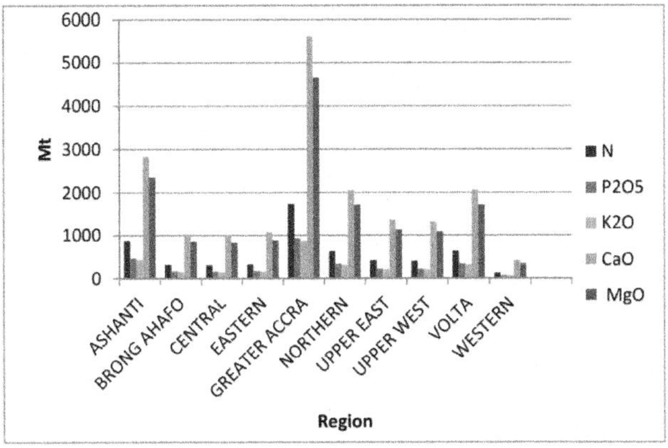

Figure 4: Nutrient equivalent of poultry manure for the various regions

Nutrient equivalent of poultry manure is presented in Figure 4. Calcium and magnesium content of poultry manure is high. Addition of calcium enriched food materials into poultry feeds especially layers is the most possible reason. Poultry manure is a very good material that may even improve the pH of the soil when applied in large quantities. The manure contains all the major nutrients. These materials can be used to improve the productivity of soils in the forest and coastal Savanna zones.

Figure 5 shows nutrient equivalent of cow dung. Cow dung contains all the major nutrients while the urine is very rich in N and K. Provision of beddings for the cattle results in an improved material since the urine is absorbed by the beddings. Cow dung will increase crop yield significantly especially the very poor soils in the Savanna zone.

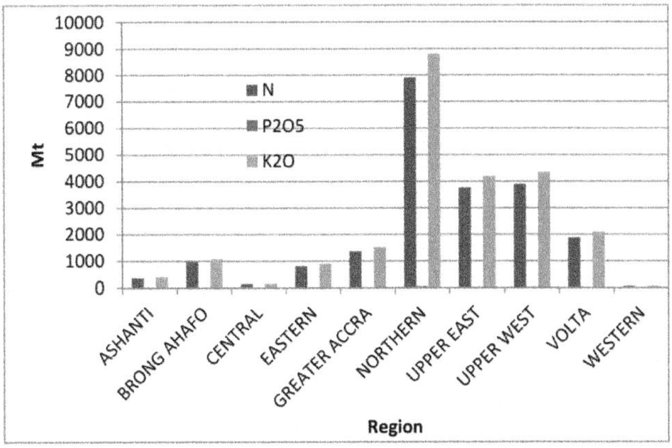

Figure 5: Nutrient equivalent of cow dung and urine for the various regions

Sheep and goats are reared throughout the country. Northern and Volta regions rear a lot of sheep and goats while in the Upper West region a lot of sheep are reared. Generally these animals are reared throughout the country. Large amount of sheep and goat manures are produced annually (Figure 6). Quantity of manure produced is directly related to number of animals in these regions. Sheep and goat manure contain high amount of nitrogen. Manure produced by these animals is normally mixed with the urine resulting in the relatively high amount of nitrogen in their manure.

Generally about 3.2 million Mt of manure was produced by animals in 2007 (Table 6). This amount of manure can reduce the amount of money spent on mineral fertilizer if most of it is used in crop production.

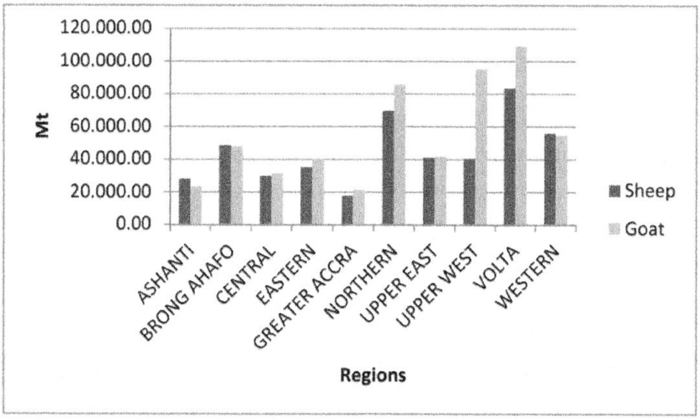

Figure 6: Quantities of sheep and goat manure for the various regions

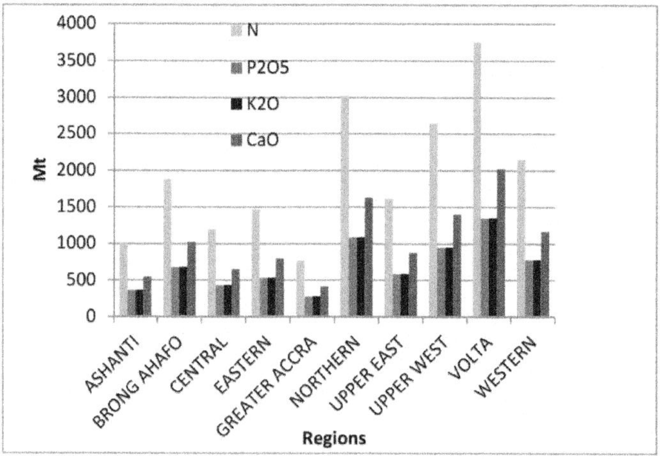

Figure 7: Combined nutrient equivalent of sheep and goat manure

Table 6a: Quantity and fertilizer equivalent of animal manures

Type	Q'ty of manure/yr Mt	Nutrient content (Mt)				
		N	P_2O_5	K_2O	CaO	MgO
Poultry	222,228.0	5,777.9	3,111.2	2,911.2	18,778.3	15,556.0
Cattle	1,848,798.0	20,706.5	31,614.4	8,874.2	14,420.6	127,012.4
Sheep	449,388.0	8,763.1	3,145.7	3,145.7	4,988.2	-
Goat	551,354.4	10,751.4	3,859.5	3,859.5	5,513.5	
Pig	118,281.9	2,696.8	2,152.7	2,129.1	378.5	-
TOTAL	3,190,050.3	48,695.7	43,883.5	20,919.7	44,079.1	142,568.4

Table 6b: Urine and total nutrient content

Region	Q'ty of urine/yr Mt	N	P_2O_5	K_2O
Cattle	1,746,087.0	21,127.65	174.61	23,572.17
Sheep	855.0	12.569	0.428	16.758
Goat	1049.0	15.42	0.525	20.560
Pig	294.6	1.119	0.295	2.917
TOTAL	1,748,285.6	21,156.8	175.9	23,612.4

A study of quantity of human excreta produced in parts of Accra and Kumasi show that large amount of the material is available and a very good source of amendment for soil improvement (Table 7). Treatment and management of human excreta is generally poor and require some attention.

Table 7: Potential available nutrients from human excreta in parts of Accra and Kumasi in 2007

Location	Quantity Disposed/yr* (Mt)	Nitrogen content/yr (Mt)	P_2O_5 content/yr (Mt)
Accra	292,000	586.04	17.23
Kumasi	54,750	109.88	3.23
Total	346,750	695.93	20.46

Issaka et al 2010

AGRO-MINERALS AND ROCKS

Large deposit of limestone and dolomite are found in several parts of the country (Table 8). Mining of some of these minerals will support the agriculture sector of the economy. The very low levels of basic cations need to be significantly raised for the application of Nitrogen, Phosphorus and Potassium fertilizers to be effective and efficient. Basic slag is a good source of P (Table 9) and can be harnessed to improve available P in our soils for better crop production.

Table 8: Lime resources of Ghana

Deposit	Type	Deposit (million tons)	CaO and MgO contents (%)	Lime-Potential quantity (million tons)
Akuse region	Shells	1.3	53.5 (0.65)	0.7
Tano	Limestone	1000		
Nauli area	Limestone	400	48.7-52 (0.57-1.84)	194.8
Nauli deposit	Limestone	23	48.9 (1.04)	11.3
Bongo Da	Limestone	15	40.0-49.3 (1.2-4.2)	6.0
Bongo Da	Dolomite	23-30	30.8-36 (12.1-19.0)	7.1
Bupei Upper horizon	Limestone	6		
Bupei Lower horizon	Dolomite	144	29.3-31 (17.5-19.2)	42.2
Oterpolu	Limestone	8-10	38.3 (3.65)	3.1
Po river	Limestone	1.5		
Daboya	Dolomite	0.7		
Total				>265.2

Adapted from Kesse (1985, 1988) cited by Owusu-Bennoah, 1997. Content of MgO in parenthesis.

Table 9: Annual production of basic slag in Ghana

Steel Industries	Capacity Mt Steel/yr	Actual Production Mt Steel/yr	Basic Slag Produced/yr* Mt	Quantity of P_2O_5 Produced/yr Mt
Tema Steel Complex	30,000	4500	675	60.75
Ferro-aibric Limited	15000	9000	1350	121.50
Total	45,000	13,500	2,025	182.2

*Adapted from: Owusu-Bennoah, 1997. * Basic slag is 15% of steel produced and contains 9% P2O5*

USES OF INDIGENOUS MATERIALS IN THE VARIOUS ECOLOGICAL ZONES

Plant Material: Even though there are many sources of plant materials emphasis will be on rice straw, rice husk and saw dust. Rice straw is generated on rice farms and is usually burnt during or before land preparation. About 40 percent of the nitrogen (N), 30 - 35 percent of the phosphorus (P), 80 - 85 percent of the potassium (K), and 40 - 50 percent of the sulfur (S) taken up by rice remains in vegetative plant parts at crop maturity (Dobermann and Fairhurst, 2002). Burning causes almost complete N loss, P losses of about 25%, K losses of 20%, and S losses of 50 – 60%. The amount of nutrients lost depends on the method used to burn the straw (Dobermann and Fairhurst, 2002). More losses occur when straw is burnt under windy conditions since most of the ash will be blown away. Thus, huge amount of nutrients are lost annually due to poor management of rice straw. Proper management of this material will greatly improve soil fertility especially in the savanna areas where organic matter is very low. Rice husk may be used as poultry and pig feed but is hardly used as amendment hence all the nutrients in the materials is lost. In Ghana saw dust is normally treated as waste. The material may be used as landfills, deposited in the out skirt of the towns/cities and burnt or left to decompose. Generally saw dust decomposes very slowly and may pose a problem when applied directly on the field.

Animal manure: Poultry manure is on high demand and farmers are unable to get enough of it. Formerly huge piles of poultry manure could be seen by the side of roads close to poultry farms. Farmers could go for the material free of charge. The story is now completely opposite. Farmers have to pay in advance (including cost of transportation) before the manure is conveyed to the field. The effect of poultry manure in the farming industry is only significant in the forest zone (especially Ashanti and Eastern regions) and the coastal savanna zone (GT Accra).

Cow dung is a material highly cherished by rural dwellers in the Upper East region. Cow dung is used for crop production, as a binding agent for plastering houses, for cooking (fuel) and to a lesser extent for trapping termites to feed chicks. Unlike poultry manure the material is not sold and therefore not easy to obtain. In the Upper West region cow dung is not valued. Farmers consider the material to be too heavy and bulky to carry to the field (farms are generally far from home). The story is mixed in the Northern region were farms are also located far from the homes. However, some areas in this region have started making good use of this material. This is in contrast to the Upper East region where most farms are close to the house. In the forest zone cow dung is scarcely used for crop production. Some of it is, however, used in crop production in the coastal savanna zone.

Sheep and goat manures are used widely in the Upper East region. Both the dung and urine are normally mixed resulting in high nitrogen content of the manure. Inclusion of litter (grasses or plant residue) results in the production of farm yard manure. In other parts of the savanna zone the manure is scarcely used. Some amount of it is used in the Volta and Western regions.

Human excreta are used in parts of the Northern region. Farmers may hire trucks to collect the effluent which is deposited on their fields in its raw form. This practice is rather unhygienic. In parts of Kumasi, near Kwame Nkrumah University of Science and Technology, human excreta is allowed to decompose. Farmers are then given the option to collect the decomposed material to their fields. Patronage is however low.

MANAGEMENT OF INDIGENOUS MATERIALS

Plant materials (saw dust, rice straw, rice husk etc) can be decomposed, charred or even ashed so as to make the material more user friendly and the nutrients more available to plants. Within the farm it should be possible to decompose rice straw before application.

Rice husk which is normally out of the farmers reach can be carbonized (material becomes easier to transport in this form) before application. The material can also be ashed (this may be environmentally unfriendly) before transporting to field. Ashed rice husk or straw has a pH of over 11.0 and hence can be used to improve the rather acidic soils that are common in Ghana. Decomposed saw dust is generally a very good material. It takes more than 6 months to get the material well decomposed. The material, however, can easily be charred or ashed.

CONCLUSION AND RECOMMENDATIONS

A wealth of indigenous materials are available in Ghana. These materials can be used to significantly improve the fertility of the soil. Some of these materials include rice straw and husk; saw dust, poultry manure, cow dung, goat and sheep manure, and human excreta. It is strongly recommended that materials of plant origin should be composted or charred before usage. Human excreta need to be composted before it is used.

ACKNOWLEDGEMENTS

The authors are grateful for the fund received from the Ministry of Agriculture, Forestry and Fisheries (MAFF), Japan, through the Japan International Research Center for the Agricultural Sciences (JIRCAS), as an out source contract for the study on "Improvement of Soil Fertility with Use of Indigenous Resources in Rice Systems of the Sub-Sahara Africa" with Dr.Satoshi Tobita as Project Leader. We also wish to thank technicians and all staff at Soil Research Institute laboratory for the analysis of both soil and plant materials.

REFERENCES

1. Adjei-Gyapong, T. and Asiamah, R. D. 2000. The interim Ghana soil classification system and its relation with the World Reference Base for Soil Resources. FAO Report on Soil Resources No. 98.
2. Brammer, H. 1962. Soils: In Agriculture and Land use in Ghana. J.
3. B. Wills ed., Oxford University Press London, Accra, New York. pp 88-126. ISSS/ISRIC?FAO. 1998. World Reference Base for Soil Resources. World Soil Resource Report 84, FAO, Rome.
4. Issaka, R. N., Buri, M. M. Adjei, O. E. and Ababio, F. O. 2010. Characterization of indigenous resources in Ghana: Locally available soil amendments and fertilizers for improving and sustaining rice production. CSIR-SRI/CR/RNI/2010/01.
5. Buri, M. M., Iassaka, R. N., Fujii , H. and Wakatsuki, T. 2009. Comparison of Soil Nutrient status of some Rice growing Environments in the major Agro-ecological zones of Ghana. International Journal of Food, Agriculture & Environment Vol. 8 (1): 384- 388
6. Buri, M. M., Issaka, R. N. T. Wakatsuki and E. Otoo 2004 Soil organic amendments and mineral fertilizers: Options for sustainable lowland rice production in the Forest agro-ecology of Ghana. Agriculture and Food Science Journal of Ghana. Vol., 3, 237-248
7. Dobermann. A. and Fairhurst, T. H 2002. Rice Straw Management. Better Crops International Vol. 16, Special Supplement.
8. MOFA 2009 Agriculture in Ghana. Facts and Figures. SRID, Ministry of Food and Agriculture

Chapter 12

INTEGRATED SOIL FERTILITY MANAGEMENT IN BEAN-BASED CROPPING SYSTEMS OF EASTERN, CENTRAL AND SOUTHERN AFRICA

Lubanga Lunze et al.

Institut National pour l'Etude et la Recherche Agronomiques (INERA), Kinshasa Democratic Republic of Congo

INTRODUCTION

Common bean (*Phaseolus vulgaris* L.) is an important grain legume in Eastern, Central and Southern Africa (ECSA), where it is grown on over 3.7 million of hectares every year. In this region, bean consumption per capita exceeding 50 kg a year is perhaps the highest in the world, reaching over 66 kg in densely populated western Kenya (Wortmann et al., 1998). Bean provides the essential dietary protein, fiber and income to at least 100 million people in Africa (Kimani et al., 2001). Beans are grown primarily as a food crop and to generate income by smallholder, resource-poor farmers, in holdings that rarely exceed 1.5 hectares. They are grown in pure stands or in association with maize, bananas, or root or tuber crops, and in recent years, between rows of fruit crops, banana and coffee, especially in the early establishment phase of these traditional cash crops. About 22% of the production area is sole-cropped, 43% is in association with maize, 15% with bananas, 13% with root and tuber crops, and 7% with other crops (Wortmann et al., 1998). In southern Africa, beans are either grown in pure stands (42%) or in association with maize (47%), or to a lesser extent, with root crops (6%) or other crops (5%). Production is mainly rain-fed, except in Mauritius and the Nile Valley of Sudan where beans are grown as an irrigated crop. In lowland areas of Madagascar, Malawi, Mozambique, and DR Congo, beans are sown after another crop in order to use residual moisture and to take advantage of lower temperatures during the winter months.

Bean is produced essentially in the Eastern and Central African highlands, where the population density is the highest (Wortmann et al., 1998). It is produced by smallholder farmers with few resources to allocate to soil improvement. In the Great Lakes region for example, Dreschsel et al. (1996) report extremely low use of mineral fertilizer, only about 0.4 kg ha^{-1}. Moreover, because of the high population density, farmers are faced with rapid soil fertility decline as a result of continuous cropping and inappropriate cropping systems with very little or no external nutrient input to replenish soil fertility. Therefore, bean yield is generally low in most regions and is most likely to decline because of the ever increasing population density. In fact, under current farming systems in small holders' fields, soil nutrient balances are negative (Bationo et al. 2006), except in banana based systems (Wortmann and Kaizzi, 1998).Although bean grain yields are variable across countries and regions, they generally vary from 200 $kgha^{-1}$ in less favorable environments to 700 kg ha^{-1} in more favorable environments when grown in pure stands, and about half of this when intercropped (Kimani et al., 2001).

In ECSA, low soil fertility is the most important yield-limiting factor in most of the bean-producing regions. The major soil fertility related problems are found to be low available phosphorus (P) and nitrogen (N), and soil acidity, which is associated aluminum (Al) and manganese (Mn) toxicity. According to the Atlas of common bean production in Africa (Wortmann et al., 1998), P is deficient in 65 to 80% of soils and N in 60% of soils in bean production areas of Eastern and Southern Africa, while about 45 to 50% of soils are acidic with a pH less than 5.2, containing high levels of either Al or Mn. The details on the importance of the edaphic stresses for common bean production are presented in table 1. Low soil fertility causes considerable losses in productivity as the estimated bean production losses due to edaphic stresses in the ECSA are about 1,128 million tons per year (Table 1).

Bean grows well in deep well drained, sandy loam, sandy clay loam or clay loam with clay content of between 15 and 35% with no nutrient deficiencies (Thung and Rao, 1999). The optimum soil pH range is 5.8 to 6.5 and Al saturation is below 10% (Lunze, 1994). It will not grow well in soils that are compacted, too alkaline or poorly drained.

Table 1: Major soil related production constraints and bean yield losses in Africa (Wortmann et al. 1998)

Constraints	Eastern and Central Africa	Southern Africa	Eastern and Central Africa	Southern Africa	Sub Saharan Africa
	% Total bean area affected		Annual losses (t)		
N deficiency	50	60	263,600	125,200	389,900
P deficiency	65	80	243,200	120,400	355,900
Acidity	45- 50	45 - 50	152,700	65,800	220,000
Al/Mn toxicity	52	42	97,500	60,300	163,900
Total			757,000	371,700	1129,700

In light of these constraints, the Pan African Bean Research Alliance (PABRA) has undertaken regional efforts to develop integrated soil fertility management (ISFM) technologies that improve sustainability, productivity and quality of the bean crop in various environments across ECSA. Over the past two decades, there have been research initiatives from both national programs and bean research networks Eastern and Central Africa Bean Research Network (ECABREN) and the Southern Africa Bean Research Network (SABRN), both of which belong to PABRA, to develop strategies and technologies to improve sustainably bean crop productivity and production in various production environments in ECSA. A range of technologies have been evaluated and developed to address the regional constraints, and efforts were made to promote the promising technologies widely.

Several technologies have been developed through collaborative research efforts within PABRA. The objective has been developing strategies and technologies that enhance resilience to environmental stresses and improve bean productivity and product quality. These include: (i) development of diagnostic tools for soil fertility assessment that are adapted to local conditions; (ii) replenishing soil nutrient pools, maximizing on-farm recycling of nutrients, and reducing nutrient losses to the environment; and (iii) improving the efficiency of external inputs. As common bean can derive part of its N from the atmosphere under low input agriculture (Giller *et al.*, 1998), improving biological nitrogen fixation using seeds inoculation with appropriate rhizobacteria and soil management was considered. Currently recommended ISFM options in bean based cropping systems include farmyard manure, compost, biomass transfer, green manure and cover crops, liming, phosphate rock (PR) and mineral fertilizers in different combinations with organic resources. The soil management options are complemented by utilization of resilient bean germplasm that perform well under low soil fertility conditions.

This paper reviews ISFM options developed by the ECABREN and SBRN and the approaches for effective and efficient delivery of these technologies to farmers.

INTEGRATED SOIL FERTILITY MANAGEMENT (ISFM) OPTIONS

Concept of ISFM

Integrated soil fertility management (ISFM) is an approach that stresses sustainable and cost-effective management of soil fertility (Sanginga and Woolmer, 2009; Vanlauwe et al., 2010). This soil fertility strategy relies on a holistic approach that embraces the full range of driving factors and consequences of soil degradation- biological, chemical, physical, social, economic, health, nutrition and political (Bationo et al., 2006). ISFM attempts to make the best use of inherent soil nutrient stocks, locally available soil amendment resources and mineral fertilizers to increase land productivity while maintaining or enhancing soil fertility. Vanlauwe et al. (2010) define ISFM as 'A set of soil fertility management practices that necessarily include the use of fertilizer, organic inputs, and improved germplasm combined with the knowledge on how to adapt these practices to local conditions, aiming at maximizing agronomic use efficiency of the applied nutrients and improving crop productivity. A conceptual diagram is shown in Figure 1.

Local Soil Fertility Diagnostic Tools

Soil fertility can vary drastically from one end of field to the other (Vanlauwe, 2006). Therefore, response to applied soil management and external inputs can vary accordingly and no single recommendation can be made for a whole farm. As ISFM technologies are generally complex, labour intensive and costly, a more accurate intervention and recommendation system is crucial to improve the chance of adoption of these technologies by farmers. To assure farmers get the maximum return from the investment in inputs for soil improvement, it is important that recommendations well target specific local conditions. Hence the indicator of soil quality is important for local communities to better manage their soil resources though better decision making (Barrios et al., 2001). These have to be simple enough for use by farmers and extension personnel. Dominant plant species on a farmland have the potential to integrate changes in soil quality, reflecting changes in the physical, chemical and biological characteristics of the soil (Pankhurst et al., 1997).

There have been research initiatives from both national programs and PABRA to integrate farmers' perceptions of soil fertility in simple soil fertility assessment. In South Kivu province of D.R. Congo, Ngongo and Lunze (2000) tested bean response to fertilizer application on soils with varying levels of weed infestation. Field trials were conducted for two seasons to test the effect of compost applied at 20 t ha^{-1} rate. Where *Gallisonga parviflora* is the dominant weed species, soil nutrient levels were high, bean (*Phaseolus vulgaris L.*) yield was high and did not increase further compost application. Where *Pennisetum polystachia* is the dominant species, soil fertility and bean productivity were low. Bean yield on these fields was increased considerably with the application of compost. The results (Table 2) confirm the farmer's perception that *Pennisetum polystachia* is an indicator of low soil fertility. *Conyza sumatrensis* and *Bidens pilosa* indicate intermediate level of soil fertility and response to compost was also intermediate.

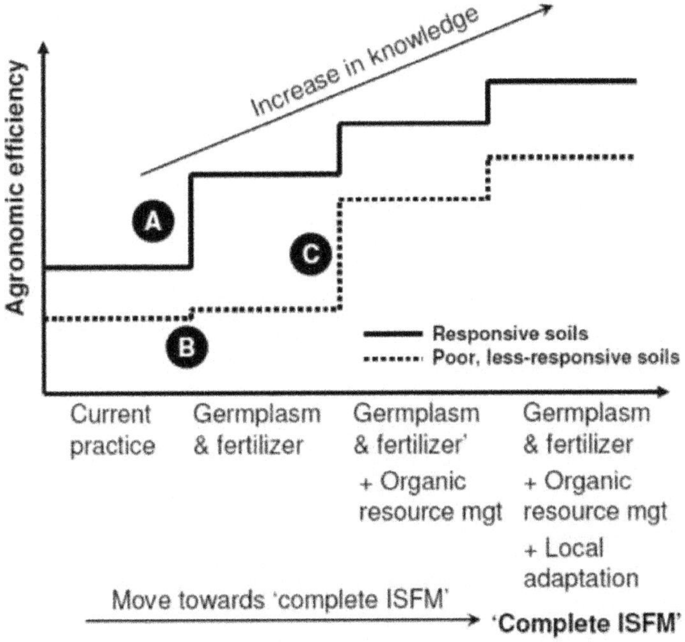

Figure 1: Conceptual relationship between the agronomic efficiency of fertilizers and organic resource and the implementation of various components of ISFM, (Vanlauwe et al., 2010)

Similar results were reported by Ugen and Wortmann (2006) in Uganda. Relative densities of *Digitaria scalarum* (blue couch), *Eleusine indica (L.)*

Gaertn. (goosegrass), *Euphorbia hirta* (garden spurge), *Cyperus spp*., *Oxalis latifolia H.B.K.* and *Sorghum halepense (L.) Pers. (johnsongrass)* varied more with nutrient supply than did other species. Soil properties had less effect on the distribution of *Ageratum conyzoides* L. (*tropic ageratum), Bidens pilosa* (*hairy beggarticks), Commelina benghalensis L. (tropical spiderwort)*, and *Galinsoga parviflora Cav. (smallflower galinsonga)*. High relative densities of *Digitaria scalarum* and *Euphorbia hirta* were generally associated with low soil nutrient levels. *Eleusine indica (L.) Gaertn.* (*Goosegrass),*

Sorghum halepense (L.) Pers. (*johnsongrass*), and *Oxalis latifolia* were associated with higher nutrient levels in soil.

A more detailed assessment of soil chemical properties based on weed species density was done in Uganda by Ugen, et al. (1999). In this study, they made observations on 39 fields with annual crops over 4 locations in Eastern and Central Uganda. Densities of weed species relative to the total weed population were determined and surface soil samples were collected and analyzed for organic carbon (OC), soil pH, available P, K, Ca and Mg and total N, P and K levels. The results are presented in table 3.

Table 2: Bean response to compost on fields with different weed species in South-Kivu, D.R. Congo (Ngongo and Lunze, 2000)

Dominant weeds	Bean yield (kg ha^{-1})			
	1994B		1995A	
	+C	-C	+C	-C
Gallinsoga parviflora	995.3a	1000.0a	1713.3a	1700.0a
Pennisetum polystachia	427.7cd	170.7cd	696.7cd	41.7e
Conyza sumatrensis	720.1b	627.8b	827.7c	419.7d
Bidens pilosa	1012.0a	950.0a	933.3c	873.3c
Digitaria vestida var scalarum	892.2a	457.7c	1335.3b	1251.0b
Mean (kg ha^{-1})	809.5	641.2	1101.3	857.1
C.V. (%)	25.9	25.9	21.2	21.2

+C: with compost and -C: without compost

Table 3: Positive (+) and negative (-) relationships of the relative densities of weeds species with soil properties which may be useful in diagnosis of soil fertility status

Species	OC	pH	P	K	Ca	Mg	N total	P total	K total
Eleusine			+	+	+	+		+	+
Euphorbia			-		-	-		-	
Sorghum	+	+		+	+		+		
Oxalis	+	+	+	+					
Nutsedge	-					-			

Source: Ugen, M., Wien A. D., Wortmann C.. CIAT Annual Report 1999

Oxalis latifolia was correlated with OC content, as this weed was unlikely to be present when percentage C was less than 2.3, and increased in density as soil C increased. Soil pH was well correlated with both *Sorghum halepense* and *Oxalis latifolia*. *Sorghum halepense* was found to be a good indicator for soil pH, being generally absent when soil pH was less than 5.8, while *Oxalis latifolia* increased in importance when soil pH increased. *Eleusine indica* was not likely to occur when soil P availability was less than 13 mg kg^{-1} and increased with increased P availability, while the total P was positively correlated with *Eleusine indica*, but negatively correlated with *Euphorbia hirta* densities. *Sorghum halepense* occurred infrequently when exchangeable K was less than 0.6 cmol(+)kg^{-1}. *Eleusine indica* and oxalis increased as exchangeable K; *Eleusine indica* and *Bidens pilosa* tended to increase as total K increased. Relative densities of *Digitaria scalarum*, *Euphorbia hirta* and *Cyperus esculentus L.* (yellow nutsedge) were negatively related to exchangeable Mg and accounted for 62% of variation in exchangeable Mg. Soil fertility assessment using natural weeds is the most convenient tool available to smallholder farmers and extension workers because it requires minimal training. Therefore, weed flora was used to develop a decision guide to aid farmers in the identification of areas in their fields with severe nutrient deficiencies in Uganda and Eastern DR Congo. Thus, farmers and extension workers have at their reach a quick and inexpensive way to assess the soil fertility levels and make decisions concerning soil management.

GENETIC APPROACHES

Genetic variation for abiotic stress tolerance exists within common bean germplasms in 15 national and regional bean programs (Singh, 1991; Aggarwaal, 1994; Lynch and Beebe, 1995; Rachier et al., 1999; Rao 2001; Beebe et al., 2009), and significant number of new lines bearing these traits were developed at CIAT headquarters in Cali, at CIAT Africa regional breeding program hosted by the University of Nairobi, Kenya and the national bean programs. Singh (2001) stated that development of high yielding cultivars adapted to low soil fertility and low input sustainable systems is essential to maximize yield of common bean to enhance food security, reduce production costs and generate income. PABRA considers development of low soil fertility-tolerant bean varieties as an option to increase bean yield at no additional cost (Kankwatsa et al., 2008). Thus, bean improvement for low soil fertility adaptation has become an important component of ISFM strategy for optimum bean production in low input systems of smallholder African farmers.

Screening common bean cultivars for low soil fertility tolerance is done under field conditions following a harmonized protocol (CIAT, 1994). The methodology consists of screening the same sets of bean genotypes at several locations under a single stress and combining results across sites with different stresses (Wortmann et al., 1995). Screening is done at two stress levels: moderate stress and no stress. The moderate stress corresponds to the stress level at which a well-adapted control variety under stress performs at 40 to 50% of its normal unstressed performance. The criterion for selection is essentially bean yield, but farmers' preferences and market preference were also considered. Five most popular market classes in ECSA are red kidney, red mottled, small red, white navy and pintos (Kimani et al., 2001). Such intentional market class choice was intended to link production to well-established markets and favor adoption by farmers. The potential genotypes identified are evaluated with farmers following the participatory varietal selection (PVS), a strategy developed within PABRA for heterogeneous environments where farmers have a range of preferences (Sperling et al., 1993). The selected varieties are then promoted through other participatory methods in on-farm trials. The sites selected in different countries represent a range of soils characteristics and agro-ecologies. Soil chemical characteristics at the experimental sites in different countries are presented in Table 4.

To date, a total of 1,400 beans lines have been evaluated through BILA (Bean Improvement for Low fertility soils in Africa) for their relative tolerance to the stresses under consideration, particularly low N, low P and soil acidity, which is associated Al and/or Mn toxicity. Considerable genetic variability in germplasm was detected and several varieties identified with specific single or multiple edaphic stress tolerance. The initial screening did not consider market factors, but allowed selection of tolerant cultivars from national and regional bean programs, already well adapted to local environments. The following cultivars were identified as tolerant lines to different stresses from the evaluation at different BILFA sites (Lunze, 2002).

Table 4: Characteristics of surface soils (0-20 cm) at test sites in several African countries

Sites and stress	Country	Altitude (m)	pH H_2O	Organic C (%)	Bray-I P (mg kg^{-1})	Exch. Ca (cmol(+) kg^{-1})	Exch. Mg (cmol(+) kg^{-1})	Exch. K (cmol(+) kg^{-1})	Exch. Al (cmol(+) kg^{-1})	Al saturation (%)
Nyamunyunye (Al)	DRC	1730	4.7	2.3	0.5	2.6	1.1	0.07	2.6	41
Gikongoro (Al)	Rwanda	1900	4.8	0.5	2.2	1.4	0.3	0.01	2.6	60
Antsirabe (P &Al)	Madagascar		4.6	3.7	1.2	0.45	0.65	0.28	2.43	59
Kakamega (low P)	Kenya	1550	4.9	2.6	7.3	1.6		2.3	na	
Mulungu 2 (low N)	Uganda	1690	5.8	2.4	9.6	5.5	3.1	0.56	na	
Selian (low N)	Tanzania		6.5	1.8	18.2	9.2	1.6	6.2	na	
Kawanda (low N)	Uganda	1190	5.5	2.1	20.0	2		0.5	na	
Kawanda (low K)	Uganda	1190	5.3	3.3	10.0	2		0.2	na	
Ikulwe (Mn)	Uganda	1200	5.2	2.8	3.0	4		0.8	na	

- Low N: RWR 382, RAO 55, ACC 433, UBR(92)25 and BAT 85
- Low P: Carioca, BAT 25, RAO 55, XAN 76 and MMS 224, ACC 433 and Ikinimba
- Low K: EMP 84, ICA Pijao, RAO 52 and BAT 1220
- Al toxicity : ACC 7/4, Ubusosera
- Mn : MCM 5001, XAN 76 and Urugezi

These cultivars were integrated into national bean breeding programs, and several tolerant varieties have been adopted by farmers and are among the released varieties (Ikinimba, XAN 76, Ubusosera, ACC 7/4, RAO 55, UBR(92)25, MwaMafutala, Ntekerabasilimu) in some countries including Madagascar, DR Congo, Malawi and Uganda, MLB 4089A , RWR 1092 in Kenya

Low Soil Fertility Adapted Bean Lines or Varieties of Various Market Classes

The most popular bean types grouped in major market classes were screened and lines tolerant to edaphic stresses were selected to allow potential users to choose for their own market. Bean lines or varieties belonging to various market classes and types, grouped by their tolerance to low N, low P and low pH conditions are presented in table 5. Bean genotypes were evaluated for low N adaptation at two sites, Mulungu in DR Congo and Selian in Tanzania. The genotypes varied significantly in their grain yield under N deficient conditions and in their response to applied N. Without applied N, the yield varied from 695 to 1,789 kg ha^{-1} while with added N at the rate of 30 kg ha^{-1}, the grain yield varied from 1,258 to 3,139 kg ha^{-1} at Mulungu. Without N, most lines gave significantly higher yield than the local sensitive check Kirundo, and previously selected tolerant variety, MwaMafutala.

The performance of the best lines selected for their tolerance to low P was evaluated at Kakamega and Antsirabe in Madagascar. Several lines were outstanding compared with the local released bean variety GLP 585. Considerable yield advantage of up to 80% was observed compared with the local check. The outstanding genotypes tolerant to low P are AFR 619, XAN 96, ARA 4, AFR 708 and RWR 1873.The screening of genotypes for low pH was done at Mulungu, DR Congo; Gikongoro, Rwanda; and Antsirabe, Madagascar (in acidic soils that demonstrate positive response to liming) revealed that the genotypes selected as Al resistant at three low pH sites were generally the same and performed consistently across sites. However, the slight variability noticed across sites could be attributed to the difference in adaptation to local environments (Wortmann et al., 1995). At Mulungu, with no lime, most test lines gave significantly higher yield than the sensitive check variety Kirundo, while only two lines VTTT 923-6-1, HM 21-7 and AFR 593-1 outperformed the tolerant check MwaSole (Table 6). The yield advantage of Al-resistant genotypes could be as high as 300% compared with the check.

At Antsirabe, Madagascar, several lines outperformed the local check Soafianarana and improved check Goiano Precoce. Most genotypes selected as resistant to Al toxicity at Antsirabe showed the same at Mulungu. The following lines gave consistently better yield under acid Al-toxic soil at the two low pH sites: AND 1056-1, AND 932-A-1, ARA 8-1B, BZ 12984-C-1 and VTTT 920-26. At Gikongoro, Rwanda many lines gave higher yield than the Al-resistant checks 7/4, Acc and RAB 478. However AND 93-A-1, BZ 12894-C-1 and AFR 593-1 were also identified as resistant to Al at the other two low pH sites, Mulungu and Antsirabe (Lunze et al., 2007).

The overall evaluation across sites and stresses allowed selection of bean lines that had consistently high yield under different stresses and across sites, with identified tolerance to one, two or even all three soil constraints considered. BZ 12894-C-1, AND 932-A-1, DRK 137-1, Nm 12806-2A are outstanding across sites and in response to all stresses - high aluminum toxicity, low N and low P availability. Several other lines have manifested tolerance to two stresses: ARA 8-B-1, AFR 709-1, AFR 703-1 and AND 1055-1 are tolerant to low P and low pH: RWK 10, ARA 8-5-1, and T 842-6F11-6A-1 tolerant to low N and low pH. These appear to have multiple tolerances to edaphic stresses and good adaptation at all environments. The results show existence of tolerance to low soil fertility in market class bean types. The promising varieties identified provide opportunities for higher bean productivity on acid soils and those with limited N and P supply (Lunze et al., 2007).

Table 5: Promising bean lines in different market classes and types, grouped according to their tolerance to stressful soil conditions- low N, low and pH (Lunze et al., 2002)

Low N			Low P			Low pH		
Line code	Seed size (100 seed weight, g)	Seed color	Line code	Seed size (100 seed weight, g)	Seed color	Line code	Seed size (100 seed weight, g)	Seed color
A 286	17	Carioca	AFR 619	34	Red	37/66/6	23	Tan
AFR 675	24	Navy	AFR 675	24	Navy	A 286	17	Carioca
AFR 699	40	Red	AFR 708	44	Calima	A 344	27	Cream
AFR 714	23	Navy	AFR 714	23	Navy	AFR 708	44	Calima
AND 871	35	Calima	AND 871	35	Calima	AFR 714	23	Navy
CAL 143	50	Calima	ARA 4	21	Cream	ARA 4	21	Cream
CAL 150	50	Calima	CIM 9314-36	41	Calima	BRB 119	31	Calima
CIM 9314-33	42	Red	CIM 9314-37	34	Calima	DB 201/77/1	19	Navy
CIM 9314-36	41	Calima	CIM 9331-1	25	Red	CIM 9314-3	37	Calima
CIM 9315-1	24	Pink	CIM 9331-2	29	Pink	CIM 9331-1	31	Red
CIM 9315-3	27	Calima	CIM 9331-3	23	Red	CIM 9415	38	Calima
CIM 9318-4	27	Calima	DB 196	20	Navy	CNF 5520	44	Calima
CIM 9331-3	23	Red	DOR 663	17	Black	DFA 53	28	White
DB 196	20	Navy	FEB 192	19	Cream	FEB 197	22	Black
DOR 715	18	Red	FEB 196	20	Carioca	G 12489	44	Calima
FEB 192	19	Cream	G 2858	21	Tan	G 2910	21	Calima
FEB 196	20	Carioca	G 5889	15	Cream	G 3480	15	Black
G 5889	15	Cream	LSA 32	32	Carioca	G 5889	15	Cream
LSA 32	32	Carioca	MORE 92018	49	Tan	HM 21-7	45	Red
MORE 92018	40	Tan	PAN 150	24	Carioca	LRK 34	45	Pink
PAN 150	24	Carioca	RWR 1873	35	Calima	LSA 144	22	Red
PRELON	20	Navy	RWR 2075	44	Red	PAN 150	24	Carioca
RAB 482	17	Red	RWR 2091	37	Red	RAB 482	17	Red
REN 22	21	Navy	SDDT 49	20	Carioca	RWR 1742	22	Red
RWK 10	40	White speckled	SDDT 54-C5	31	Pink	RWR 1873	35	Calima
SDDT 55-C4	42	Calima	UBR(92)24/11	15	Navy	UBR(92)11	18	Carioca
UBR(92)25	18	Navy	VEF 88(40)L1PYT6	24	Red	VEF88(40)L1 PYT6	24	Red
VEF88(4O)L1 PYT6	24	Red	XAN 76	18	Calima	XAN 76	18	Cream
			ZAA 5/2	33	Cream	ZAA 5/2	33	Calima

Table 6: Bean lines and varieties resistant to Al toxicity in soil with a yield advantage over the check

Line	Bean grain yield		Loss due toxicity (%)	% Yield advantage over check
	Stress (kg ha-1)	Non-stress (kg ha-1)		
VTTT 923-6-1	1494	1588	5.9	310.6
HM 21-7	1317	1455	9.5	273.8
AFR 593-1	1040	1801	42.2	216.2
MwaSole	999	1264	20.9	207.7
ARA-8-5-1	957	1394	31.3	199.2
AND 932-A-1	932	1206	22.7	193.8
BZ 12984-C-1	866	874	0.9	180.0
Mwamafutala	625	1510	58.6	129.9
Kirundo (control)	481	988	51.3	100

Adapted from Lunze et al., 2007

Dissemination and Impact

Efforts have been made to promote all the promising and potential cultivars in all regions with problem soils as their yield advantage in similar environments has been confirmed. The benefit of low soil fertility adapted bean varieties has been demonstrated in Eastern DR Congo (Njingulula, 2003; Mastaki, 2006), Malawi, Kenya and Uganda (Kankwatsa et al., 2008). In Eastern DR Congo, with very low fertility soils, Ubusosera and RWR 382 have become the main varieties grown by farmers, replacing previously grown local or improved varieties. Kimani (2005) reported that different countries have selected and widely promoted bean genotypes that are adapted to their own environment (table 7). More than 10 years later, most farmers still kept BILFA varieties because of their adaptation to marginal soil conditions (Musungayi et al., 2008). Njingulula (2003) conducted a study to assess the impact of two low soil fertility adapted varieties in Eastern Congo and has indicated that farmers who have adopted those two varieties (Ubusosera named MwaSole, resistant to Al and RWR 382, locally named MwaMafutala adapted to low N) have noticeably improved their socio-economic conditions. In this study, (1) 32% of respondents noted that bean quantity for consumption and sale was increased, (2) 28.4% of respondents said that bean become more permanent throughout the year, (3) 30% of respondents mentioned that nutrition was improved as they could eat twice a day compared to only once previously, (4) and other farmers responded they purchased livestock and various household items such as bicycles, radio, etc and paid medical care and school fees.

Table 7: Bean lines/varieties adapted to low soil fertility in different countries

Lines	Countries where adopted
RWR 1873	Uganda, Kenya, DRC
RWR 1946	Uganda, Kenya, DRC
RWR 2075	Uganda, Kenya, DRC
UBR(92)25	DRC, Malawi, Tanzania, Uganda
RWR 719	Ethiopia, Rwanda and Kenya
ACC 7/4, Ubusosera	Rwanda, DRC
DFA 54, DFA 53, DOR 633, DOR 715, AFR 708	Uganda, Kenya

Source: Kimani, 2005

USE OF ORGANIC RESOURCES

Organic matter based soil nutrient management is a traditional practice that continues on smallholder farms. Among the organic resources used are animal manure, compost, crop residues for soil incorporation, natural fallowing, improved fallows, relay or intercropping of legumes, and biomass transfer (Place *et al.,* 2003). Organic manure, compost and farmyard manure are the most common inputs used to improve soil fertility by small scale farmers (Musungayi et al. 1990; Kankwatsa et al., 2008). The need for both organic and mineral inputs to sustain soil health and crop production through ISFM has been highlighted due to their positive interactions (Vanlauwe et al. 2010).

Compost and Farmyard Manure

FYM use is the only possible practice for many resource poor farmers to improve crop and soil productivity. FYM and compost application are the most common practices used to improve soil fertility on smallholder farms although the quantities available are inadequate to meet crop nutrient demand. Organic manure available on farm estimates on dry matter basis varied from to 3.1 to 18.9 t ha^{-1} in Central Kenya (Opala, 2011) and 0.6 t ha^{-1} in Eastern DR Congo, sufficient to cover only 25% of their land under crops every season (Musungayi et al, 1990).Thus, scarcity of organic materials on farm limits their use at recommended rates. Because of scarcity of these resources, most farmers have developed localized application rather than broadcasting. This involves applying FYM or compost in the furrow or planting hole and covering it with soil before placing bean seeds, to avoid direct contact of manure with seeds. Considerable yield increase is achieved using compost alone, and the increase is dependent on initial fertility level in soil (Table 8).

Table 8: Effect of compost application to fields dominated with different weeds on bean yield

Soil dominated by	Bean yield (kg ha^{-1})		
	Without compost	With compost	Yield increase
Pennisetum polystachia	41.7	697	655
Conyza sumatrensis	418	828	408
Bidens pilosa	876	933	60
Digitaria sestida	1251	1335	84
Mean	857	1101	244
CV (%)	24.2	21.2	

Adapted from Ngongo, 2001

The potential of organic resources, such as FYM and compost to improve bean productivity, either as source of nutrients or by improving mineral fertilizer efficiency is well established in the region and various fertilization recommendations have been formulated. However, any substantial effect of organic amendments requires very large quantities that are not readily accessible to the majority of bean growers who are smallholder farmers (Thung and Rao, 1999).

Biomass Transfer

Search for alternative sources of organics and more economical resources have always been a concern of the national bean programs. Many shrubs and trees, such as *Lantana camara, Tephrosia vogelii* and *Tithonia divesrifolia* which are common on smallholder's farms in ECSA have been studied in different countries.. Among all shrubs, *Tithonia* is the most common shrub with substantial biomass production in most countries of the region. It is grown for land stabilization along the road and for erosion control on cropland, and as plot or compound boundary, and as an ornamental plant. *Tithonia* use and popularization as soil improving resource has gained interest with the increasing need of intensifying bean-based production system with the recognition of the potential of this species to accumulate nutrients. Jama et al., (2000) estimated *Tithonia* biomass available after nine months for transfer to fields at 2 t ha^{-1} kg of dry matter in Western Kenya. *Tithonia* has very high shoot vigour with relatively high nutrient concentrations in its biomass. Concentration of N ranges from 3.0 to 4.1%, P from 0.24 to 0.56% and K from 2.7 to 4.0% (Jama et al. 2000). Besides this advantage in nutrient concentration, the biomass of *Tithonia* is also known to be rapidly decomposing (Buresh and Tian, 1998) due to proper balance in lignins, polyphenols and N (Palm et al., 1997). In fact, it has the ability to extract relatively high amounts of nutrients from the soil, a property which may make the practice not sustainable in the long run due to nutrient depletion. On soils with high natural nutrient stock however, such as soils of volcanic origin of the Great Lakes region, *Tithonia* use has shown considerable potential in improving crop production. In most cases, as in the productions systems of a growing number of small scale farmers without livestock, *Tithonia* remains the only and essential source of organic material available on the farms (Rabary, 2001).

Tithonia biomass transfer is a practice extensively studied in Rwanda, Kenya, Tanzania and DR Congo for its integration into bean-based production systems. In Rwanda and DR Congo, considerable bean yield increase is obtained, 227% in Rwanda (Ruganzu and Nabahungu, 2002) and 68% in DR Congo (Ngongo, 2002).

Rate Application of TITHONIA Biomass

Determining the appropriate rate of biomass transfer is essential for integration of *Tithoni*a biomass use in farmers' cropping systems, either to beans grown as monocrop or as a mixed crop. The effect of the rate of application on bean yield of monocrop was evaluated in DR Congo and Rwanda (Ngongo, 2002; Ruganzu and Nabahungu, 2002; Nabahungu and Ruganzu, 2001). In Rwanda, the study was conducted on soils with pH varying from 3.8 to 4.9. The rate of application of *Tithonia* biomass varied from 0 to 12 t ha^{-1} on dry matter basis (Table 9). The optimum rate of application was determined to be 6 t ha^{-1} in Rwanda, and 4 to 6 t ha^{-1} depending on the initial soil fertility in DR Congo (Table 10). This slight difference is due to the difference in soil fertility at experimental sites, Rubona in Rwanda and Mulungu in DR Congo. In light of these results, the appropriate rate of application of *Tithonia* biomass on climbing bean may vary between 4 and 6 t ha^{-1}, which is slightly lower than the recommendation on maize of 7.5 t ha^{-1} in Tanzania (Ikerra et al. 2007). In the later study, they found that the optimum rate of application on maize could be set at 5 t ha^{-1} in the long rainy season while in short rainy season, the recommendation could be higher at 7.5 t ha^{-1}.

Table 9: Climbing bean yield response to applied *Tithonia* biomass in Rwanda (Ruganzu and Nabahungu, 2002)

Tithonia rate of application ($t\ ha^{-1}$)	Bean yield ($kg\ ha^{-1}$)
0	917
2	1314
4	1714
6	2086
8	1486
10	942.9
12	1271

Table 10: Climbing bean yield response to applied *Tithonia* biomass in DRCongo

Rate of application *Tithonia* ($t\ ha^{-1}$)	Bean yield ($kg\ ha^{-1}$) Soil fertility level		
	Low	Medium	High
0	230 d	949 c	1777 b
2	533 c	1358 b	1841 ab
4	838 b	1788 a	1901 b
6	1565 a	1788 a	2034 a
CV %	39.2	39.2	39.2
Mean yield (kg ha-1)	791.5	1470.7	1887.7

The rate of application in bean- maize mixed cropping, a more common cropping system in farmers' field, was evaluated by Ngongo (2002) in DR Congo (Table 11). In this system, the best yield of both crop either sole crop or intercropped was obtained at the rate of 8 $t.ha^{-1}$ofbiomass.

Table 11: Bean and maize yield response to *Tithonia* biomass in sole cropping and intercropping systems in DR Congo

Tithonia Biomass (t/ha)	Bean yield (kg. ha^{-1})		Maize yield (kg. ha^{-1})	
	Bean sole crop	Bean- Maize intercrop	Maize sole crop	Maize- bean intercrop
0 t ha^{-1}	703.6 *c*	748.0 *c*	1250 *c*	2250 *b*
4 t ha^{-1}	887.0 c	900.1 *c*	1930 *bc*	2450 *b*
8 t ha^{-1}	1410.0 *a*	1139.0 *b*	2220 *b*	3470 *a*
C.V. (%)	16.34	16.34	21.72	21.72
Mean (kg ha^{-1})	1000.2	929.03	1800	2720

Figures followed by the same letter are not significantly different at 0.05 probability. Source: Ngongo, 2002

Method of Application of TITHONIA Biomass

The technique of *Tithonia* use was fine-tuned by Ngongo (2002), who studied the appropriate time for biomass application. The time of application prior to bean sowing varied from 0 to 7 days. The best bean yield was obtained when biomass application was made at least 4 days prior to sowing. It was concluded that there was no reason to apply *Tithonia* biomass 7 days before planting (Table 12). The mode of application involved applying the fresh biomass on soil surface and incorporating it into soil 24 hours later, leaving it to decompose for 4 to 7 days before sowing bean.

Table 12: Effect of different application of *Tithonia diversifolia* on bean yield

Biomass application 6 t ha^{-1}	Bean yield (kg ha^{-1}) 2000A	2000B
Control	563 fg	375 g
At sowing	1,253 bc	985 cde
At 7 days before sowing	1,327 bc	1,174 bcd
At 4 days before sowing	1,703 a	1,273 bc
At 2 days before sowing	1,423 ab	1,105 bcde
Surface applied and worked in at sowing	776 ef	854 def
C.V. (%)	19.4	
Mean yield (kg ha^{-1})	1,174	

Figures followed by the same letters are not significantly different at 0.05 %

Green Manure and Cover Crops

Research and extension efforts were made to identify legumes that present farmers with as many options as possible so that they can choose cropping systems that are best suited to their needs. Many plant species have been selected and their potential benefit demonstrated using both on-station and on-farm experiments. More detailed field investigations have been done in Uganda and DRC. The green manure and cover crops are primarily used to enhance land productivity by improving N economy in the production systems. Several legume species have been evaluated and found to be well adapted to the region and produce substantial amounts of biomass. Among the potential green manure species identified were *Crotalaria ochroleuca*, *Mucuna pruriens* (L.) DC, *Lablab purpureus* (L), and *Tephrosia vogelii* Hook, which were tested extensively in on-farm experiments for their potential to improve productivity and to obtain feedback based on farmers' perception (Wortmann et al., 1994; Esilaba et al., 2001). High N concentration in *Crotalaria* biomass grown under N-limiting conditions indicated that large quantities of N were biologically fixed. Mean maize grain yields following *Crotalaria* sole crop were 180% and 240% of maize grain yields following maize in two on-station trials and nine on-farm trials, respectively. In the study done by Fishler and Wortmann (2001), grain yields of maize and bean following one season of *Crotalaria* fallow were 41% and 43%, respectively, and the yields were more than following a two-season weedy fallow. Grain yields of maize following a one-season fallow with *Mucuna* and *Lablab* were 60% and 50% higher, respectively, compared to maize following maize. Maize and bean yield were more, although effects were small, during the second and third subsequent seasons, indicating additional benefits from the residual effects of the green manures.

In spite of the demonstrated advantages of green manure and cover crops, the adoption by farmers has always been very poor, hampered by several factors such as availability of seeds, difficulties in planting and establishing the crop, and labor intense practice such as mowing and incorporating the cover crop (Esilaba et al. 2005). Alternative methods of green manure integration on farm were studied, particularly simultaneous cultivation of green manure with main crop, bean or another crop. In fact, in densely populated regions, leaving land under green manure for the whole season is not always acceptable to farmers. Therefore, intercropping with main crop has been developed. In Eastern Uganda, *Mucuna* and *Lablab* were successfully produced by intersowing into maize at three weeks after sowing maize, although the yields of the associated maize crop were reduced by 24% to 28%, and farmers estimated the labor requirements for *Mucuna* and *Lablab* to be less than for *Crotalaria* (Fishler and Wortmann, 2001). In Uganda, the yields of the green manure species were reduced by 40–70% when intercropped with a food crop as compared to sole crop production and that yields of food crops were reduced by 61–87% when intercropped with *Crotalaria ochroleuca* G. Don. In contrast, maize grain yield response in the first season following sole-crop of green manure production ranged from 0 to 240% (Fishler and Wortmann, 2001). Production of the green manure by intercropping *Crotalaria* with either maize or beans was found to be feasible with little reduction in food crop yield with a mean Land Equivalent Ratio of 1.3.

Additional benefits of cover crops reported in Uganda are weed suppression, soil erosion control and control of mole rats (Fishler and Wortmann, 2001). Results from these trials also indicated that *Mucuna pruriens* (L.) and *Lablab purpureus* (L) were best for weed suppression and to control soil erosion. Also the requirements for tillage and weeding could be lower following green manure and cover crops. In addition, maize can often be planted directly in the holes left from uprooting *Mucuna* and *Lablab*, thus reducing labor requirements during the following season. *Tephrosia vogelii* Hook. f. was effective in controlling mole rats and there was significant adoption by neighboring farmers.

Climbing Bean Rotation Effects

Climbing bean is by far the bean type with high biomass production and probably with high BNF, and therefore considerable green manure effect is expected. Its integration in the production system can improve productivity and sustainability of bean-based cropping systems. In fact, climbing bean develops extensive nodulation, up to three times more than bush bean (Van Schoonhoven and Pastor-Corrales, 1994). This is an indication of higher capacity for N fixation. It has been suggested that the contribution from N fixing legume in rotation could be responsible for most of the beneficial rotation effect observed in the subsequent maize crop (Sanginga, 2003; Bado, 2002; Balldock et al, 1981). Evidence of net positive soil N balance by climbing bean has been reported (Kumarasinghe et al. 1992). In this study, they reported that at the late pod-filling stage the climbing bean had accumulated 119 kg N ha^{-1}, 84% being derived from fixation, 16% from soil, and only 0.2% from the ^{15}N fertilizer. In a long-term experiment, Wortmann (2001) reported improved sorghum yield in the rotation with climbing bean and estimated that N derived from the atmosphere was 40% to 57% of plant N, depending on the calculation used. It is clear that evidence of benefit of climbing bean cultivation exists, either as rotational effects or improved N nutrition.

On-station and on-farm farmer participatory trials were conducted by Lunze and Ngongo (2011) to assess the beneficial effects of the climbing bean on the subsequent maize crop in rotation, compared to bush bean and continuous maize cropping systems. The estimate of N contribution from the climbing bean to the system estimated as N fertilizer replacement values (Bado, 2002), and these values varied from 15 kg N ha^{-1} to 42 kg N ha^{-1} in the first season of rotation. Note that this method is believed to overestimate the amount of N supplied by the legumes in the systems (Bullock, 1992). Without applied N fertilizer, the average maize grain yield increase over three cropping seasons in response to the preceding climbing bean effect were 489 kg ha^{-1} and 812 kg ha^{-1} compared to bush bean and maize as preceding crops, respectively, which is 17.5% and 33.8% increase. However better yield advantage of climbing bean over continuous maize was obtained in the long rains cropping season (Season A), 43.2% compared to 24.2% in short rains season (Season B) (Table 13).

Table 13: Effects of preceding crop and fertilizer N on maize grain yield over 3 seasons at Mulungu, DR Congo (Lunze and Ngongo, 2011)

Preceding crop	N Rate (kg/ha)	Maize grain Yield (kg/ha)		
		Season 2002A	Season 2002 B	Season 2003 A
Bush bean	0	2911	2735	2625
	33	4542	3738	3656
	66	5297	4206	4003
Climbing bean	0	3821	2847	3072
	33	4869	4115	4000
	66	5019	4375	4278
Maize	0	2724	2477	2102
	33	4141	3556	2872
	66	4836	3931	3897
Mean Yield LSD (0.05) CV (%)		4540 334.9 6.5	3558 707.1 9.8	3336 379.2 10.0

These studies in Uganda and DR Congo provide evidence for beneficial effects of climbing bean to improve N nutrition of the following maize crop. It is presumed that climbing bean promotion is an appropriate strategy for higher productivity and sustainability on smallholder farms, as substantial gain of N will be achieved by proper integration of climbing bean in the bean-based production systems in ECSA. Additional effects of growing climbing bean are expected, such as reducing erosion because it provides more extensive soil cover through its canopy compared to bush bean. These beneficial effects are considerable on relatively medium fertility soils where climbing bean effects were highest, because of the large quantity of biomass produced.

BIOLOGICAL NITROGEN FIXATION IN BEAN

The enhancement of the capacity of beans for biological fixation through symbiosis with *Rhizobium phaseoli* was recognized by biological nitrogen fixation(BNF) working group within PABRA as an important option to improve the productivity of bean crop on farms (Nyabienda, 1988). This was particularly true for regions where farmers have limited access to N fertilizers. However, improvement of bean BNF requires a multiplidisciplinary approach that will require the plant breeder to increase the host capacity to fix N_2 as bean is considered as a poor N_2 fixing legume (Giller, 2001), and selection of effective *Rhizobium* strains that can compete for nodulation with native populations of bacteria present in most soils.

N_2 Fixation Variability in Bean Germplasm

BNF is among the strategies extensively investigated as one of the avenues to improve bean productivity. Matheson (1997) measured N_2 fixation at Kawanda and Namulonge, Uganda using two common methods: ^{15}N abundance method and N-difference method using non-nodulating beans lines INIAP 404 and EXRICO. N_2 fixed was lower at Namulonge than at Kawanda. It ranged from 10.6 to 35.1 kg Nha^{-1} and from 0.7 to 20.4 kg N/ha, respectively, at Kawanda and Namulonge. This study found that the genotypes identified as tolerant to low soil N had greater N fixing capacity. Table 14 summarizes the genotypic variation in N_2 fixation of common bean at Kawanda, Rwanda.

The same author calculated the net N balance and found a negative net N balance for all varieties, ranging from - 23.6 kg ha^{-1} to - 5.3 kg ha^{-1}, indicating that more N had been exported in grain at harvest than had been fixed and retained in soil. The yield was negatively related to net N balance: the higher the yield, the more negative the net N balances. However, unlike bush bean, climbing beans exhibit greater N_2 fixation capacities (Wortmann, 2001; Lunze and Ngongo, 2011). Lunze and Ngongo (2011) estimated N_2 fixation by climbing bean r from 16 to 42 kg ha^{-1} per season. N fixation is further enhanced by several cultural practices and agronomic management, i.e. inoculation, P fertilization and liming, that are discussed below.

Inoculation

BNF is extensively investigated as one of the avenues to improve bean productivity. Rhizobial inoculation is a common practice to promote BNF in bean, given that the *Rhizobium* strain is well adapted to local environments. Strain CIAT 899 is currently used for bean seed inoculation in all regions of ECSA. Inoculation with this strain in Burundi showed a yield advantage up to 59% in Gitega (Ruraduma, 2002). Evaluation by the same author of local *Rhizobium* strains so far has failed to identify a more efficient strain.

Liming and P Fertilization Effects

The efficiency of BNF is further improved by liming and P fertilization. This is not always affordable for smallholder farmers. Considering these factors, bean seed pelleting with lime and phosphate was evaluated in Burundi and Malawi. On-farm trials at Ikulwe, Iganga District of Uganda, the %N_2 fixation was estimated for a well-adapted high yielding variety MCM 5001, considering the effects of inoculation and P application on N_2 fixation. Neither inoculation nor 100 kg ha^{-1} of TSP (triple super phosphate) application increased bean yield, but 100 kg ha^{-1} TSP with inoculation yielded significantly higher yields than these treatments individually. Inoculation had therefore greater effects on bean yield at higher P levels. The estimate of % N_2 fixation without addition of P fertilizer and inoculation ranged from 5% to 32%, while the addition of 50 kg $ha^{-1}$of TSP increased N_2 fixation by 50%. However, they did not observe any significant differences ($P<0.05$) in %N_2 fixation between treatments and farms.

In Malawi, Chilimba and Kapapa (2002) evaluated bean seed pelleting with dolomitic lime. On-farm trials conducted in Dedza where soils are strongly acidic showed significant responses of four bean varieties to inoculation and seed pelleting with lime on BNF and grain yield. Seed pelleting and inoculation showed beneficial effects in increasing plant N content, nodule numbers and even grain yield. In acidic soils, both inoculation plus seed pelleting significantly improved BNF and grain yield whereas in a normal non-acidic soil, inoculation only or pelleting only significantly improved BNF and grain yield. The on-farm evaluation was conducted in Bembeke Extension Planning Area where soil pH is 4.5 and the soil is fine, kaolinitic, thermic, Kandiudafic Eutrudox while at Chitedze research station the soil pH is 5.9 and the soil is classified as fine, kaolinitic, thermic, Udic Kandhaplustalfs.The results on the evaluation on soils with varying soil pH are presented in Table 15.

Table 14: Nitrogen fixation, biomass and yield of 30 varieties grown under low N conditions at Kawanda (1996)

Variety	Biomass kg/ha R9	Yield (kg/ha)	Total N (kg/ha)	% N fixed	N2 Fixed (kg/ha) 15N method	N2 fixed (kg/ha) Difference Method
BAT 1297	218,4	1033	63,9	43	26,8	28,6
BAT 308	217,9	969	51,5	34,4	17,6	16,2
INIAP 404	136	601	31,1	-12,5	-3,7	-4,2
CNF 5513	220,6	1413	61,5	35,6	22,1	26,2
H2 MULATHINO	230	1061	61,5	39,7	23,1	26,5
IBHBN 69	241,9	1161	57	45,4	25,2	21,7
IBR (92B) 43	184	890	52,6	31,2	16,1	17,3
MCM 1015	201,7	1090	52,9	41,4	22	17,6
MCM 1016	223,8	1015	60	35,1	22,2	24,7
MLB645689A	202,4	1279	55,8	42,1	21,7	20,5
MMS 243	230,5	1000	69,4	36,2	23,3	34,1
MMS 250	207	1175	56,3	38,7	21,9	21
MMS 253	176,4	1080	50,2	51	25,9	14,9
MORE 90040	230	1318	62,1	55,1	33,9	26,8
MUS 97	217,7	1465	58,8	44,4	25,9	23,5
EXRICO	199,1	545	39,5	12,5	5,1	4,2
RWK 5	186,9	1326	51,5	39,7	19,9	16,2
RWR 109	209,8	1025	59,7	44,8	27,4	24,2
RWR 382	267,9	956	68,7	50,9	35,1	33,4
UBR (92) 09	198,3	1304	57,8	58,4	33,7	22,5
UBR (92) 10	233,2	941	66,2	33,1	21,3	30,9
UBR (92) 11	197,5	1155	53,9	49,5	26,1	18,6
UBR (92) 12	209,7	1209	58,7	29,4	10,6	23,4
UBR (92) 17	180,4	888	52,3	49,4	25,4	17
UBR (92) 20	258,2	889	69,9	42	28,8	34,6
UBR (92) 25	202	1132	61,9	47,6	29,4	26,6
UBR (92) 38	208,8	788	65,3	38,9	52,1	30
XAN 76	185,3	1416	54	50,6	26,1	18,7
CAL 96	201,4	962	59,9	39,2	23,4	24,6
MCM 5001	171,7	1004	52,8	49,6	26,1	17,5
Mean	208,3	1070	57,2	39,9	22,9	21,9
s.e.d.	27,37	129	8,77	10,93	6,129	8,77

INIAP 404 and EXRICO: non-nodulating reference lines; CAL and MCM 5001: check varieties

Table 15: Effect of inoculation and seed pelleting on % N, nodule number and the yield of beans at three locations (Marieta, Ester and and Chitedze) in Malawi

	Marieta			Ester			Chitedze		
Treatments	% N	Nodule No.	Grain yield (kg ha-1)	% N	Nodule No.	Grain yield (kg ha-1)	% N	Nodule	Grain yield (kg ha-1)
Control	3.12	15.5	1056	2.05	21.6	389	2.30	10,8	1438
Inoculation	3.64	82.7	1599	3.48	85.5	1119	3.56	78,8	1937
Pelleting	2.18	18.3	1043	2.68	28.0	911	2.01	12,3	2410
Inoc & Pell	4.52	115.1	1981	4.37	117.6	1612	4.40	94,4	2184
Mean	3.35	57.9	1419,75	3.12	63.2	1007,7	3.07	49.0	1992.25
SE	0.483	4.839	231.2	0.428	5.07	230.4	0.4677	2,66	156.1
CV%	21.3	28.97	28.3	12.8	27.83	39.7	58.9	18,8	26.13

Adapted from Chilimba and Kapapa, 2002

The results of the above studies confirm that both lime and P fertilizer application rate can be considerably reduced by coating bean seeds to achieve relatively higher yield. This technology is considered a good ISFM strategy.

MINERAL FERTILIZERS

Most smallholder farmers in the ECSA regions are well aware of the value of mineral fertilizers, but the rate of application remains low, below the recommended rates, except for commercial farmers. The essential reason for this is the high cost of mineral fertilizers, and low profitability. In most regions however, bean comes after cereals, which are commonly fertilized so that bean benefits from residual fertilizer effects. Nonetheless, fertilizer recommendations have been developed in several countries for bean, particularly in Kenya, Tanzania, Burundi and South Africa.

On Ferrarsols (Oxisols) of Burundi, mineral fertilizer applied alone on bean was found to be non-economical, and under certain conditions bean yield was depressed. The recommended rate (kg ha^{-1}) for Burundi was 15 – 30 for N, 50 – 60 for P_2O_5 and 30 for K_2O, with 2 to 5 t organic manure (Ruraduma, 2002). The ISFM strategy suggests options that combine organic resources and mineral fertilizers to achieve higher yield, economically. Fertilizer application to bean crop is developed for those grown as a sole crop or intercropped with maize. In addition to determining the rate of application, the proper time of application is important to maximize N_2 fixation, as early application of N can inhibit nodulation of bean. In Uganda, Wortmann, (1998) recommended applying a small amount of N at the planting time and the major part of fertilizer N at the second weeding time, to favor bean productivity and N2 fixation. That is an application of 5 kg N ha^{-1} and 20 kg P ha^{-1} at sowing time and applying 35 kg N ha^{-1} at second weeding time for both maize and bean that are intercropped. In on-farm trials with the objective of developing appropriate fertilization practice on snap bean, Ugen et al. (2009) evaluated rates and timing of the mineral fertilizers CAN, NPK 17-17-17 and DAP (diammonium phosphate), which are the common fertilizers in the marketplace. In this study, application of DAP at planting and topdressing with NPK-17:17:17 at 21 days after emergence appeared to be appropriate for bean production. More complete mineral fertilizer recommendations used in South Africa are presented in Tables 16, 17 and 18 (Liebenberg, 2002).

Nitrogen

According to Liebenberg (2002), inoculation of dry bean seed is regarded as ineffective in South Africa. Consequently, dry beans should be considered as incapable of satisfying all of their N requirements through N_2-fixation. The application of all the N fertiliser at planting time is recommended, particularly where non-decomposed material has been ploughed in before planting.

Table 16: Guidelines for nitrogen application

Yield potential (t ha^{-1})	1,5	2,0	2,5
N fertilisation (kg ha^{-1})	15,0	30,0	45,0

Phosphorus

Commercial production in South Africa showed modest yield responses to P fertiliser application in dry beans and P is not normally a yield-restrictive factor. Under subsistence production, where small quantities of fertilizer are applied, P can be a yield-limiting factor. Where the P content of the soil is lower than 20 μg g^{-1} (Bray 1), it is recommended that TSP be broadcasted and ploughed into the soil to a depth of 15 to 20 cm before planting. P fertiliser must still be band-placed at the time of planting. In low pH soils, P can be utilised efficiently by band-placing 3.5 cm to the side of the row and 5 cm below the seed.

Table 17: Guidelines for phosphorus fertilisation

Soil analysis Bray 1 (mg kg^{-1})	P application for potential (t ha^{-1})		
	1,5	2,0	2,5
	P fertilization (kg/ha)		
13	16	22	28
20	12	16	20
27	10	13	16
34	9	12	15
> 55	5	5	5

Potassium (K)

When dry beans are grown on soils with high clay content, K is not normally a limiting factor. Deficiencies are most likely to occur on sandy soils with an analysis of less than 50 mg kg^{-1} K. The optimum leaf K content is 2% potassium.

Table 18: Guidelines for potassium fertilisation

Soil analysis NH_4OAc, pH 7 K (mg kg^{-1})	K application for potential (t ha^{-1})		
	1,5	2,0	2,5
	P fertilization (kg ha^{-1})		
40	22	27	32
59	19	24	29
78	17	21	26
98	15	19	24
> 98	0	0	0

Effect of Cropping Systems

Wortmann et al. (1996) studied the relationship between maize monoculture to fertilizer N and P and the intercrop response from 62 fertilizer response trials conducted in Kenya. Their results indicate that intercrop was more productive than the monoculture with no fertilizers applied, but overall responses of the systems to applied nutrients did not differ. Maize both in monoculture and intercropped responded more frequently to applied N than did the intercrop bean. Frequency of response to applied P was similar for both crops and both production systems.

Combination with Organic Manure

It is well established that combining mineral fertilizer with organic resources improves fertilizer use efficiency. In Western Kenya's Vihiga District, the recommendation is 50 N kg ha^{-1} and 50 P kg ha^{-1} with 5 t ha^{-1} of farmyard manure (Rachier et al., 2001). The N and P rates are reduced by 50% in the following season. If money or credit is not available, in addition to 5 t ha^{-1} FYM, only 25 kg ha^{-1} N and P are recommended. The following season, the N rate is reduced by 25% and no P is applied. If FYM is available on farm, 5 t ha^{-1} FYM together with 25 kg ha^{-1} N and P are recommended. This rate is reduced by 50% in the second season. If the green manure is grown, then N, P and FYM rates are reduced by 50% in the first season, and in the second season only 25 kg P ha^{-1} is needed. In the study conducted by Ngongo (2002) in DRC, the aim was to reduce the quantities of both mineral fertilizer and *Tithonia* biomass recommended. The author evaluated bean response to application of 4 t ha^{-1} *Tithonia* biomass with varying rates of mineral fertilizer (Table 19).

Table 19: Effects of *Tithonia* biomass in combination with mineral fertilizers on bean yield in DR Congo.

	Bean grain yield (kg ha^{-1})	
Treatments	1998B	1999 A
Control	551 d	1066 cd
4 t/ha *Tithonia*	1132 cd	1744 a
4 t/ha *Tithonia* + 20 kg NPKha^{-1}	1341 bc	2201 ab
4 t/ha *Tithonia* + 40 kg NPKha^{-1}	1313 bc	2296 a
4 t/ha *Tithonia* + 60 kg NPKha^{-1}	1402 bc	2390 a
60 kg/ha NPKha^{-1}	1405 bc	
C.V. (%)	23,41	23.41
Mean (kg. ha^{-1})	1190,67	2004

The results confirmed that less fertilizer could be recommended when *Tithonia* is applied. The application of 4 t ha^{-1} *Tithonia* with 20 kg ha^{-1} of NPK fertilizer was as efficient as the recommended mineral fertilizer rate of 60 kg ha^{-1}. Thus a 3-fold reduction of inorganic fertilizer is possible when *Tithonia is* applied.

MIJINGU PHOSPHATE ROCK

Bean growing environments are well described in all ECSA countries and P deficiency is widely reported as the most important bean production constraint (Wortmann et al., 1998). The low P availability to plants is explained by the nature of the soils that are highly weathered with low total P and/or high P fixing capacity (Rao et al., 1999). Rock phosphate (RP) which is a resource relatively common throughout ECSA countries is recognized as a fertilizer with high potential to improve bean crop productivity having low cost, compared to conventional mineral P fertilizers. The deposits of the RP are reported in many countries and exploited to some extent in Tanzania (Mijingu), Malawi (Tundulu), Zambia (Isoka), South Africa, DR Congo (Kanzi) and Burundi (Matongo). Minjingu Phosphate Rock (MPR), a sedimentary biogenic deposit which contains about 13% total P and 3% neutral ammonium citrate soluble P is reported to be highly reactive (Jama and Van Straaten, 2006).

On acidic soils of Tonga in Rwanda, Nabahungu et al. (2002) studied the effects MPR associated with limestone and green manures (GM) (*Tithonia* and *Tephrosia* biomass) on P uptake and on maize yield. This study showed that MRP significantly contributed to P increase in the soil and resulted in increased P uptake and maize yield. Green manures in combination with MPR increased P uptake significantly. The results indicated that using a combination of limestone, MPR and GM is the best strategy in improving maize productivity on acid soils (Table 20). The effect of RP alone or GM alone is low, whereas the combination of MPR with organic resources improved its effect.

Table 20: Maize yield response to Mijingu rock phosphate and fertilizers (adapted from Nabahungu, 2007)

Treatment	Maize Yield (kg/ha)
Control	148
MPR	3267
Tithonia	3349
Tephrosia	1452
Tithonia + MPR	3993
Tephrosia + MPR	4554
Tithonia + Lime + MRP	5594
Tephrosia + Lime + MRP	5907
CV %	11.1

A combination of RP and GM produces a demonstrated yield advantage over control or each of these treatments alone. The recommendation for acidic soils of Rwanda is to combine organic resources with RP and lime (ground limestone or burnt lime) allow maize yields to increase up to approximately 6 t ha^{-1}. The same recommendation is applied for bean. In Northern Tanzania, the recommended rate of application is 250 kg ha^{-1} MRP combined with FYM at 5 to 10 t ha^{-1} or a GM grown in the previous season (*Mucuna pruriens or Vernonia subligera*).

Other practices for MRP utilization have been developed in Kenya and Tanzania. The RP-fortified compost technology is well accepted by farmers in Western Kenya (Odera and Okalebo, 2009). Another option which combines PR, liming, *Rhizobium* inoculation and bean seeds, developed by a Phosphate rock evaluation project (PREP), known as PREP-PAC. This product comprises PR (2 kg), urea (0.2 kg), legume seed, rhizobial seed inoculants, seed adhesive and lime pellet, packed to fertilize 25 m^2 of land. In Western Kenya the use of PREP-PAC and climbing bean package increased maize and bean yield by 0.72 and 0.25 t ha^{-1} respectively, resulting in a 161% return in investment (Nekesa et al., 1999). PREP-PAC use on bean for the two seasons in seven districts in Easter Uganda showed significant increases of 881 kg ha^{-1} in bean yield, increasing from 1,316 to 2,197 kg ha^{-1} (Esilaba et al., 2005). This product is commercialized, and makes the PR more accessible to smallholder farmers, thus offers opportunity for easier handling and use for increased bean productivity. Another PR from Uganda, Busumbu RP (BRP) was studied by Nabahungu et al. (2007) who reported that composting increased P availability and P recovery from BRP to the extent similar to that of TSP, as well as the availability of Ca and Mg.

LIMING

Lime application is a common soil improvement practice generally recommended on acidic soils. The rate of application is based on exchangeable soil Al concentration. For tropical soils, lime recommendation is 1.5 times the exchangeable Al in t ha^{-1}, the rate which is sufficient to neutralize toxic Al. One of the first studies to determine the optimum rate of lime for bean was done in Malawi by Aggarwal et al. (1994). In this study, bean response to applied calcitic lime at the rate of 0, 25, 50, 75 and 100% of the exchangeable Al was evaluated in 1992-93 cropping season using 15 varieties and in 1993 -94 season using 8 varieties. Additional treatment was lime applied at 100% of exchangeable Al plus P since the soil was P- deficient. The results indicated that although the performance of the varieties was poor due to low fertility, liming caused a linear increase in nodule number, nodule weight, and grain weight up to 75% level of Al neutralization (Table 21). The yield declined at higher rate of lime application due to nutrient imbalance, which might have been induced by lime. Lime recommendation is therefore soundly based on neutralizing the exchangeable soil Al concentration, rather than increasing soil pH.

Table 21: Effect of lime application on seed yield, nodule number, root weight and shoot weight of common bean varieties grown during 2 seasons at Bembeke, Malawi

Lime level to neutralize (%) exch. Al	Seed Yield ($kgha^{-1}$)	Nodule/plant	Nodule weight ($mgplant^{-1}$)	Root weight ($gplant^{-1}$)	Shoot weight ($gplant^{-1}$)
0	147	0.85	7.59	0.41	1.59
25	161	1.52	13.98	0.42	1.63
50	153	2.43	19.11	0.40	1.41
75	266	5.01	37.29	0.43	1.78
100	235	3.95	32.46	0.43	1.64
100 + P	253	4.84	44.93	0.44	1.71
SE	194	0.92	6.88	0.04	0.16

Finely powdered ground limestone, cheaper than lime (neutralizing value of 90 to 100) evaluated in Rwanda, was comparable to agricultural lime found in the marketplace. Thus lime recommendation was devised for ground limestone. Table 22 presents the results of bean response to lime (agricultural lime) and limestone in two seasons (Nabahungu and Ruganzu, 2001).

Table 22: Bean response to agricultural lime and ground limestone at Rubona, Rwanda

Treatments	Bean yield (kg.ha-1)	
	2000 A	2000 B
Control	733	456
Lime 1 t ha^{-1}	1283	295
Limestone 0.5 t ha^{-1}	1300	209
limestone 1 t ha^{-1}	1817	247
Limestone 1.5 t ha^{-1}	1467	333
LSD (0.05)	831	241

Bean yield was very low without lime and doubled with lime application, but was still considered low. For better yield, lime application should be associated with organic resource application, which according to Palm et al. (1997) acid soils by its interactions with the mineral soil in complexing toxic cations and reducing the P sorption capacity of the soil. Examples from DR Congo, Rwanda and Burundi indicate that combining lime with organic manure results in considerable bean yield increase. Rutunga et al. (1998) studied crop rotation system of maize and beans, established at Rubona

(Rwanda) from 1984 to 1992. They evaluated the effects of different types and rates of fertilizers in improving the productivity of acidic Oxisols. Continuous cropping of maize followed by beans for a period of 8 years gave no yield in control plots. A single application of 2 t ha^{-1} of lime increased significantly (p = 0.01) the soil pH, Ca^{2+} content, cationic exchange capacity, and decreased the level of the exchangeable Al. However, this quantity of lime, when applied every two years for a period of eight years, led to overliming. The application of more than 8 t ha^{-1} of FYM annually, combined with 300 kg ha^{-1} of NPK (17:17:17) fertilizer (every six months) significantly improved soil organic C and crop production at Rubona, Rwanda. The high rate 35 t ha^{-1} of FYM or the combination of "lime, FYM and NPK fertilizers" gave the best crop performance.

DECISION SUPPORT TOOLS

Several technologies have been developed and widely tested with successful results, resulting in development of soil fertility management decision guide, particularly in Uganda (Farley, 1998;Esilaba et al. 2001), Kenya (Rachier et al., 2001) and Eastern Congo (Ngongo, 2001). Soil fertility management research results have been translated into recommendations that take into account the soil physical conditions as well as farmers' socio-economic status. Decision support systems (DSS) to the use of inorganic and organic sources have been developed for different environments and socio-economic conditions in bean based production systems in the region at different locations. These guides are efficient tools for realizing better bean crop yield in areas where they are developed. However, the extrapolation to other sites and location with different soil conditions, climate and population density and socio-economic conditions is not guaranteed with acceptable results as they are site specific (Esilaba et al., 2001). A model decision tool for Uganda is presented in Table 23 while for Eastern DR Congo the tool was developed using soil fertility assessment based on farmers' perception (Table 24). Other DSS are similar, but the local conditions and available resources vary.

PROMOTION AND USE OF ISFM OPTIONS

PABRA has developed and disseminated different technologies to address farmers' production constraints and to increase bean productivity in sustainable ways (Kimani et al., 2001). These comprise simple technologies such as new bean varieties and more complex, knowledge intensive ones such as integrated pest and disease management (IPDM) or ISFM. The technologies and their dissemination methods were based on low cost options that were in many cases developed through collaboration with multiple partners, and engaged and empowered end users in participation and adoption (Kankwatsa et al., 2008). The spread and adoption of new bean varieties have been very impressive; reaching several million farming families, while the spread and adoption of ISFM technologies have been slow. Scaling up and disseminating information on recommended technologics was through Farmer Research Groups using a modified farmer field school approach. Because farmers were driving the experimentation and dissemination processes, all aspects of local culture were taken into consideration (Kankwatsa et al., 2008) and they often made recommendation for improvement to suit their production circumstances, both in the social and the technological aspects: this led to the development of the ownership by the farming communities and a high rate of adoption among participating farmers.

Table 23: Tentative guide to fertilizer use for maize and bean in Uganda

Conditions	Maize, sole crop	Bean, sole crop	Maize-bean intercrop
Adequate money or credit available	Apply 50 kg ha^{-1} TSP and 25 kg ha^{-1} urea at sowing; apply 50 kg ha^{-1} urea at second weeding	Apply 100 kg ha^{-1} TSP and 20 kg ha^{-1} urea at sowing	Apply 100 kg ha^{-1} and 20 kg ha^{-1} urea at sowing; apply 50 kg ha^{-1} urea at 2nd weeding
Money or credit is inadequate	Apply 50 kg ha^{-1} urea at first weeding	Apply 50 kg ha^{-1} TSP and 20 kg ha^{-1} urea at sowing	Apply 50 kg ha^{-1} TSP and 20 kg ha^{-1} urea at sowing; apply 50 kg ha^{-1} urea at 2nd weeding
Green manure was produced the previous season	Do not apply inorganic fertilizer	Do not apply inorganic fertilizer	Do not apply inorganic fertilizer
Lantana, etc is available	Reduce application of urea at 2nd weeding by 30% for each ton of fresh leafy material applied	Do not reduce fertilizer rate	Reduce application of urea at 2nd weeding by 30% for each ton of fresh leafy material applied
Sowing is delayed until after 15 march or 15 September	Reduce fertilizer rate by 50%	Do not reduce fertilizer rate	Reduce fertilizer rate by 50%
Sowing is delayed until after 30 march or 30 September	Do not use fertilizer at sowing; top-dress urea at 50% rate if conditions are promising	Do not reduce fertilizer rate	Apply 50% of TSP at sowing; top dress urea at 50% rate if conditions are promising
Farm yard manure in available	Reduce fertilizer by 25% for each ton/ha of dry FYM applied	Reduce fertilizer by 40% for each ton of dry FYM applied	Reduce fertilizer by 20% for each ton/ha dry FYM applied
FYM was applied last season	Reduce fertilizer by 15% for each ton/ha of dry FYM applied	Reduce fertilizer by 30% for each ton of dry FYM applied	Reduce fertilizer by 10% for each ton/ha dry FYM applied
Land was rotated from banana or fallow within last one season	Apply N at 2nd weeding, but only if maize is yellowing	Do not apply fertilizer	Apply N at 2nd weeding, but only if maize is yellowing

Top-dress with urea if the crop is well established, the season appears promising, and especially if the lower leaves are yellowish-green in color.

Table 24: Decision guide for ISFM in a Bean-based cropping system in Eastern DR Congo

Farmer Conditions		Soil type according to farmer criteria*	Bean type	Cropping system	Soil fertility management Recommendations
Finances	**Organic resources**				
No money, no Credit	Compost not available	Plot dominated by *Gallinsoga parviflora*	Bush	Bush beans sole crop	No fertilizers
				Bush beans-maize intercropped	Apply 20 t ha⁻¹ Kitchen ash
			Climbing	Climbing beans sole crop	Apply 4 t ha⁻¹ *Tithonia* fresh biomass + 20 t ha⁻¹ ash
				Climbing beans- maize intercropped	Apply 6 t ha⁻¹ *Tithonia* fresh biomass + 20 t ha⁻¹ ash
		Plot dominated by *Pennisetum polystachia* or *Bidens pilosa* (poor soils)	Bush	Bush beans sole crop	6 t ha⁻¹ *Tithonia* + 20 t ha⁻¹ ash
				Bush beans-maize intercropped	8 t ha⁻¹ *Tithonia* + 20 t ha⁻¹ ash
			Climbing	Climbing beans sole crop	10 t ha⁻¹ *Tithonia* + 20 t ha⁻¹ ash
				Climbing beans- maize intercropped	10 t ha⁻¹ *Tithonia* + 30 t ha⁻¹ ash
	Compost available	Plot dominated by *Gallinsoga parviflora*	Bush	Bush beans sole crop	No of fertilizers
				Bush beans-maize intercropped	20 t ha⁻¹ ash
			Climbing	Climbing beans sole crop	6 t ha⁻¹ *Tithonia* or 20 t ha⁻¹ compost
				Climbing beans- maize intercropped	8 t ha⁻¹ *Tithonia* or 20 t ha⁻¹ Compost + 20 t ha⁻¹ ash
		Plot dominated by *Pennisetum polystachia* (poor soils)	Bush	Bush beans sole crop	4 t ha⁻¹ *Tithonia* or 10 t ha⁻¹ compost
				Bush beans-maize intercropped	(6 t ha⁻¹ *Tithonia* or 20 t ha⁻¹ compost)
			Climbing	Climbing beans sole crop	6 t/ha *Tithonia* or 20 t ha⁻¹ compost
				Climbing beans- maize intercropped	(8 t ha⁻¹ *Tithonia* or 30 t ha⁻¹ compost) + 30 t ha⁻¹ ash
Money or credit Available	Compost not available	Plot dominated by *Gallinsoga parviflora*	Bush	Bush beans sole crop	30 kg/ha (P,K)
				Bush beans-maize intercropped	4 t ha⁻¹ *Tithonia* + 50 kg ha⁻¹ (P,K)
			Climbing	Climbing beans sole crop	6 t ha⁻¹ *Tithonia* + 50 kg ha⁻¹ (P, K)
				Climbing beans- maize intercropped	6 t ha⁻¹ *Tithonia* + 50 kg/ha (N.P,K)
		Plot dominated by *Pennisetum polystachia* or *Bidens pilosa* (poor soils)	Bush	Bush beans sole crop	6 t/ha *Tithonia* + 50 kg/ha (N,P,K)
				Bush beans-maize intercropped	6 t/ha *Tithonia* + 75 kg/ha (N,P,K)
			Climbing	Climbing beans sole crop	8 t ha⁻¹ *Tithonia* + 50 kg ha⁻¹ (N,P,K)
				Climbing beans- maize intercropped	8 t ha⁻¹ *Tithonia* + 75 kg ha⁻¹ (N,P,K)

	Compost available	Plot dominated by *Gallinsoga parviflora*	Bush	Bush beans sole crop	4-6 t ha^{-1} *Tithonia* or 20 t ha^{-1} compost
				Bush beans-maize intercropped	(8 t ha^{-1} *Tithonia* or 30 t ha^{-1} compost) + 50 kg/ha (N,P,K)
			Climbing	Climbing beans sole crop	(8 t ha^{-1} *Tithonia* or 20 t/ha compost) + 50 kg ha^{-1} (P,K)
				Climbing beans- maize intercropped	(8 T t ha^{-1} *Tithonia* or 20 t ha^{-1} compost) + 75 kg ha^{-1} (N,P, K)
		Plot dominated by *Pennisetum polystachia* or *Bidens pilosa* (poor soils)	Bush	Bush beans sole crop	(4 t ha^{-1} *Tithonia* or 10 t ha^{-1} compost) + 50 kg ha^{-1}(P,K)
				Bush beans-maize intercropped	(6 t ha^{-1} *Tithonia* or 20 t ha^{-1} compost) + 50 kg ha^{-1} (N,P,K)
			Climbing	Climbing beans sole crop	(8-10 t ha^{-1} *Tithonia* or 20 t ha^{-1} compost) + 50 kg ha^{-1}(N,P,K)
				Climbing beans- maize intercropped	(10 t ha^{-1} *Tithonia* or 30 t ha^{-1} compost) + 75 kg ha^{-1} (N,P,K)

Awareness and adoption of ISFM technologies by farmers is found to be essential factor to their eventual widespread adoption in ECSA. Kankwatsa et al. (2008) found that farmer's awareness varied significantly with their participation to the technology promotional activities of farmers. They reported that participating farmers were more rational in their choice of technologies to solve specific constraints and more participating farmers adopted the IPDM and ISFM technologies than the non-participating farmers. To favor farmers' capacity to experiment and eventually to adapt technological options to their biophysical and socioeconomic conditions, they were exposed to a whole range of new soil fertility management options through farmers' field schools and dialogue with researchers in the process of farmers' participatory research. They have continued to experiment and ended to suggest modifications to the technologies. In Bushumba, Estern DRC, Njingulula and Ngongo (2007) reported the results of participatory evaluation of the rate of *Tithonia* application. They found that although the rate recommended by research was not much modified, farmers found the better manner to express the rate of application using a common volume measure. Farmers' recommendation of *Tithonia* biomass application was the rate of 3 to 4 basins per 4 m x 4 m portion of land, which is equivalent to 5 to 7.5 t ha^{-1}, in perfect agreement with the recommendation by researchers. Going through this process increased their confidence in their own ability to find solutions to different problems, and improved the rate of adoption of suitable technology. The impact studies were conducted by Kankwatsa et al (2008) in PABRA member countries to evaluate the fate of promoted ISFM technologies.

CONCLUSIONS

African smallholder farming conditions are worsened by declining soil fertility as a consequence of population pressure on a limited landbase. PABRA, as an African research for development program has made outstanding efforts in establishing partnership with numerous stakeholders, farmers, rural communities, non-governmental and governmental organizations, private sector, traders and research organizations (national, regional and international). In spite of complexity of ISFM technologies developed, bean productivity has been improved under low soil fertility conditions by developing strategies and technologies that enhance resilience to environmental stresses and by enhancing farmers' access to adapt and use cost effective integrated environmental stress management options (PABRA, 2008). The strategy of PABRA on ISFM has two broad avenues for achieving the objective. These include soil fertility management and deployment of resilient bean germplasm.

Genotypes with improved performance on low fertility soils is valued by farmers and readily adopted, and have made it possible to improve bean production in regions where they are adopted at no additional cost. Identification and use of cultivars tolerant to mineral deficiencies and toxicities are essential for reducing production costs and dependence of farmers on soil amendment inputs. The resilient bean cultivars have been shown superior to existing popular varieties developed without consideration of tolerance to soil infertility tolerance. Many bean genotypes have been selected with tolerance to single or multiple edaphic stresses (Wortmann et al., 1995; Rao, 2001). Therefore, as stated by Lynch et Beebe (1995), efficient cultivars would have several benefits, including increased food production, increased farm income, increased bean availability and consumption, and thereby improved nutrition for low income producers and consumers. Indeed, improved variety is considered as a fast adoption technology (Kankwantsa et al., 2008). The nurseries of these genotypes are maintained with the PABRA regional networks with characteristics of market preferences, besides tolerance to low soil fertility.

Bean productivity is further enhanced by various soil management options developed in participatory manner. The use of farmer participatory approach to technology dissemination enabled farmers to be familiar with, understand, experiment and adapt ISFM technologies. Through capacity building through farmers' groups and farmer research groups, farmers have been empowered to obtain various technologies and services from appropriate partners. Various locally available resources are used, as limestone and phosphate rock, alone or in combination with mineral fertilizers, and the effect of organic resources in improving nutrient use efficiency is well demonstrated. Considerable and beneficial bean yield increase is achieved. However, the low production of organic resources on farms calls for biomass transfer. *Tithonia* biomass is readily available and adopted by farmers and contributes considerably to bean productivity enhancement as an essential source of nutrients and organic amendment. Maximal use of locally available nutrients through low-external input technologies, techniques, combined with optimal use of external nutrients appears to be the most appropriate strategy in the existing economic environment (de Jager, et al., 2004).

Based on research data on soil fertility management in major bean growing areas, a comprehensive decision support system (DSS) was developed for use by farmers and extension workers in different agro-ecological zones and under diverse socio-economic scenarios. In contrast to standard guidelines, the DSS provides farmers with ISFM options under different scenarios, thereby allowing farmers the ability to choose the option(s) that best suit their needs and socio-economic conditions. Soil fertility conditions are assessed using criteria and indicators that are easy to measure, such as dominant weed species. The desired outcome is that the tool will be used by researchers, extension workers, and farmers for assessing and implementing options of using scarce resources for maintaining soil fertility and improving crop yields in bean-based cropping systems.

PABRA places special emphasis on increased access to cost effective and environmentally friendly integrated bean production, especially by female bean growers. Currently, two approaches are used to facilitate and increase farmer's access to ISFM technologies: a deliberate promotion and delivery of improved varieties and ISFM technologies as a single package; and policy advocacy and harnessing of enabling policies (including input support system) to deliver ISFM technologies to bean farmers. The Alliance (PABRA) feels confident that these approaches will greatly enhance the adoption of ISFM technologies by bean growers and overall bean production in eastern, central and southern Africa.

REFERENCES

1. Aggarwal, V. D., Mughogho, S. K., Chirwa, R. & Mbvundula, A. D. (1994). Results of testing for tolerance to a low pH complex in Malawi. *Proceedings of a Working group meeting. Bean improvement for a low fertility soils in Africa*. CIAT Africa Workshop Series No. 26, Kampala, Uganda, January 23 – 26, 1994
2. Bado, B.V. (2002). Rôle des légumineuses sur la fertilité des sols ferrugineux tropicaux des zones guinéenne et soudanienne du Burkina Faso. *Thèse de Philosophiae Doctor (Ph. D.)*, Département des Sols et de Génie Agroalimentaire, Faculté des Sciences del'Agriculture et de l'Alimentation Université Laval, Québec, Décembre 2002
3. Barrios, E., Bekunda, M., Delve, R., Esilaba, A. & Mowo, J. (2001).

Identifying and classifying local indicators of soil quality: Methodology for decision making in natural resource management. Eastern Africa version. *CIAT, SWNM, TSBF, AHI.* 2001

4. Bationo, A., Hartemink, O., Lungu, M., Naimi, P., Okoth, E., Smaling& Thiombiano L.(2006). African soils: Their productivity and profitability of fertilizer use. *Background paper prepared for the African Fertilizer Summit,* Abuja June 9 – 13, 2006
5. Beebe, S., Rao, I. M., Blair, M. W. & Butare L. (2009). Breeding for abiotic stress tolerance in common bean: present and future challenges. In: *Proc 14th Australian Plant Breeding & 11th SABRAO Conference*, 10 to 14 August, 2009, Brisbane, Australia.
6. Bullock DG (1992) Crop rotation. Crit Rev Plant Sci 11(4):309–326
7. Chilimba, A.D.C. & Kapapa, C. (2002). On-Farm Evaluation of Rhizobium inoculation and seed pelleting and Inoculation of Field Beans (*Phaseolus vulgaris*) in acid soils to enhance nodulation and grain yield. *Soil fertility management and bean improvement for low fertility soils in Africa "BILFA" Working Group Meeting,* Awassa, Ethiopia, November 12 – 16, 2002
8. Dreschel, P., Gyiele, L., Kunze, D., & Coffie, O. (2001). Population density, soil nutrient depletion and economic growth in Sub-Saharan Africa. *Ecological Economics*, Volume 23, pp. 152 – 258
9. Esilaba, A.O., Byalebeka, J.B., Delve, R.J., Okalebo, J.R., Ssenyange, D., Mbalule, M., & Ssali H. (2005). On farm testing of integrated nutrient management strategies in eastern Uganda. *Agricultural Systems*, Volume 86, pp. 144–165
10. Esilaba, A.O., Byalebeka, J.B., Nakiganda, A., Mubiru, S., Delve, R.J., Ssali, H. & Mbalule M. (2001). Integrated Nutrient Management strategies in Eastern Uganda. CIAT, Kampala. *CIAT Africa Occasional Publication Series*, No. 35, pp. 71
11. Fishler, M. & Wortmann, C.S. (2008). Green manures for maize-bean systems in Eastern Uganda: Agronomic performance and farmers' perceptions. *Agroforestry Systems*, Volume 47, Number 1 – 3, pp. 123 – 138
12. Giller, K. E. (2001). Nitrogen fixation in tropical cropping systems. 2 ed. CABI, Wallingford, Oxon UK.
13. Giller K.E., Amijee F., Brodrick S.J.& Edje O.T. 1998. Environmental constraints to nodulation and nitrogen fixation of Phaseolus vulgaris L. in Tanzania. II. Response to N and P fertilizers and inoculation with Rhizobium. African Crop Science Journal 6(2):171-178

14. Ikerra, S. T., Semu, E. & Mrema, J. P. (2007). Combining tithonia diversifolia and Mijingu phosphate rock for improvement of P availability and maize grain yield on a Chromic Acrisol in Morogoro, Tanzania, In: *Advances in integrated soil fertility management in sub_Saharan Africa: Challenges and Opportunities*, Bationo, A., Waswa, B., Kihara, J. & Kimetu J. pp. 333 – 344, Springer, Dordrecht, The Netherlands

15. Jama, B. & Van Straaten, P. (2006). Potential of East African phosphate rock deposits in integrated nutrient management strategies. *Anais da Academia Brasileira de Ciencias*, Volume 78, Number 4, pp. 781-790

16. Kumarasinghe KS, Danso SKA,& Zapata F (1992) Field evaluation of N-2 fixation and N-partitioning in climbing bean (Phaseolus vulgaris L.) using N-15. *Biol Fertil Soils* 13:142–146

17. Kankwatsa P., Ampofo K., Kasambala C. & Mukankusi C. (2008). Analysis of the socio-economic validity of new bean based integrated soil fertility management and integrated pest and disease management technologies: Uganda, Malawi, Swaziland and Eastern DRC. CIAT, Kampala

18. Kimani, P. M., Buruchara, R., Ampofo, K., Pyndji, M., Chirwa, R. M., & Kirkby, R. (2001). Breeding Beans for Smallholder Farmers in Eastern, Central, and Southern Africa: Constraints, Achievements, and Potential. *PABRA Millennium Workshop*, Arusha, Tanzania, 28 May – 1 June 2001

19. Liebenberg, A.J. (2002). Dry bean production. Directorate Agricultural Information Services, Department of Agriculture in cooperation with ARC-Grain Crops Institute, South Africa

20. Lunze, L. (1994). Screening for tolerance to aluminum toxicity in bean. In. *Bean Improvement for Low Fertility in Soils Africa: Proceedings of a Working Group Meeting*. Kampala, Uganda, May, 23-26, 1994

21. Lunze, L. & Ngongo, M. (2011). Potential Nitrogen Contribution of Climbing Bean to Subsequent Maize Crop in Rotation in South Kivu Province of Democratic Republic of Congo. In: *Innovations as Key to the Green Revolution in Africa*, A. Bationo et al. (eds.), 677 – 681, Springer Science, Dordrecht, The Netherlands . In press

22. Lunze, L., Kimani, P.M., Ndakidemi, P., Rabary, B., Rachier, G.O., Ugen, M.M. & Nabahungu, L. (2002). Selection of bean lines tolerant to low soil fertility conditions in Africa. *Bean Improvement Cooperative, BIC* Volume 45, pp. 182–183

23. Lunze L., Kimani P.M., Ngatoluwa R., Rabary B., Rachier G.O., Ugen M.M., Ruganza V. & Awad elkarim E.E. 2007. Bean improvement for low soil fertility in adaptation in Eastern and Central Africa. In: *Advances in*

*integrated soil fertility management in sub_Saharan Africa: Challenges and Opportunities,*Bationo, A., Waswa, B., Kihara, J. & Kimetu J. pp. 325 – 332. Springer, Dordrecht, The Netherlands

24. Lynch, J.P., & Beebe, S.E. (1995). Adaptation of beans (Phaseolus vulgaris L.) to low phosphorous availability. *HortScience*, Volume 30, pp. 1165-1171
25. Mastaki, N. J. L. 2006 Le rôle des goulots d'étranglement de la commercialisation dans l'adoption des innovations agricoles chez les producteurs vivriers du Sud-Kivu (Est de la R.D.Congo). *Dissertation originale présentée en vue de l'obtention du grade de docteur en sciences agronomiques et ingénierie biologique. Filière : Economie et Développement Rural.* Faculté Universitaire des Sciences Agronomiques de GEMBLOUX
26. Mauyo, L.M., Okalebo, J.R., Kirkby, R.A., Buruchara, R., Ugen, M. & Maritim, H.K. (2007). Spacial pricing efficiency and regional market integration of cross-border bean (Phaseolus vulgaris L.) marketing in Rest Africa: The case of Western Kenya and Eastern Uganda. In: *Advances in Intergrated Soil Fertility Management in Sub-Saharan Africa: Challenges and Opportunities*, A. Bationo et al. (eds), pp. 1027 – 1033, Springer
27. Musungayi, T., Njingulula, M., Mbikayi, N., & Lunze, L. 2006. Expansion des variétés biofortifiées de haricot et de patate douce à chair orange dans les provinces du Sud Kivu et Nord Kivu en R. D. Congo. Rapport d'Enquête. INERA (Unpublished)
28. Musungayi, T. Sperling, L. Graf, W. & Lunze, L. 1990. Enquêtes diagnostiques de la zone de Walungu, zone d'action de la Femme Solidaire pour le développement du Bushi, Rapport d'enquêtes (Inédit).
29. Nabahungu L. et Ruganzu, V. 2001. Effet du Travertin et de Tithonia diversifolia sur la Productivité du Haricot Volubile en Sols Acides du Rwanda. *PABRA Millenium Symposium*. Arusha, Tanzania, 28 May – 1 June, 2001
30. Nabahungu, N.L, Semoka, J.M.R &. Zaongo, C. (2007). Limestone, Minjingu Phosphate Rock and Green Manure Application on Improvement of Acid Soils in Rwanda.
31. *Advances in Integrated Soil Fertility Management in sub-Saharan Africa: Challenges and Opportunities.*Springer, pp 703-712
32. Nekesa P., Maritim H.K., Okalebo J.R. & Woomer P.L. 1999. Economic analysis of maize-bean production using a soil fertility replenishment

product (PREP-PAC) in Western Kenya. *African Crop Science. J.* Vol 7. No. 4, pp 585-590

33. Ngongo, M. & Lunze, L. (2000). Espèces d'herbe dominante comme indice de la productivité de sol et de la réponse du haricot commun à l'application du compost. *African Crop Science J.* Volume 8, Number 3 pp. 251–261
34. Ngongo, M. (2001). Management of soil fertility in South Kivu with green manure of tithonia. In: PABRA Millenium Symposium..Arusha, Tanzania, 28 May – 1 June, 2001
35. Njingulula, M. (2003). Etude de l'impact socio-économique des variétés améliorées du haricot dans le système de culture paysan au Congo, République Démocratique. Proceedings of a Regional Workshop to Develop a Work Plan for Impact Assessment Studies and Monitoring and Evaluation.Kampala, Uganda. PABRA, CIAT. pp 83–87.
36. Njingulula, P. M. & Ngongo, M. (2007). Initiating rural farmers to participatory research: case of soil fertilization in Bushumba, East of DR Congo. In. *Advances in integrated soil fertility management in Sub_ Saharan Africa: Challenges and Opportunities.* Bationo, A., B. Waswa, Kihara, J. & Kimetu J. (Eds) pp. 1051-1059, Springer
37. Nyabienda, P. & Hakizimana, A. (1988). Atelier sur la fixation biologique de l'azote du haricot en Afrique, CIAT African Workshop Series, No. 8, Rubona, Rwanda, 27 – 29 Octobre 1988
38. Opala PA. 2001. Management of organic inputs in East Africa: A review of current knowledge and future challenges. Archives of Applied Science Research, 2011, 3 (1): 65-76. Available from: www.scholarsresearchlibrary.com
39. PABRA. 2008. Supporting nutrition and health, food security, environmental stresses and market challenges that will contribute to improve the livelihood and create income of resource poor small holder families in Sub-Saharan Africa. PABRA, CIAT, Kampala, Uganda
40. Palm, C.A., Myers R. J. K. & Nandwa, S. M. (1997) Combined use of organic and inorganic nutrient sources for soil fertility maintenance and replenishment. In: *Replenishing Soil Fertility in Africa*, Buresh RJ, Sanchez PA & Calhoun F (eds), pp 193–217. SSSA Special Publication 51. SSSA, ASA, Wadington
41. Pankhurst, C., Doube, B.M. & Gupta, V.V.S.R. (Eds.) (1997). Biological Indicators of SoilHealth. CAB International, Wallingford.
42. Rabary, B. 2001. Participatory research for improved agroecosystemmanagement: a mean for rural community development.

PABRA Millenium Symposium. Arusha, Tanzania, 28 May – 1 June 2001

43. Rachier G.O., Wortmann C.S., Tenywa J.S. and Osiru D.S.O. 1999. Phosphorus acquisition and utilization in common beans: mechanisms and genotype differences. African Crop Science Journal Vol. 4: 187-193

44. Rachier, G.O., Salaya, B.D. & Wortmann, C.S. (2001). Verification of recommended rates of inorganic fertilizers and farmyard manures in maize intercrop under contrasting soil types. In: *PABRA Millenium Symposium*. Arusha, Tanzania, 28 May – 1 June 2001

45. Rao, I. M. (2001). Role of physiology in improving crop adaptation to abiotic stresses in the tropics: The case of common bean and tropical forages. In: *Handbook of Plant and Crop Physiology*, Pessarakli M, (ed). Marcel Dekker, Inc, New York, USA, p. 583–613.

46. Rao, I. M., Friesen, D. K. & Osaki, M. (1999). Plant adaptation to phosphorus-limited tropical soils. In: M. Pessarakli (ed.) Handbook of Plant and Crop Stress. Marcel Dekker, Inc., New York, USA, pp. 61-96

47. Ruraduma, C. (2002). Improving biological nitrogen fixation of beans. *Soil fertility management and bean improvement for low fertility soils in Africa "BILFA" Working Group Meeting*. November 12 – 16, Awassa, Ethiopia.

48. Rutunga, V, Steiner, K.G., Karanja, N.K., Gachene, C.K.K. & Nzabonihankuye, G. (1998). Continuous fertilization on non-humiferous acid Oxisols in Rwanda "Plateau Central": Soil chemical changes and plant production. *Biotechnol. Agron. Soc. Environ*. Volume 2, Number 2, pp. 135–142

49. Rutunga, V. (1997). Sols acides de la région d'altitude de la Crête Zaïre-Nil (Rwanda): Potentialités agricoles et forestières. Lengo Publishers, Nairobi, 68pp.

50. Sanginga, N. (2003). Role of biological nitrogen fixation in legume based cropping systems; a case study of West Africa farming systems. *Plant and Soil,* Volume 252, Number 1, pp. 25-39 (15)

51. Sanginga, N., & Woolmer, P. L, eds. (2009). Integrated Soil Fertility Management in Africa:*Principles, Practices and Developmental Process. Tropical Soil Biology and Fertility Institute of the International Centre for Tropical Agriculture.* Nairobi. 263 p.

52. Singh, S.P. (2001). Broadening the genetic base of common bean cultivars: a review. *Crop Science* Volume 41, pp. 1659–1675

53. Sperling, L., Scheidegger, U., Buruchara, R., Nyabienda, P. & Munyanesa. (1993). Intensifying production among smallholder farmers: The impact

of improved climbing bean in Rwanda. *CIAT African Occasional Publication Series* No. 12. CIAT/RESAPAC, Butare, Rwanda, p13

54. Thung, M. & Rao, I. M. (1999). Integrated management of abiotic stresses. In: *Common Bean Improvemen in the Twenty-First Century*.S. P. Singh (ed.),. Kluwer Academic Publishers, Dordrecht, The Netherlands, pp. 331-370.
55. Tumuhaire, J.B., Rwakaikara, S.M.C., Muwnga, S. & Natigo, S. (2007). Screening legume green manure for climatic adaptability and farmer acceptance in the semi-arid agro-ecological zone of Uganda. In. *Advances in Integrated Soil Fertility Management in Sub-Saharan Africa: Challenges and Opportunities*, A. Bationo et al., pp. 255 – 259, Springer,Dordrecht, The Netherlands
56. Ugen, M. & Wortmann, C. S. (2001). Weed Flora and Soil Properties in Subhumid Tropical Uganda, *Weed Technology*, Vol. 15, Number. 3, pp. 535-543
57. Ugen, M. A., Ndegwa, A. M. Nderitu, J. H., Musoni, A. & Ngulu, F. (2009). Enhancing competitiveness of snap beans for domestic and export markets. ASARECA CGS Revised Full Proposal Document. Entebbe, Uganda
58. Vanlauwe, B., Bationo, A., Chianu, J., Giller, K. E., Merckx, R., Mokwunye, U., Ohiokpehai, O., Pypers, P., Tabo, R., Shepherd, K., Smaling, E., Woomer, P. L. & Sanginga, N. (2010) Integratedsoil fertility management: Operational definition and consequencesfor implementation and dissemination. *Outl onAgric* 39, 17–24
59. Vanlauwe B., Chianu J., Giller K.E., Merckx R , Mokwunye U., Pypers P., Shepherd K., Smaling E, Woomer P.L. & Sanginga N. 2010. Integrated soil fertility management: operational definition and consequences for implementation and dissemination. *19th World Congress of Soil Science, Soil Solutions for a Changing World.* 1 – 6 August 2010, Brisbane, Australia.
60. Vanlauwe B, Tittonell P, &Mukalama J (2006) Within-farm soil fertility gradients affect response of maize tofertilizer application in western Kenya. *Nutrient Cycling in Agroecosystems* 76,171-182.
61. van Schoonhoven, A. & Pastor-Corrales, M.A. (1994) . Système standard pour l'évaluation du germoplasme du haricot. Publication CIAT, Cali, Colombia, No. 207
62. Wortmann, C.S.(2001). Nutrient dynamics in a climbing bean and sorghum crop rotation in the Central Africa Highlands, *Nutrient Cycling in Agroecosystems*, Volume 61, Number 3, pp. 267-272

63. Wortmann, C. S. & Kaizzi, C. K. (1998). Nutrient balances and expected effects of alternative practices in the farming systems of Uganda. *Agriculture, Ecosystems and Environment* Volume 71, pp. 117-131

64. Wortmann, C. S. & Ssali, H. (2001) . Integrated nutrient management for resource-poor farming systems: A case study of adaptive research and technology dissemination in Uganda. *American Journal of Alternative Agriculture*, Volume 16, pp. 161-167

65. Wortmann, C.S., Isabirye, M. & Musa, S. (1994). Crotalariaochroleucaas a Green Manure Crop in Uganda. *African Crop Science J.*, Volume 2, Number 1, pp. 55-61

66. Wortmann, C.S., Kirkby, R.A., Aledu, C.A. & Allen, J.D. (1998). Atlas of common bean (Phaseolus vulgaris L.) Production in Africa, CIAT Cali, Colombia

67. Wortmann, C.S., Lunze, L., Ochwoh, V.A. & Lynch, J.P. (1995). Bean Improvement for Low Fertility Soils in Africa. *African Crop Science J*, Volume 3, Number 4, pp. 469-477

68. Wortmann, C.S., Schnier, H.F. & Muriuki A.W. (1996). Estimation of the fertilize response of maize and bean intercropping using sole crop response equations. *African Crop Science Journal*, Vol. 4. No.1, pp. 51-55, 1996

CITATION

CHAPTER 1

Elmira Saljnikov, Dragan Cakmak and Saule Rahimgalieva (2013). Soil Organic Matter Stability as Affected by Land Management in Steppe Ecosystems, Soil Processes and Current Trends in Quality Assessment, Dr. Maria C. Hernandez Soriano (Ed.), ISBN: 978-953-51-1029-3, InTech, DOI: 10.5772/53557.

CHAPTER 2

C.A. Igwe and S.E. Obalum (2013). Microaggregate Stability of Tropical Soils and its Roles on Soil Erosion Hazard Prediction, Advances in Agrophysical Research, Prof. Stanisław Grundas (Ed.), ISBN: 978-953-51-1184-9, InTech, DOI: 10.5772/52473.

CHAPTER 3

M.C. Hernandez-Soriano, A. Sevilla-Perea, B. Kerré and M.D. Mingorance (2013). Stability of Organic Matter in Anthropic Soils: A Spectroscopic Approach, Soil Processes and Current Trends in Quality Assessment, Dr. Maria C. Hernandez Soriano (Ed.), ISBN: 978-953-51-1029-3, InTech, DOI: 10.5772/55632.

CHAPTER 4

Shinya Funakawa, Hiroshi Yoshida, Tetsuhiro Watanabe, Soh Sugihara, Method Kilasara and Takashi Kosaki (2012). Soil Fertility Status and Its Determining Factors in Tanzania, Soil Health and Land Use Management, Dr. Maria C. Hernandez Soriano (Ed.), ISBN: 978-953-307-614-0, InTech, DOI: 10.5772/29199.

CHAPTER 5

Manoj K. Jha (2012). Quantifying Soil Moisture Distribution at a Watershed Scale, Soil Health and Land Use Management, Dr. Maria C. Hernandez Soriano (Ed.), ISBN: 978-953-307-614-0, InTech, DOI: 10.5772/29990.

CHAPTER 6

Andrea Rubenacker, Paola Campitelli, Manuel Velasco and Silvia Ceppi (2012). Fire Impact on Several Chemical and Physicochemical Parameters in a Forest Soil, Soil Health and Land Use Management, Dr. Maria C. Hernandez Soriano (Ed.), ISBN: 978-953-307-614-0, InTech, DOI: 10.5772/30460.

CHAPTER 7

Bart Minten and Claude Randrianarisoa (2012). Forest Preservation, Flooding and Soil Fertility: Evidence from Madagascar, Soil Health and Land Use Management, Dr. Maria C. Hernandez Soriano (Ed.), ISBN: 978-953- 307-614-0, InTech,

CHAPTER 8

Celia Maria Maganhotto de Souza Silva and Elisabeth Francisconi Fay (2012). Effect of Salinity on Soil Microorganisms, Soil Health and Land Use Management, Dr. Maria C. Hernandez Soriano (Ed.), ISBN: 978- 953-307-614-0, InTech,

CHAPTER 9

Alberto C. de Campos Bernardi, Patrícia P. A. Oliveira and Odo Primavesi (2012). Soil Fertility of Tropical Intensively Managed Forage System for Grazing Cattle in Brazil, Soil Fertility Improvement and Integrated Nutrient Management - A Global Perspective, Dr. Joann Whalen (Ed.), ISBN: 978-953-307-945-5, InTech,

CHAPTER 10

Ren Wan-Jun, Huang Yun and Yang Wen-Yu (2012). Effects on Soil Fertility and Microbial Populations of Broadcast-Transplanting Rice Seedlings in High Standing-Stubble Under No-Tillage in Paddy Fields, Soil Fertility Improvement and Integrated Nutrient Management - A Global Perspective, Dr. Joann Whalen (Ed.), ISBN: 978-953-307-945-5, InTech,

CHAPTER 11

Roland Nuhu Issaka, Moro Mohammed Buri, Satoshi Tobita, Satoshi Nakamura and Eric Owusu-Adjei (2012). Indigenous Fertilizing Materials to Enhance Soil Productivity in Ghana, Soil Fertility Improvement and Integrated Nutrient Management - A Global Perspective, Dr. Joann Whalen (Ed.), ISBN: 978-953-307-945-5, InTech,

CHAPTER 12

Lubanga Lunze, Mathew M. Abang, Robin Buruchara, Michael A. Ugen, Nsharwasi Léon Nabahungu, Gideon O. Rachier, Mulangwa Ngongo and Idupulapati Rao (2012). Integrated Soil Fertility Management in Bean-Based Cropping Systems of Eastern, Central and Southern Africa, Soil Fertility Improvement and Integrated Nutrient Management - A Global Perspective, Dr. Joann Whalen (Ed.), ISBN: 978-953-307-945-5, InTech,

INDEX